Intelligent Networks

IFIP – The International Federation for Information Processing

IFIP was founded in 1960 under the auspices of UNESCO, following the First World Computer Congress held in Paris the previous year. An umbrella organization for societies working in information processing, IFIP's aim is two-fold: to support information processing within its member countries and to encourage technology transfer to developing nations. As its mission statement clearly states,

> IFIP's mission is to be the leading, truly international, apolitical organization which encourages and assists in the development, exploitation and application of information technology for the benefit of all people.

IFIP is a non-profitmaking organization, run almost solely by 2500 volunteers. It operates through a number of technical committees, which organize events and publications. IFIP's events range from an international congress to local seminars, but the most important are:

- the IFIP World Computer Congress, held every second year;
- open conferences;
- working conferences.

The flagship event is the IFIP World Computer Congress, at which both invited and contributed papers are presented. Contributed papers are rigorously refereed and the rejection rate is high.

As with the Congress, participation in the open conferences is open to all and papers may be invited or submitted. Again, submitted papers are stringently refereed.

The working conferences are structured differently. They are usually run by a working group and attendance is small and by invitation only. Their purpose is to create an atmosphere conducive to innovation and development. Refereeing is less rigorous and papers are subjected to extensive group discussion.

Publications arising from IFIP events vary. The papers presented at the IFIP World Computer Congress and at open conferences are published as conference proceedings, while the results of the working conferences are often published as collections of selected and edited papers.

Any national society whose primary activity is in information may apply to become a full member of IFIP, although full membership is restricted to one society per country. Full members are entitled to vote at the annual General Assembly, National societies preferring a less committed involvement may apply for associate or corresponding membership. Associate members enjoy the same benefits as full members, but without voting rights. Corresponding members are not represented in IFIP bodies. Affiliated membership is open to non-national societies, and individual and honorary membership schemes are also offered.

Intelligent Networks

Proceedings of the IFIP workshop on intelligent networks 1994

Edited by

Jarmo Harju, Tapani Karttunen

Lappeenranta University of Technology
Lappeenranta
Finland

and

Olli Martikainen

Telecom Finland Ltd
Helsinki
Finland

Published by Chapman & Hall on behalf of the International Federation for Information Processing (IFIP)

CHAPMAN & HALL

London · Glasgow · Weinheim · New York · Tokyo · Melbourne · Madras

Published by Chapman & Hall, 2–6 Boundary Row, London SE1 8HN, UK

Chapman & Hall, 2–6 Boundary Row, London SE1 8HN, UK

Blackie Academic & Professional, Wester Cleddens Road, Bishopbriggs, Glasgow G64 2NZ, UK

Chapman & Hall GmbH, Pappelallee 3, 69469 Weinheim, Germany

Chapman & Hall USA, 115 Fifth Avenue, New York, NY 10003, USA

Chapman & Hall Japan, ITP-Japan, Kyowa Building, 3F, 2-2-1 Hirakawacho, Chiyoda-ku, Tokyo 102, Japan

Chapman & Hall Australia, 102 Dodds Street, South Melbourne, Victoria 3205, Australia

Chapman & Hall India, R. Seshadri, 32 Second Main Road, CIT East, Madras 600 035, India

---

First edition 1995

Printed in Great Britain by TJ Press, Padstow, Cornwall

ISBN 0 412 71730 1

A catalogue record for this book is available from the British Library

∞ Printed on permanent acid-free text paper, manufactured in accordance with ANSI/NISO Z39.48-1992 and ANSI/NISO Z39.48-1984 (Permanence of Paper).

CONTENTS

Preface v

1 Introduction to intelligent networks
T. Karttunen, O. Harju and O. Martikainen vii

2 Intelligent networks for personal communications
H. Bisseling and J. den Hartog 34

3 Service creation from IN to mobile and broadband
C. Knight 47

4 Nokia's IN solution for fixed and cellular networks
P. Lehtinen and M. Warsta 61

5 Service creation environment as a software development platform
A. Bihain and J. White 68

6 IN service specification using the KANNEL language
K. Granö, J. Harju, T. Järvinen, T. Karttunen, T. Larikka and J. Paakki 77

7 Performance analyzer for Intelligent Network
J. Jormakka 97

8 Service prototyping in the OVOPS environment
P. Puro and J. Sonninen 111

9 The local management for a service control point
T. Haitao and E. Kärkkäinen 126

10 An advanced management architecture for IN
D. Gaïti 137

11 Secure IN internetworking
A. Herrigel and X. Lai 149

12 Charging generalised: a generic role model for charging and billing services in telecommunications
D. Ranasinghe and J. Nørgaard 159

13 Database access in intelligent networks
K.E.E. Raatikainen 173

14 Problem classes in intelligent network database design
J. Taina 194

15 Customer value creation in value added telecommunication services
J. Takala, O. Martikainen and J. Ruusunen 208

16 Value creation for multimedia services on broadband networks
S. Luukkainen and C. Englund 225

17 Use of INtelligent networks in the Universal Mobile Telecommunications System (UMTS)
H. Mitts 236

18 On location of service control
H. Hämmäinen and P. Lahtinen 246

19 Comparison of broadband intelligent network signalling architecture
O. Martikainen, T. Karttunen, V. Naoumov and K. Samouylov 253

20 Broadband intelligent network project
O. Martikainen and K. Molin 270

21 Which intelligence for future networks?
D. Gaïti and G. Pujolle 280

Index of contributors 291

Preface

Substantial new breakthroughs are going on in telecommunications technology. The introduction of cellular radio networks and mobility is probably the most influential one in the next few years. The broadband transmission and switching technology is also maturing and will provide a cost effective platform for service provision. Interactive business and consumer services based on video and multimedia will become possible, after the broadband customer access is in the market. Common to all these developments will be the computer controlled structure of modern telecommunications, where protocols, application technology and resource management are key factors.

This book contains the papers presented in the Workshop on Intelligent Networks, sponsored by IFIP Technical Committee 6 (Communication Systems). The Workshop was organized as part of the Third Summer School on Telecommunications at the Lappeenranta University of Technology, Lappeenranta, Finland on August 8–9, 1994. The call for papers was the first in the short history of the Summer School, yet we were able to select 20 high quality papers out of the submitted contributions. One can distinguish three main streams in the topics of the papers, reflecting the trends mentioned above: service creation and management platforms, use of software tools and methods for the IN specification, analysis and database design, and mobility and broadband aspects of future Intelligent Networks.

We wish to thank the members of the program committee of the Workshop for their help in the reviewing process and for chairing the Workshop sessions: James Aitken (United Kingdom), Maria J. B. Almeida (Brazil), Dominique Gaiti (USA), Peter Delgado (United Kingdom), Heinz Dibold (Germany), Roberto Kung (France), Kari Lautanala (Finland), Valeri Naoumov (Russia), Jørgen Nørgaad (Denmark), Guy Pujolle (France), Konstantin Samouylov (Russia) and Lennart Söderberg (Sweden). Telecom Finland sponsored this Workshop by providing us with the excellent Workshop Secretary, Ms. Ansa Laakkonen. Local arrangements were carried out by the Centre of Continuing Education of the Lappeenranta University of Technology. The Summer School Secretaries, Ms. Päivi Pönni and Ms. Minna-Maija Mäenpää deserve our gratitude for their careful and enthusiastic work in carrying out the registration process and dealing with the practical affairs of more than 120 Summer School participants.

Jarmo Harju
Tapani Karttunen
Olli Martikainen

1

Introduction to intelligent networks

J. Harju, T. Karttunen and O. Martikainen

Lappeenranta University of Technology,
Datacommunications Laboratory,
P.O.Box 20,
FIN-53851 Lappeenranta, Finland
Tel. +358 53 574 3613, Fax. +358 53 574 3650

Abstract

The development of telecommunications techniques and the need for more advanced services has created projects on standardization of international Intelligent Networks (*IN*). The standards of Intelligent Networks define IN in an abstract point of view, so it leaves the service providers the decisions on their own implementations. The first standard sets of IN are Bellcore's AIN release 1 and the CCITT's Capability Set 1 (*CS1*). They define the basic services of IN, additional features such as rapid service introduction and a flexible architecture that provides future expansion to further IN Capability Sets. The standardization organisations, such as CCITT and ETSI, work hard to help the service providers to implement their IN architecture in order to be able to provide international IN services. This kind of architecture is better known as global Intelligent Network architecture and it should be taken into consideration already in the early implementations of IN. This paper presents some history of telecommunications technology, an overview of IN and its services.

1. TURNING-POINTS IN TELECOMMUNICATIONS

Several turning-points can be found in the history of telecommunications technology (marked as circles in the figure) (Figure 1).

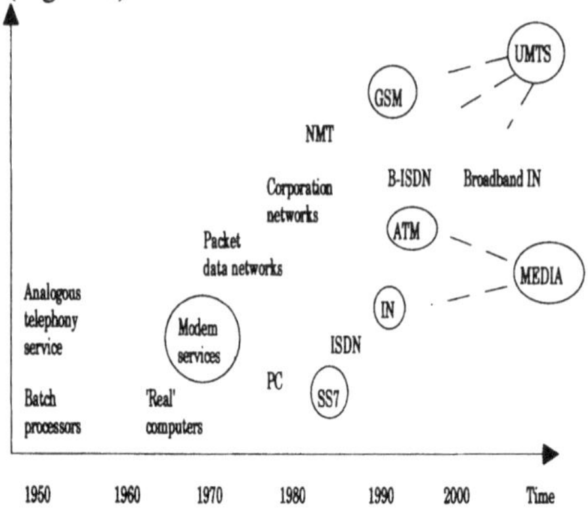

Figure 1. The development of telecommunications.

First, the beginning of data transfer by the use of analogous telephony service was an important stage in the history. This service was not good for use in corporations because of its low data transfer speed. Then, there was a need for a data transfer service that used billing by data amount while the expences of the analogous telephony service consisted mainly of the data transfer time. The packet switched data networks were developed especially for corporations use. Second, CCITT (*Consultative Committee for International Telephone and Telegraphy*) introduced its SS7 protocol stack to replace the analogous signalling system. This was the corner-stone for the digital telecommunications technology that is used, for instance, in ISDN (*Integrated Services Digital Network*). In the late 1980's radio signalling technology was advanced enough to provide digital telephony service. The GSM (*Global System for Mobile communications*) mobile phone technology, introduced into use 1991, is also suitable for low-speed data transfer. The Intelligent Network is an architecture capable to integrate all the telecommunications services mentioned in a flexible way.

The telecommunications networks and wide area networks used PDH (*Plesiochronous Digital Hierarchy*) technology in the physical data transfer. At the introduction of CCITT's SDH (*Synchronous Digital Hierarchy*) technology the physical data transfer rates increased remarkably. A new technology, ATM (*Asynchronous Transfer Mode*), was introduced to use the available bandwidth efficiently in the 1992. By the introduction of ATM it was possible to imagine of such concepts as B-ISDN (*Broadband Integrated Services Digital Network*), broadband mobility and broadband IN. Broadband infrastructure will make it possible to introduce advanced value added, mobile and media services (Figure 2).

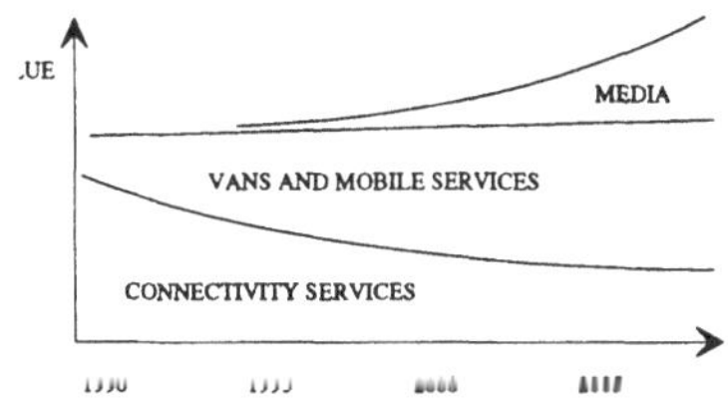

Figure 2. Turnover Value of Service Types

In modern telecommunications deeply influential changes are taking place, caused by the emerging competitive media services market and the new technological breakthroughs. The market changes are due to the integration of telecommunications and information technology, which brings interactive real time video and multimedia services available to users. Examples of these services are digital interactive TV, video on demand services for banking, shopping and leisure, electronic press and publishing. The technological requirements for these services are cost effective broadband transmission and access technologies, flexible computer based management and control of networks, switching and service applications and the support of mobility.

1.1 UMTS

UMTS (*Universal Mobile Telecommunications System*) is intended to be an international standard for global telecommunication system. It is a third generation mobile telecommunications system which integrates several second generation mobile systems like cordless telephones (CT2 (*Cordless Telephone 2*) and DECT (*Digital European Cordless Telecommunications*)), mobile telecommunications systems (GSM and PCN) and radio message systems (ERMES (*European Radio Message System*)).

UMTS is researched in RACE and financed by EC (*European Community*) and ETSI's group SGM5. UMTS defines a mobile communications system where a mobile phone could be used at home, office and elsewhere. UMTS is an open system which is based on TMN and IN concepts. The system supports ISDN services and could be at some degree compatible with B-ISDN with ATM-switching and possible broadband mobile access. This system is a very advanced telecommunications system that supports global mobility and Intelligent Network services and is not expected to be introduced before the year 2000.

2. COMPUTER CONTROLLED TELECOMMUNICATIONS

2.1 CCITT Signalling System No. 7

With the introduction of electronic processors in switching systems came the possibility of providing Common Channel Signalling (*CCS*). This is an out-of-band signalling method in which a common data channel is used to convey signalling information related to a number of trunks. CCITT published this new signalling protocol stack SS7 (*Signalling System No. 7*) based on CCITT OSI (*Open Systems Interconnection*) Reference Model (*OSIRM*) in 1980.

SS7 is fully digital and SS7 protocol stack corresponds to the seven layers of the OSIRM and includes the Application Services and User Parts *(UP)* (Figure 3). The signalling network structure component of SS7 is the Network Service Part (*NSP*), and it consists of the Message Transfer Part (*MTP*) and the Signalling Connection Control Part (*SCCP*). The OSIRM layers 4 - 6 are provided by Intermediate Service Part *(ISP)* and each User Part.

SS7 is quite an advanced protocol stack. It includes capabilities for congestion control and overload control. It also includes features for avoiding congestion by alternative routing or capacity expansion when heavy load is detected. With congestion is ment, generally, shortage of resources, which is caused by an excessive amount of load, or a failure that reduces the installed capacity of a network element. SS7 also includes capabilities for sending congestion and overload indications to the adjacent exchanges or traffic sources. [Lehti93]

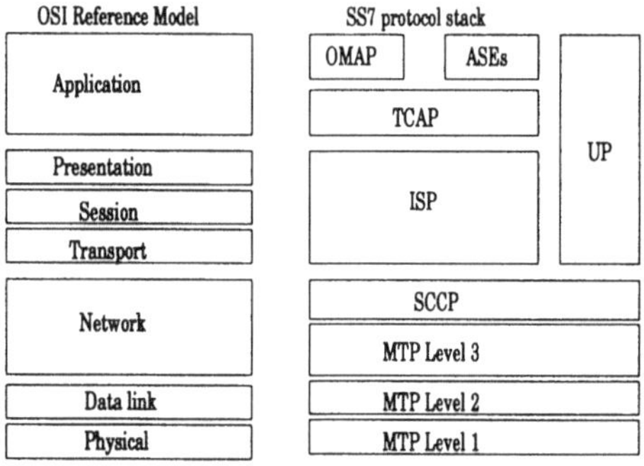

Figure 3. SS7 protocol architecture.

2.1.1 Network Services Part

MTP consists of levels 1-3 of the SS7 protocol stack and it provides a connectionless message transfer system that enables signalling information to be transferred across the network to its desired destination. Functions are included in MTP that allow system failures to occur in the network without adversely affecting the transfer of signalling information. So the overall purpose of MTP is to provide a reliable transfer and delivery of signalling information across the signalling network and to have the ability to react and take necessary actions in response to system and network failures to ensure that reliable transfer is maintained. The first level of MTP presents the signalling data link functions. A signalling data link functon is a bidirectional transmission path for signalling, consisting of two data channel operating together in opposite directions at the same data rate. It fully complies with the OSI's definition of the physical layer. Level 2 of MTP presents the signalling link functions. The signalling link functions correspond to the OSI's data link layer. Together with a signalling data link, the signalling link functions provide a signalling link for the reliable transfer of signalling messages between two directly connected signalling points. The third level of MTP presents the signalling network functions. They correspond to the lower half of the OSI's network layer, and they provide the functions and procedures for the transfer of messages between signalling points, which are the nodes of the signalling network. [Modar90]

SCCP provides additional functions to MTP for both connectionless and connection-oriented network services. SCCP enhances the services of the MTP to provide the functional equivalent of OSI's network layer. The addressing capability of MTP is limited to delivering a message to a node and using a four-bit service indicator to distribute messages within the node. SCCP supplements this capability by providing an addressing capability that uses DPCs (*Destination Point Code*) plus Subsystem Numbers (*SSN*). The SSN is local addressing information used by SCCP to identify each of the SCCP users at a node.

2.1.2 User Part

The User Part forms the most upper layer of the SS7 protocol stack that use the services provided by the lower layers SCCP and MTP. User Part functions are ISDN-UP, TCAP (*Transaction Capabilities Application Part*) and OMAP (*Operations, Maintenance, and Administration Part*). The ISDN-UP is not discussed in this paper. TCAP refers to the set of protocols and functions used by a set of widely distributed applications in a network to communicate with each other. TCAP directly uses the service of SCCP. Essentially, TCAP provides a set of tools in a connectionless environment that can be used by an application at a node to invoke execution of a procedure at another node and exchange the results of such invocation. As such, it includes protocols and services to perform remote operations. It is closely related to the OSI Remote Operations Service Element (*ROSE*). The OMAP of the SS7 protocol stack provides the applications protocols and procedures to monitor, coordinate, and control all the network resource that make communications based on SS7 possible.

2.1.3 Signalling network structure

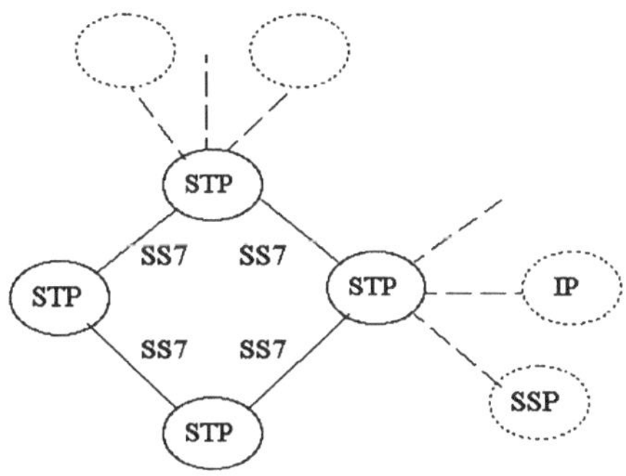

Figure 4. CCITT SS7 network structure.

Signalling networks consist of signalling points and signalling links connecting the signalling points together. (Figure 4) As alluded to earlier, a signalling point that transfers messages from one signalling link to another at level 3 is said to be a STP (*Signalling Transfer Point*). Signalling points that are STP's can also provide functions higher than level 3, such as SCCP and other level 4 functions like ISDN-UP. When signalling point has an STP capability and also provides level 4 functions, it is commonly said to have an integrated STP functionality. When the signalling point provides only STP capability, or STP and SCCP capabilities, it is commonly called a stand-alone STP. Signalling links, STP's (stand-alone and integrated), and signalling points with level 4 protocol functionality can be combined in many different ways to form a signalling network. The SS7 Network Services Part protocol is specified independent of the underlying signalling network structure. However, to meet the stringent availability requirements given below (e.g., signalling route set unavailability must not exceed ten minutes per year), it is clear that any network structure must provide redundancies for the signalling links, which have unavailabilities measured in many hours per year. In most cases the STP's must also have backups.

The worldwide signalling network is intended to be structured into two functionally independent levels: the national and international levels. This allows numbering plans network management of the international and the different national network to be independent of one another. A signalling point can be a national signalling point, an international signalling point, or both. If it serves both, it is identified by a specific signalling point code in each of the signalling networks.

2.2 Intelligent Network

2.2.1 The need for IN

In the past few years the development of telecommunications networks has been rapid. Earlier, the telecommunications network functions before were controlled mainly by operators. The desire to share data and distribute application processing among network elements, the need for standard interfaces between them and user demands for more sophisticated telecommunications services has changed the controlling of network elements notably. The telecommunications network elements today are controlled by the network operator, the service provider or the customer himself. To integrate the control and management of different services inside the operator, or to be able to provide third party control and management services, control and management interfaces with software support are needed.

The development of IN architecture was initiated by Bellcore in USA almost ten years ago in order to help the Regional Bell Operating Companies to become more competitive in the deregulated telecommunications environment. The original goal was to provide network operators with the ability to introduce, control and manage services more effectively by using a centralized database in a Service Control Point (*SCP*) for controlling and managing the various network services. [Lauta93]

The objective of IN is to allow the inclusion of additional capabilities to facilitate provisioning of service, independent of the service or network implementation in a multi-vendor environment. Service implementation independence allows service providers to define their own services independent of service specific developments by equipment vendors [Q1201].

Network implementation independence allows network and service operators to allocate functionality and resources within their networks and to efficiently manage their networks independent of network implementation specific developments by equipment vendors.

The network architectures, so far, have developed almost independently of each other. This point of view, of course, causes the network operators and service providers to provide independently implemented service to customers. The basic idea of IN has been that it facilitates the provisioning of services independently from the telecommunications networks and equipment vendors. So, the IN acts as a distributing and centralizing framework of the telecommunications services. With this framework, it is possible to introduce advanced customer oriented services rapidly and cost effectively.

2.2.2 *Definition of Intelligent Network*

Intelligent Network (IN) is an architectural concept for the operation and provision of new services which is characterized by [Q1201]:

- extensive use of information processing techniques;
- efficient use of network resources;
- modularization and reusability of network functions;
- integrated service creations and implementation by means of the modularized reusable network functions;
- flexible allocation of network functions to physical entities;
- portability of network functions among physical entities;
- standardized communication between network functions via service independent interfaces;
- service subscriber 1) control of some subscriber-specific service attributes;
- service user 2) control of some user-specific service attributes;
- standardized management of service logic.

IN is applicable to a wide variety of networks, including but not limited to: public switched telephone network (PSTN) mobile, packet switched public data network (PSPDN) and integrated services digital network (ISDN) - both narrowband-ISDN (N-ISDN) and broadband-ISDN (B-ISDN).

IN supports a wide variety of services, including supplementary services, and utilizes existing and future bearer services (e.g. as those defined in N-ISDN and B-ISDN contexts).

3. INTELLIGENT NETWORK ARCHITECTURE

3.1 Origins of IN

The Intelligent Networks is a telecommunications network services control and management architecture. In February 1985, Regional Bell Operating Companies (*RBOC*) submitted a Request For Information (*RFI*) for a Feature Node concept with the following objectives:

- support the rapid introduction of new services in the network,
- help establish equipment and interface standards to give the RBOCs the widest possible choice of vendor products and
- create opportunities for non-RBOC service vendors to offer services that stimulate network usage.

As with the past telecommunications technology, it was not desirable to introduce short term services, because of the long implementation and development period. Now, with IN technology it is possible to introduce new services rapidly without affecting the available services. IN defines a large set of standards that describe the interfaces between different network control points. With only specifying the interfaces IN makes it possible for vendor systems to provide with different products and, of course, for operators to use any of these products in their network configuration. IN includes also capabilities for other than operators to introduce new services into the telecommunications network.

The IN's main advantage is the ability to orchestrate exchange service execution from a small set of Intelligent Network nodes known as Service Control Points (*SCP*). SCPs are connected to the network exchanges (known as Service Switching Points) via a standardized interface; CCITT Signalling System No. 7. The SS7 will facilitate a multi-vendor SCP and SSP marketplace, and the standardization of application interfaces allows a multi-vendor software marketplace for SCP applications (that is, the service control logic and its related data) (Figure 5).

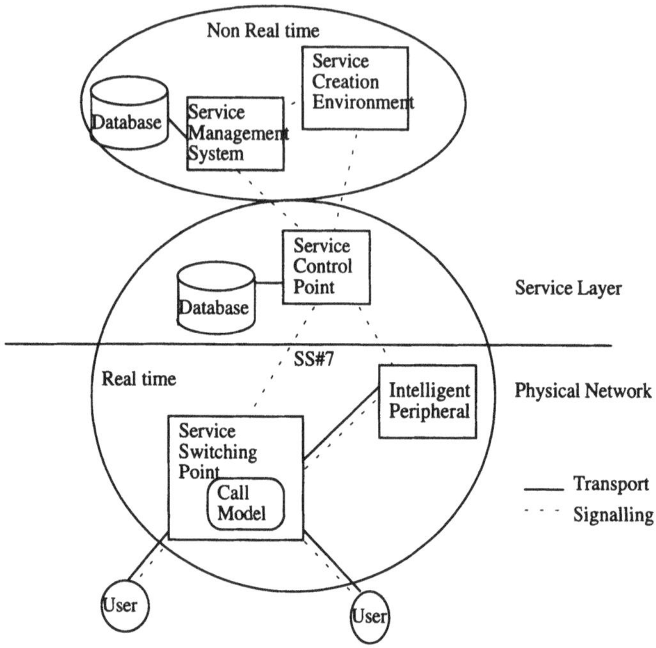

Figure 5. Intelligent Network overview.

The IN's long term goal is the ability to introduce new services, or change existing services quickly, without having to adapt SSP software (only parameters or trigger updates). The adaptation will be confined to the SCP where parameters or stimuli are updated. This goal was first planned by Bellcore to be achieved in two stages: IN/1 and IN/2. IN/1 definitions introduced the term Intelligent Network in 1986 and in 1987 IN/2 definitions were introduced. In 1988 IN/2 was delayed and IN/1+ was introduced instead. In 1989 Bellcore abandoned IN/1+ for several reasons, some being problems in the technology and lack of multivendor involvement. Instead a MultiVendor Initiative (MVI) was started in 1989 to define Advanced Intelligent Network (AIN). At the same time CCITT and ETSI started work on IN. The IN basic concepts for a service independent architecture were introduced already in IN/1. The AIN concepts were essentially those of IN/2 defining a fully service independent architecture with total separation of service logic from the underlying seitching system. These principles were accepted also by CCITT and ETSI work. The AIN Release 1 and CCITT CS1 were published in 1993.

Table 1: IN/1 outlines

Table 2: IN/2 outlines

An Intelligent Network is able to separate the specification, creation, and control of telephony services from physical switching networks. The key benefit of this capability is that exchange carriers will be able to rapidly engineer new revenue-producing services, in response to market opportunities, without having to rely on lenghty cycles for implementing them entirely on switching fabric. Ultimately, service creation, or at least service customization, can be extended to subscribers [Homa92]

The original IN concepts IN/1 and IN/2 were not considered sufficient to support vendor independence and open interfaces, and extensive standardization activities were started in 1989's. The first available publications were the Advanced Intelligent Network (AIN), and after that CCITT and ETSI provided their first draft recommendations. Our presentation here is mainly based on the CCITT, presently ITU-T, recommendations.

3.2 IN standardization

3.2.1. IN standards bodies

The IN standards are defined by ETSI and CCITT. Also, in the USA, the work is being done by Bellcore, which is not a standards body but provides the major input to the American National Standards Institute committee TS.1. [Roger90]

ETSI

ETSI was created in 1988 and its members are the European Telcos (*Telecommunications Operating Company*), manufacturers, user representatives and research bodies. ETSI has two purposes. IN belongs to the latter category.

- ETSI is to achieve workable versions of international standards for the European environment.
- ETSI is to define European standards in areas where quick response is required for technical development.

CCITT

Work on international standards for IN began at CCITT in 1989. Study Group XI.4 is responsible of the standartization. CCITT expects that the specification and deployment of IN will continue over a number of study periods. CCITT name has changed to ITU (*International Telecommunications Union*) and there the Special Interest Group (*SIG*) is T (ITU-T). Its approach to the development of IN standards assumes that it is necessary to start with a minimum set of criteria which are sufficiently open that they can evolve to meet the needs of the long-term concept as this becomes a practical reality.

Both ETSI and ANSI are keen to ensure that CCITT recommendations agree substantially with their own activities, and collaboration between all three bodies is likely to be an important determinant in the rapid development of realistic IN standards.

3.2.2. Phased standardization

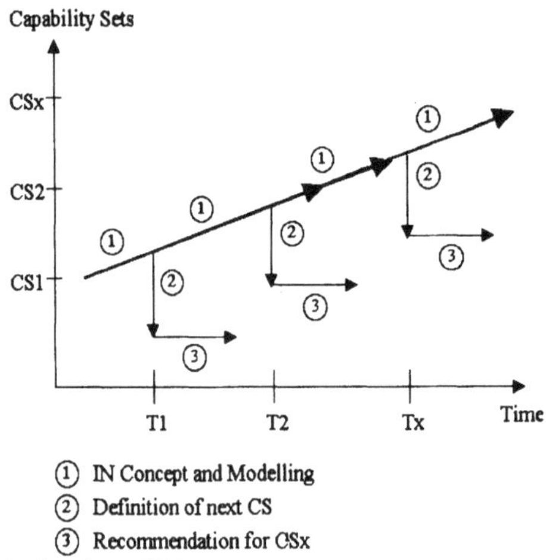

Figure 6. Phased standardization of IN.

To meet the goals and objectives, CCITT has embarked on a phased standardation process toward the target IN architecture (INA) [Q1201]. CCITT works on defining a set of capabilities for each phase and simultaneously on evolving the view of the target IN architecture called the long-term capability set (LTCS) (Figure 6) The IN subjects of standardization are called Capability Sets (*CS*). The Capability Sets involve service creation, management and interaction and also network management, service processing and network internetworking.

3.2.3. Structure of CCITT IN standards

The basic standard that defines the framework of other IN standards is Q.1200 - Q-Series Intelligent Network Recommendations Structure. The standards have been numbered so that every new CSx will have a number that begins with 12x and the description of the CSx recommendation part y will be numbered also systematically such as 12xy. (Table 3) So, the principles introduction for IN CS2 will be recommendation number Q.1221.

Table 3: IN recommendations structure.

0X - General	
1X - CS1	1 - Principles Introduction
2X - CS2	2 - Service Plane (not included for CS1)
3X - CS3	3 - Global Functional Plane
4X - CS4	4 - Distributed Functional Plane
5X - CS5	5 - Physical Plane
6X - CS6	6 - For future use
7X - CS7	7 - For future use
8X - CS8	8 - Interface Recommendations
9X - Vocabulary	9 - Intelligent Network Users Guide

3.2.4. Capability Set 1

It has been an international and european wide aim to define the first step of INA. These recommendations are gathered into a set called IN Capability Set 1 (CS1). There are two standardation organisations working on CS1: CCITT and ETSI. CCITT has gathered these recommendations into the Q.121y -series. (Table 4) CCITT's and ETSI's standards do not substantially differ from each other.

CCITT Study Group XI, Working Party XI/4 includes representatives from most of the important telecommunications network operators and equipment vendors in the world. Study Group XVIII also is involved in the initial set of IN standards, and is sharing responsibility for the Introductory Recommendations. At these meetings, there is an obvious willingness to strongly focus on achieving a realistic initial set of IN capability, which is both technically implementable and commercially deployable.

Table 4: IN CS1 recommendations.

Recommendation Q.1200	Q-Series Intelligent Network Recommendations Structure
Recommendation Q.1201	Principles of Intelligent Network Architecture

Recommendation Q.1202	Intelligent Network - Service Plane Architecture
Recommendation Q.1203	Intelligent Network - Global Functional Plane Architecture
Recommendation Q.1204	Intelligent Network - Distributed Functional Plane Architecture
Recommendation Q.1205	Intelligent Network - Physical Plane Architecture
Recommendation Q.1208	Intelligent Network - Application Protocol General Aspects
Recommendation Q.1211	Intelligent Network - Introduction to Intelligent Network Capability Set 1
Recommendation Q.1213	Intelligent Network - Global Functional Plane for CS1
Recommendation Q.1214	Intelligent Network - Distributed Functional Plane for CS1
Recommendation Q.1215	Intelligent Network - Physical Plane for CS1
Recommendation Q.1218	Intelligent Network - Intelligent Network Interface Specifications
Recommendations Q.1219	Intelligent Network Users guide for Capability Set 1

In defining IN CS1, CCITT applied the INCM (*Intelligent Network Conceptual Model*) using both "bottom-up" and "top-down" approaches. The former approach focused on modelling the capabilities of existing networks in terms of functional and physical architectures that could evolve the target IN architecture, given CCITT's objective of evolving IN from existing networks. The latter approach was service-driven and it focused on identifying a set of IN CS1 services and Service Features.

IN CS1 defines capabilities of direct use to both manufactures and network operators in support of circuit-switched voice/data services either defined or in the process of being defined by CCITT. The primary characteristic of the target set of IN CS1 services is that they apply during the setup phase of a call or during the release phase of a call. CCITT chose this *single-ended* service characteristic to limit the operational, implementation, and control complexity for IN CS1. Even with this limitation, it may be expected that equipment suppliers will support interworking of IN CS1 capabilities with existing switch-based services, including more complex services such as those that apply during the active phase of a call. For example, IN CS1 routing, charging, and user interaction capabilities may be used to customize or improve existing switch-based services to better satisfy market needs.

It is anticipated that CS1 recommendations of CCITT and ETSI will be adopted world-wide. This can help to develop open interfaces between the SSP (*Service Switching Point*) and SCP (*Service Control Point*), thus putting into effect one of the important goals of the IN, namely vendor independence. [Lauta93]

3.2.5 IN CS1 Services

Although, by nature, the IN is a service independent architecture, it is relevant to describe the general CS-1 service capabilities. The services and Service Features that are to be supported by CS-1 are fundamental to the CS-1 Service Building Blocks, call processing model and service control principles.

The target set of CS-1 defines several services (Table 5) and service features. A service is a stand-alone commercial offering, characterized by one or more core Service Features, and can be optionally enhanced by other Service Features. A Service Feature is a specific aspect of a service that can also be used in conjunction with other services/Service Features as part of a commercial offering. It is either a core part of a service or an optional part offered as an enhancement to a service. [Q1211]

Table 5: Target set of IN CS1 services.

Automatic Alternative Billing (*ABB*)	Mass Calling (*MAS*)
Abbreviated Dialling (*ABD*)	Malicious Call Identification (*MCI*)
Account Card Calling (*ACC*)	Premium Rate (*PRM*)
Credit Card Calling (*CCC*)	Security Screening (*SEC*)
Call Distribution (*CD*)	Selective Call Forward on Busy/Don't Answer (*SCF*)
Call Forwarding (*CF*)	Split Charging (*SPL*)
* Completion of Call to Busy Subsrciber (*CCBS*)	Televoting (*VOT*)
* Conference Calling (*CON*)	Terminating Call Screening (*TCS*)
Call Rerouting Distribution (*CRD*)	User-Defined Routing (*UDR*)
Destination Call Routing (*DCR*)	Universal Access Number (*UAN*)
Follow-Me-Diversion (*FMD*)	Universal Personal Telecommunications (*UPT*)
Freephone (*FPH*)	Virtual Private Network (*VPN*)

Note: The service indicated with a * may only be partially supported in CS1, because they require capabilities beyond those of type A services.

3.3 IN Functional Requirements

IN functional requirements arise as a result of the need to provide network capabilities for both customer needs (service requirements) and network operator needs (network requirements) [Q1201].

A service user is an entity external to the network that uses its services. A service is that which is offered by an administration to its customers in order to satisfy a telecommunications requirement. Part of the service used by customers may be provided/managed by other customers of the network. These are often called as third party services and their providers as 3rd party service providers.

Service requirements will assist in identifying specific services that are offered to the customer. These service capabilities are also referred to as (telecommunication) services. Network requirements span the ability to create, deploy, operate and maintain network capabilities to provide services.

Service and network requirements can be identified for the following areas of service/network capabilities: service creation, service management, network management, service processing and network interworking.

- **Service creation**: An activity whereby supplementary services are brought into being through specification phase, development phase and verification phase.
- **Service management**: An activity to support the proper operation of a service and the administration of information relating to the user/customer and/or the network operator, Service management can support the following processes: service development, service provisioning, service control, billing and service monitoring.
- **Network management**: An activity to support the proper operation of an IN-structured network.
- **Service processing** consists of basic call and supplementary service processing which are the serial and/or parallel executions of network functions in a coordinated way, such that basic and supplementary services are provided to the customers.
- **Network interworking**: A process through which several networks (IN to IN or IN to non-IN) cooperate to provide a service.

3.3.1 Service Requirements

The goal of work for IN is to define a new architectural concept that meets the needs of telecommunication service providers to rapidly, cost effectively, and vendor-independently satisfy their existing and potential market needs for services, and to improve the quality and reduce the cost of network service operations and management. In [Q1201] the following overall service requirements are given when defining the IN architecture:

- it should be possible to access services by the usual user network interface (e.g. POTS, ISDN);
- it should be possible to access services that span multiple networks;
- it should be possible to invoke a service on a call-by-call basis or for a period of time, in the latter case the service may be deactivated at the end of the period;
- it should be possible to perform some access control to a service;
- it should be easy to define and introduce services;
- it should be possible to support services involving calls between two or more parties;
- it should be possible to record service usage in the network (service supervision, tests, performance information, charging);
- it should be possible to provide services that imply the use of functions in several networks;
- it should be possible to control the interactions between different invocations of the same service.

Service requirements for service creation refer to the network capabilities that are used by network operators for the provision of service creation services to customers.

Service requirements for service management refer to the network capabilities that are necessary for the provision of service management services to customers.

Service requirements for service processing refer to the network capabilities that are necessary for the provision, from a customer's point of view, of basic and supplementary services by an IN-structured network. The IN is primarily a network concept that aims for efficient creation, deployment and management of supplementary services that enhance basic services. Hence, from a customers point of view the provision of services is transparent, the customer is unaware whether the service is provided in an IN way. Service processing requirements can be identified for service and access capabilities. The service capabilities of IN can be applied to the support of supplementary services for the following basic services [Q1201]:

- bearer services including speech, audio and data,
- teleservices as telephony, telefax and videotex,
- broadband interactive services and
- broadband distribution services.

The access capabilities of IN should be applicable to all telecommunications networks, such as Public Switched Telecommunications Networks (*PSTN*), including Integrated Services Digital Networks (*ISDN*), both narrowband and broadband, packet-switched public data networks, and mobile networks. Although, IN CS1 enables only the use of PSTN, PLMN (*Public Land Mobile Network*) and ISDN, IN should enable service providers to define their own services, independent of service-specific developments by equipment suppliers.

CS1 is intended to address services with high commercial value, focusing at addressing flexible routing, charging, and user interaction services. The list of benchmark services and features will be listed later on. Standardization of these services, however, is not CCITT's role. An important characteristic is that the services will be technologically feasible and understandable, but do not significantly impact existing deployed technology. In this context, services have been categorized by CCITT as specified in tables 6 and 7.

Table 6: CCITT Type A service features.

All type A services are invoked on behalf of and directly affect a single user. Most type A services can be invoked only during call setup or tear down and fall in the category of "single-user, single-ended (no requirements for representing end-to-end messaging or control), single point-of-control (no requirement for representing interaction points between multiple service logic programs), and single-bearer capability (one media profile)". Type A services may be used in conjunction with other services, switch-based or not, of any type, to form a more complete service package.

Table 7: CCITT Type B service features

Type B services can be invoked at any point during the call. These services may be invoked on behalf of and directly impact one or more users. Feature interaction and arbitration, and topology manipulation are capabilities that need to be addressed to deploy these services. Note that it is possible to use type A capabilities to enhance some existing type B services.

The services addressed by CS1 fall under type A services. The type A category lead to a series of advantages in the context of CS1 standardization. First, they represent a wide range of services of proven value. Second, these services depend on well-understood control relationships between network components and this represents an achievable target within required time frame of IN CS1 product deployment in 1993. Finally, complexity in the transition to rapid service delivery process is minimized both for service provider and for the equipment manufacturer.

3.3.2 Network Requirements

Overall network requirements of IN are stated in [Q1201] as follows:

- it should be possible to move cost-effectively from existing network bases to target network bases in a practical and flexible manner,
- it should be possible to reduce redundancies among network functions in physical entities,
- it should be possible to allow for the flexible allocation of network functions to physical entities,
- there is a need for communication protocols that allow flexibility in the allocation of functions,
- it should be possible to create new services from network functions in a cost and time efficient manner,
- it should be possible to quarantee the integrity of the network when new service is being introduced and
- it should be possible to manage network elements and network resources such that quality of service and network performance can be quaranteed.

Network requirements for service creation refer to the network capabilities that are necessary from a network operator point of view for the creation of new supplementary services. The service creation process consists of specification, development and verification steps.

Network requirements for service management refer to the network capabilities that are necessary from a network operator point of view to support the proper operation of services.

Network requirements for service processing refer to the network capabilities that are necessary for the provision, from a network operator point of view, of basic and supplementary services by an IN-structured network. The main network requirements for service processing stem from the inability of network operators of traditional "non-IN" networks to rapidly create and deploy new supplementary services. To overcome this inability the IN aims for:

- rapid service implementations by means of reusable network functions;
- modularization of network functions;
- standardized communication between network functions via service independent interfaces.

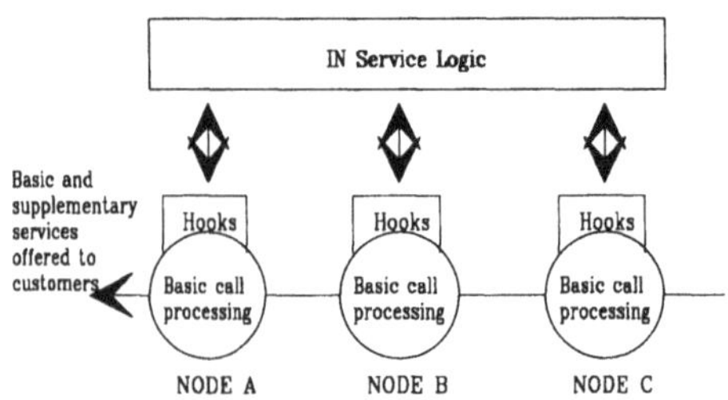

Figure 7. IN Service Processing Model.

To achieve the goal of fast service implementation, the IN Service Processing Model is introduced (Figure 7).

The three main elements of this model are: the basic call processes, the "hooks" that allow the basic call processes to interact with IN service logic, and IN service logic that can be "programmed" to implement new supplementary services. For these elements the main principles are described below:

- The basic call process should be available all over the network and is designed to support, with optimal performance, services that do not require special features. In order to achieve flexibility in service processing, the basic call process needs to be modularized into service-independent sub-processes such that these can be executed autonomously (without interference from the outside during execution).
- "Hooks" are to be added to the basic call process forming the links between the individual basic call sub-processes and the service logic. The "hooks" are able to start an interaction session with the IN service logic. For this it should continuously check the basic call process for the occurrence of conditions on which an interaction session with IN service logic should be started. During an interaction session the basic call process can be temporarily suspended.
- IN service logic uses a programmable software environment that needs to be developed to allow fast implementation of new supplementary services. New supplementary services can be created by means of "programs" containing IN service logic. The IN service logic is able, via the "hooks" functionality, to interact with the basic call process. In this way IN service logic can control the sub-processes in the basic call process and the sequencing of these sub-processes.

Thus, by changing logic at the service control point and modifying network data, a new service that uses existing network capabilities can readily be implemented.

In addition IN service logic can decide to terminate an interaction session with the basic call process. The basic call process will then resume its execution as specified by the IN service logic. In order to allow fast service implementation, the IN service logic should have a logical view of the network resources that constitute the basic call process and additional (specialised) network functions. For proper service processing, the following principles apply:

- it should be possible to distribute resources between services in a well balanced way;
- it should be possible for IN supported services to share resources with non-IN supported services;
- it should be possible to provide a different method of resource data management from the current embedded method;
- it should be possible to introduce IN supported services specific resources.

To define an IN architecture including the network elements within this architecture, there is a need for a **call model** that describes the real-time behaviour of call control capabilities for the provision of basic and supplementary services. In order to be consistent with the principles of the above-described IN service processing model, the IN call model should cover the following aspects:

- it should specify which basic services can be supported by the model;
- it should model the basic call processes (each individual basic service may require its own IN basic call process);

- it should describe **trigger mechanisms** ("hooks") that allow the IN basic call process to interact with service logic;
- it should provide a logical view (from the service logic point of view) of call processing functions and network resources, which as a consequence allows fast service implementation;
- it should specify the mechanisms according to which an IN-basic call process may interact with the service logic (e.g. single-ended interactions, simultaneous interactions, service logic initiated interactions, etc.);
- it should be evolvable from the existing technology base.

3.4 IN Conceptual Model

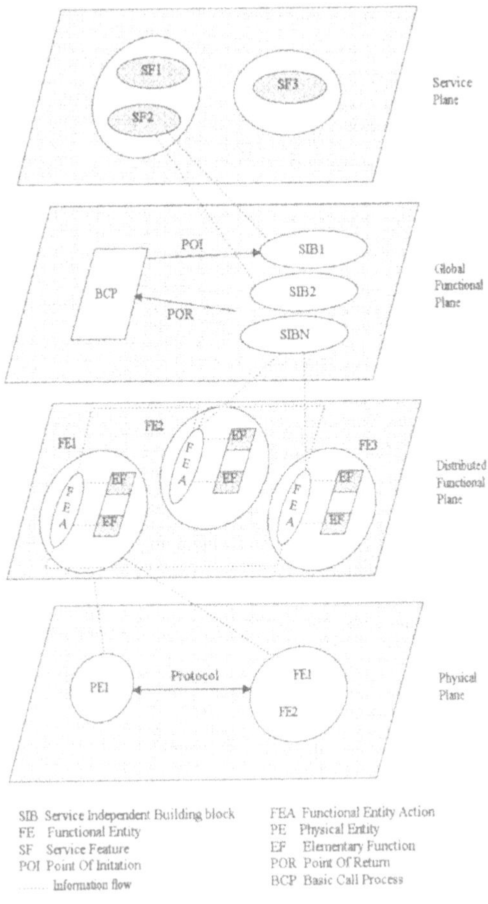

Figure 8. IN Conceptual Model

The IN Conceptual Model *(INCM)* is defined in the CCITT Recommendation Q.1201. The conceptual model is divided into four planes and it forms the basis for the standardization work. The IN conceptual Model was designed to serve as a modelling tool for the Intelligent Network. It is also a tool that can be used to design the IN architecture to meet the following main objectives [Q1201]:

- service implementation independence,
- network implementation independence and
- vendor and technology independence.

Each INCM plane represents a different abstract view of the capabilities provided by an IN-structured network. These views address service aspects, global functionality, distributes functionality and physical aspects of an IN (Figure 8).

The **Service Plane** represents an exclusively service-oriented view. This view contains no information whatsoever regarding the implementation of the services in the network, e.g. an "IN-type" implementation is not visible. All that is perceived is the network's service-related behaviour as seen, for example, by a service user. Services are composed of one or more Service Features (SFs), which are the "lowest level" of services.

The **Global Functional Plane** (GFP) models an IN-structured network as a single entity. Contained in this view is a global (network-wide) basic call processing *(BCP)* SIB, the service independent building blocks *(SIBs)*, and point of initiation *(POI)* and point of return *(POR)* between the BCP and a chain of SIBs.

The **Distributed Functional Plane** (DFP) models a distributed view of an IN-structured network. Each functional entity *(FE)* may perform a variety of functional entire actions *(FEAs)*. Any given FEA may be performed within different functional entities. However, a given FEA may not be distributed across functional entities.

Within each functional entity, various FEAs may be performed by onc or more elementary functions. The manner in which elementary functions result in FEAs is for further study.

Service-independent building blocks (SIBs) are realised in the distributed functional plane (DFP) by a sequence of particular FESs performed in the functional entities. Some of these FEAs result in information flows between functional entities. The information flows consist of messages which exhance information between functional entities. The messages comply with OSI structures and principles.

The **Physical Plane** models the physical aspects of IN-structured networks. The model identifies the different physical entities and protocols that may exist in real IN-structured networks. It also indicates which functional entities are implemented in which physical entities.

The entities contained in adjacent planes of the INCM are related to each other. The nature of the relationship is as follows (Q1201):

- Service plane to GF plane: Service features within the service plane are realised in the GF plane by a combination of global service logic and SIBs including the basic call process SIBs. This mapping is related to the service creation process.
- GF plane to distributed functional (DF) plane: Each SIB identified in the GF plane must be present in at least one FE in the DF plane. A SIB may be realised in more than one FE. Thus, cooperation of several FEs may be needed. The service logic in the GF plane maps onto one or more DSLs in the DF plane. This mapping is related to the service creation process.

- DF plane to physical plane: FEs identified in the DF plane determine the behaviour of the physical entities (PEs) onto which they are mapped. Each FE must be mapped onto one physical entity, but, each PE contains one or more FEs. Relationships between FEs, identified in the DF plane, are specified as protocols in the physical plane. DSLs may be dynamically loaded into physical entities and this mapping is related to the service management process.

3.5 Distributed Functional Plane

The global Distributed Functional Plane (*DFP*) is of primary interest to network designers and providers. It describes the functional architecture of an IN-structured network in terms of units of network functionality (Figure 9). These functionalities are referred to as Functional Entities (*FE*). The information that flows between Functional Entities are referred to as relationships. The functional entities are described independently of how the functionality is physically implemented or deployed in the network. SIB's on the global functional plane are realized on the Distributed Functional Plane by a sequence of Functional Entity Actions (*FEA*) and resulting information flows. [Q1214]

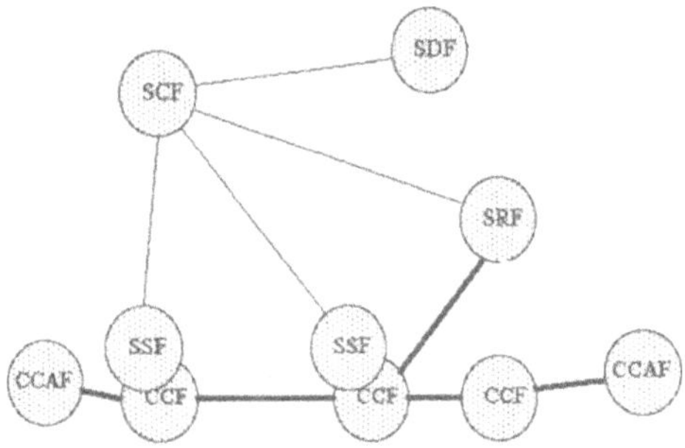

CCAF Call Control Agent Function
CCF Call Control Function
SSF Service Switching Function
SCF Service Control Function
SRF Special Resource Function
SDF Service Data Function

Figure 9. Distributed functional plane architecture.

The DFP architecture provides flexibility to support a large variety of services and facilitates the evolution of IN by organizing the functional capabilities in an open-ended and modular strtucture to achieve service independence. The DFP architecture is vendor/implementation independent, thereby providing the flexibility for multiple physical networking configuration and placing no constraints on national network architecture beyond the network and interface standards which will be developed for IN structured networks. The definition of the DFP architecture initially accomodates service execution capabilities and will accomodate service creation and service and network management capabilities when they become available.

A Functional Entity is a unique group of functions in a single location and a subset of the total set of functions required to provide a service. One or more Functional Entities can be located in the same Physical Entity. Different Functional Entities contain different functions, and may also contain one or more of the same functions. In addition, one Functional Entity cannot be split between two Physical Entities; the Functional Entity is mapped entirely within a single Physical Entity. Finally, duplicate instances of a FE can be mapped to different PEs, though not the same PE.

3.5.1 Definition of FEs

This section gives a description of the Functional Entities at the Distributed Functional Plane related to IN service execution and how they are mapped to the Physical Plane architecture.

CCAF

The CCAF is the Call Control Agent Function that provides access for users. It is the interface between user and network call control functions. It has the following characteristics: It

a) provides for user access, interacting with the user to establish, maintain, modify and release, as required, a call or instance of service;
b) accesses the service-providing capabilities of the Call Control Function, using service requests (e.g. setup, transfer, hold, etc.) for the establishment, manipulation and release of a call or instance of service;
c) receives indications relating to the call or service from the CCF and relays them to the user as required and
d) maintains call/service state information as perceived by this functional entity.

CCF

The CCF is the Call Control Function in the network that provides call/connection processing and control. It

a) establishes, manipulates and releases call/connection instances as "requested" by the CCAF;
b) provides the capability to associate and relate CCAF functional entities that are involved in a particular call and/or connection instance (that may be on SSF requests);
c) manages the relationship between CCAF functional entities involved in a call (e.g. supervises the overall perspective of the call and/or connection instance);
d) provides trigger mechanism to access IN functionality (e.g. passes events to the SSF) and
e) is managed, updated and/or otherwise administred for its IN-related functions (i.e. trigger mechanisms) by a Service Management Function.

SSF

The SSF is the Service Switching Function, which, associated with the CCF, provides the set of functions required for interaction between the CCF and Service Control Function. It

a) extends the logic of the CCF to include recognition of service control triggers and to interact with the SCF;
b) manages signalling between the CCF and the SCF;
c) modifies call/connection processing functions (in the CCF) as required to process requests for IN provided service usage under the control of the SCF and
d) is managed, updated and/or otherwise administred by an SMF.

SSF/CCF Model

The SSF/CCF model described below include the Basic Call Manager *(BCM)*, the IN-Switching Manager *(IN-SM)*, the Feature Interactions Manager *(FIM)*/Call Manager *(CM)*, the relationship of the BCM to the IN-SM, the relationship of the BCM and IN-SM to the FIM/CM, and the functional separation provided in the SSF/CCF (Figure 4-7). [Q1214]

a) BCM - The entity in the CCF that provides basic call and connection control to establish communication paths for users and interconnects such communication paths, that detects basic call and connection control events that can lead to the invocation of IN service logic instances or should be reported to active IN service logic instances, and that manages CCF resources required to support basic call and connection control. The BCM interacts with the FIM/CM as described in the FIM/CM description below.

b) IN-SM - The entity in the SSF that interacts with the SCF in the course of providing IN service features to users. It provides the SCF with an observable view of SSF/CCF call/connection processing activities, and provides the SCF with access to SSF/CCF capabilities and resources. It also detects IN call/connection processing events that should be reported to active IN service logic instances, and manages SSF resources required to support IN service logic instances. The IN-SM interacts with the FIM/CM as described below.

c) FIM/CM - The entity in the SSF that provides mechanisms to support multiple concurrent instances of IN service logic instances on a single call. In particular, the FIM/CM can prevent multiple instances of IN an non-IN service logic instances from being invoked. The ability of the FIM/CM to arbitrate between multiple instances of IN and non-IN service logic instances is for further study. The FIM/CM integrates these interactions mechanisms with the BCM and IN-FM to provide the SSF with a unified view of call/service processing internal to the SSF for a single call.

d) BCM Relationship to IN-SM - The relationship that encompasses the interaction between the BCM and the IN-SM, through the FIM/CM. The information flow related to this interaction is not externally visible and is not standardized for CS-1. However, an understanding of this subject is required to identify how basic call and connection processing and IN call/connection processing may interact.

e) BCM and IN-SM Relationships to FIM/CM - The relationships that encompass the interaction between the BCM and FIM/CM, and the IN-SM and the FIM/CM. The information flows related to these interactions are not externally visible and are not standardized for CS-1. However, an understanding of this subject is required in order to unify the BCM, IN-SM and FIM/CM.

f) Functional Separation in the SSF/CCF. The functional separation of processes and resources in the SSF/CCF that provides a means of handling service logic instance interactions for CS-1. This functional separation services to isolate single-ended service logic instances related to the calling party from single-ended service logic instances related to the called party for the same call. Within the scope of CS-1, there is no functionality in the SSF for handling service feature interactions between the separate SSF calling party processes and SSF called party processes.

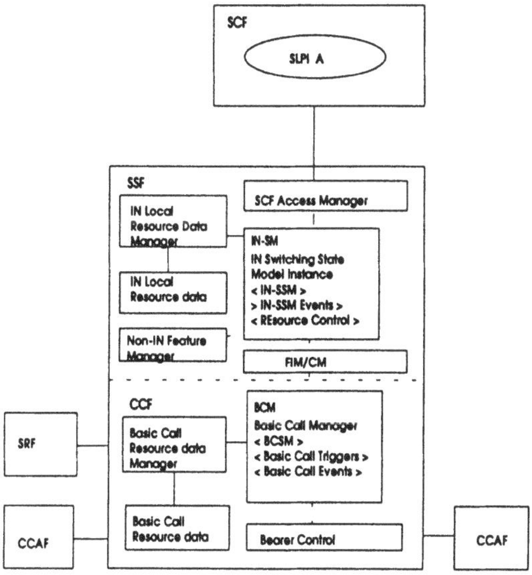

Figure 10. SSF/CCF Model

BCSM

The BCSM (Basic Call State Model) is a high-level finite state machine description of CCF activities required to establish and maintain communication paths for users. As such, it identifies a set of basic call and connection activities in a CCF and shows how these activities are joined together to process a basic call and connection (i.e., establish and maintain a communication path for a user). [Q1214]

Many aspects of the BCSM are not externally visible to IN service logic instances. However, aspects of BCSM will be the subject of standardization. As such, the BCSM is primarily an explanatory tool for providing a representation of CCF activities that can be analysed to determine which aspects of the BCSM will be visible to IN service logic instances, if any, and what level of abstraction and granularity is appropriate for this visibility.

The BCSM identifies points in basic call and connection processing when IN service logic instances are permitted to interact with basic call and connection control capabilities. In particular, it provides a framework for describing basic call and connection events that can lead to the invocation of IN service logic instances or should be reported to active IN service logic instances, for describing those points in call and connection processing at which these events are detected, and for describing those points in call and connection processing when the transfer of control can occur.

Figure 10 shows the key components that have been identified to describe a BCSM, to include: Points in Call *(PICs),* Detection Points *(DPs),* transitions, and events. PICs identify CCF activities required to complete on or more basic call/connection states of interest to IN service logic instances. DPs indicate points in basic call and connection processing at which transfer of control can occur. Transitions indicate the normal flow of basic call/connection processing from one PIC to another. Events cause transitions into and out of PICs. Information Flows [Q1214] (e.g. between SSF/CCF and SCF) corresponding to Events and PICs are represented by Operations [Q1218] and modelled as Application Service Elements (ASEs).

The BCSM for CS-1 should model existing switch processing of basic two-party calls, and should reflect the functional separation between the originating and terminating portions of calls. In addition, though CCAF functionality is not explicitly modelled in the BCSM, a mapping is required between access signalling events and BCSM events, for each access arrangement supported by CS-1.

Since the BSCM is generic, it may describe events that do not apply to certain access arrangements. It is important to understand and describe how each access arrangement applies to the BCSM.

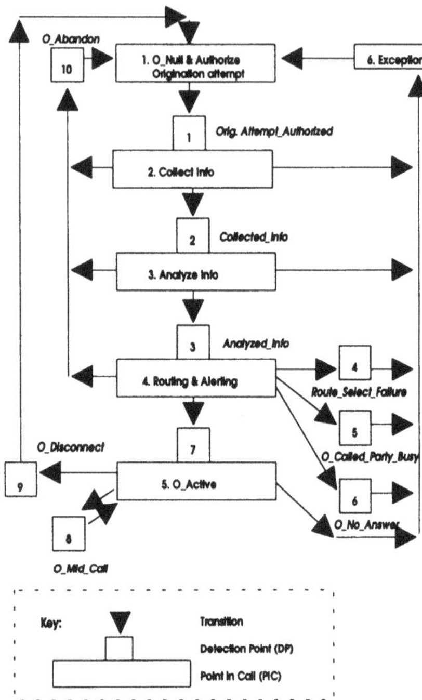

Figure 11. Originating BCSM for CS1

SCF

The SCF is a function that commands call control functions in the processing of IN provided and/or custom service requests. The SCF may interact with other functional entities to access additional logic or obtain information (service or user data) required to process a call/service logic instance. It

a) interfaces and interacts with SSF/CCF, SRF and SDF functional entities;

b) contains the logic and processing capability required to handle IN provided service attempts;

c) interfaces and interacts with other SCFs, if necessary;

d) is managed, updated and/or otherwise administered by an SMF.

SDF

The SDF contains customer and network data for real time access by the SCF in the execution of an IN provided service. It

a) interfaces and interacts with SCF as required;

b) interfaces and interacts with other SDFs, if necessary;

c) is managed, updated and/or otherwise administered by an SMF.

SRF

The SRF provides the specialized resources required for the execution of IN provided services (e.g. digit receivers, announcements, conference bridges, etc.). It

a) interfaces and interacts with SCF and SSF (and with the CCF);

b) is managed, updated and/or otherwise administered by an SMF;

c) may contain the logic and processing capability to receive/send and convert information received from users;

d) may contain functionality similar to the CCF to manage bearer connections to the specialized resources.

3.6. Global Functional Plane

The Global Functional Plane (*GFP*) is of primary interest to service designers. [Wyatt91] The Global Functional Plane models network functionality from a global, or network-wide, point of view. As such, the IN structured network is said to be viewed as a single entity in the GFP. In this plane, services and Service Features are redefined in terms of the broad network functions required to support them. These functions are neither service nor Service Feature specific and are referred to as SIB's (*Service-Independent building Block*). [Q1203]

Services identified in the service plane are decomposed into their service features, then mapped onto one or more SIBs in the GFP. Each SIB is similarly mapped onto one or more FEs in the Distributed Functional Plane [Q1203] (Figure 12).

3.6.1 SIB

IN CS1 contains 14 SIBs that include algorithm, charge, compare, translate, basic call process, among others. In principle many other services described in CCITT Recommendations Q.1211 could be specified. [Raat93] SIBs are standard reusable networkwide capabilities residing in the Global Functional Plane, used to create services. As such they are global in nature and their locations need not to be considered as the entire network is regarded as a single entity. A Service Feature is provided by a combination of one or more SIBs. SIBs have the following characteristics:

- SIBs are defined completely independent from any physical architecture considerations.
- Each SIB has a unified and stable interface, with one or more inputs an one or more outputs.
- SIBs are reusable, monolithic, building blocks, describing a single complete activity, and used by the service designer to create services.

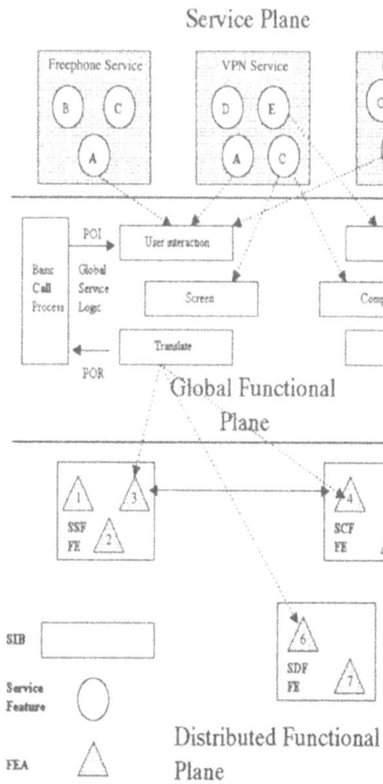

Figure 12. Service decomposition.

A SIB can exist independently, or it can coexist with other SIBs in the same network element. IN-based services can be distinguished from one another by the sequence of SIB functions and by the specific parameters within each SIB. IN CS1 describes 13 SIBs plus a specialized SIB called Basic Call Process (Table 8).

Table 8: The CS1 SIBs

Algorithm	Screen
Charge	Service Data Management
Compare	Status Notification
Distribution	Translate
Limit	User Interaction
Log Call Information	Verify
Queue	

Basic Call Process (*BCP*) identifies the normal call process from which IN services are launched, including Points Of Initiation (*POI*) and Points Of Return (*POR*) which provide the interface from the BCP to Global Service Logic (*GSL*). The GSL describes how SIBs are chained together to describe Service Features. The GSL also describes interaction between the BCP and the SIB chains. [Q1203] (Figure 13) By definition, SIBs, including the BCP, are service independent and cannot contain knowledge of subsequent SIBs. Therefore, GSL is the only element in the GFP which is specifically service dependent

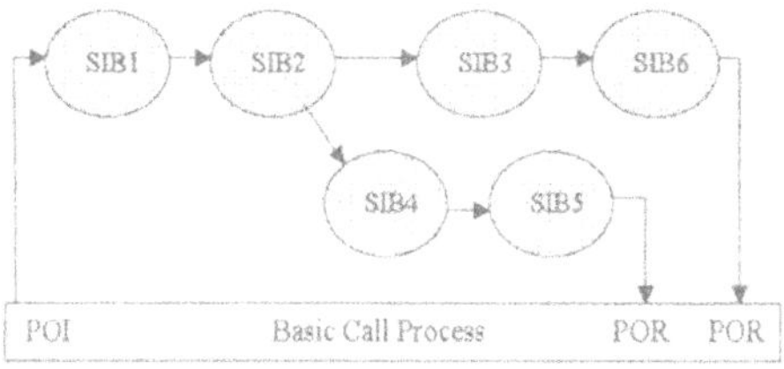

Figure 13. Modelling of Global Functional Plane.

In order to chain SIBs together, knowledge of the connection pattern, decision options, and data required by SIBs must be available. Therefore, the pattern of how SIB are chained together must be maintained within the GFP, and described in the GSL. The GSL describes subsequential SIB chaining, potential branching, and where branches rejoin. When an IN supported service is to be invoked, its GSL is laucnhed at the POI by a triggering mechanism from the BCP. At the end of chain of SIBs, the GSL also describes returning point to the BCP by indicating the specific POR. For a given service or Service Feature at least one POI is required. However, depending upon the logic required to support the service or Service Feature, multiple PORs may be defined. [Q1203]

In order to describe Service Features with these generic SIBs, some elements of service dependency is needed. Service dependency can be described using data parameters which enable a SIB to be tailored to perform the desired functionality. Data parameters are specified independently for each SIB and are made available to the SIB through GSL. Two types of data parameters are required for each SIB, dynamic parameters called Call Instance Data (*CID*) and static parameters called Service Support Data (*SSD*). [Q1203]

3.6.2 Basic Call Process

The Basic Call Process is responsible for providing basic call connectivity between parties in the network. The BCP can be viewed as a specialized SIB which then provides basic call capabilities including connecting call with appropriate disposition; disconnecting calls, with appropriate disposition; and retaining CID for further processing of that call instance [Q1203]

The need for specific POI/POR functionality is that the same chain of SIBs may represent a different service if launched from a different point in the BCP. Similarly, the same chain of SIBs launched from the same point may represent a different service if returned to the BCP at a different point. [Q1203]

3.6.3 Global Service Logic

The Global Service Logic can be defined as the "glue" that defines the order in which SIBs will be chained together to accomplish services. Each instance of global service logic is (potentially) unique to each individual call, but uses common elements, comprising specifically: BCP interaction point (POI and POR); SIBs; logical connections between SIBs, and between SIBs and BCP interaction points; input and output data parameters, service support data and call instance data defined for each SIB. [Q1203] The GSL will then chain together these elements (SIBs) to provide a specific service.

3.7. The IN-structured network

The IN concept is an extension of, rather than a replacement for, traditional service control. Since an IN primarily affects only the internal service processing of switching systems, it should have little influence on the signalling procedures of a traditional network. Therefore, we can place intelligent nodes in existing networks without affectting traditional network operations or capabilities. [Wyatt91]

The Intelligent Network consists of integrated hardware and software distributed throughout the service providers' network. Thanks to the new technologies, service providers will be able to create their own services. [Nerys91] Compared to the convenient telecommunications network architecture, IN forms an excellent and fast way of introducing services.

IN promises to change the way vendors, telephone companies, and customers run their businesses and work with one another. Today, vendors develop a product that delivers a certain service, then sell it to telecommunications operators. With IN, vendors will develop software "building blocks" [Nerys91], then deliver these to telephone companies who assemble them to create new services.

3.7.1. SCE

The Service Creation Environment capability of IN enables effective service creation. Service Creation Environments enable network and service providers to create new revenue-generating services that are independent of equipment vendor's deployment schedules. Many administrations are asking vendors of IN equipment to provide them with Service Creation Environment capabilities. This is also true of large service subscribers, who prefer to control the operation of their IN-based services. In the current Service Creation Environment, service subscribers can control services using existing capabilities or modifying parameters within these capabilities. Current Service Creation Environments are user friendly and support updates of service control points. The next generation of Service Creation Environment will also support updates of intelligent peripherals and Adjuncts. Because SIBs are being defined for the IN, it is now possible to develop a Service Creation Environment platform to support new services and direct them to appropriate Physical Entities. In addition, new SCEs must provide extensive validations for new IN-based services so they do not have an adverse effect on the overall operation of the network or the subscribers services. [Wyatt91]

The service designers are staff members of the provider's company. They have to create new services by definite and unambiguous descriptions. Such descriptions are called Service Logic Programs (*SLP*). After deployment of a new service in the network, one can buy or subscribe to such a service. [Abram92]

The services are determined by single Service Features. Following the ETSI framework this should be reflected in the service representation: each SLP should be composed from SIBs. The interface for composition of new services may differ. The interface might be an advanced specification language for the construction of SIBs and their interfaces/(inputs and outputs). However, it is possible to build a Graphical User Interface (*GUI*) on the top of the specification language and by so ease and speed up the introduction of new IN services.

3.7.2. IN Application Protocol

The IN Application Protocol (*INAP*) is intended to be used between the following four functions: SSF, SCF, SDF and SRF. The INAP in CS1 is ment to be using the SS7 protocol stack, but it does not imply that only this signalling protocol should be used. [Q1218]

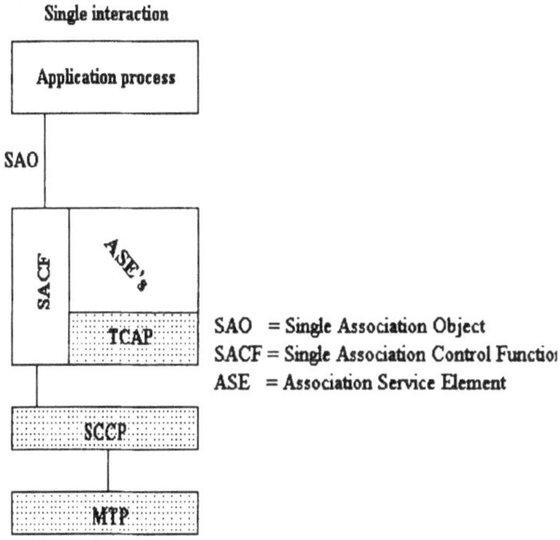

Figure 14. INAP Protocol Architecture.

The INAP protocol architecture is based on the OSI Application Layer Structure (Figure 14). A physical entity has either single interactions or multiple co-ordinated (not discussed here) interactions with other physical entities. The Single Asociation Control Function provides a co-ordination function using Application Service Elements *(ASEs)*, which includes the ordering of operations supported by ASEs (based on the order of received primitives) [Q1218]. The SAO represent the SACF plus a set of ASE's to be used over a single interaction between a pair of Physical Entities. If there were need for multiple interactions, the use of MACF (*Multiple Association Control Function*) would be acceptable. In this case, MACF would provide a co-ordinating function among several SAOs, each of which interacts with an SAO in a remote PE.

Each ASE supports one or more operations. Information flows of [Q1214] are in principle mapped one to one with operations. For example, the operations corresponding to the information flows of the Originating BCSM for CS1 (Figure 11) are the following:

- Origination Attempt Authorized
- Collect Information
- Collected Information
- Analyze Information
- Analyzed Information
- Route Select Failure
- OCalled party Busy

- O_No Answer
- ODisconnect
- OAnswer
- O_Mid Call

In the CCITT New Recommendation Q.1218 the INAP and TCAP messages are specified using the Abstract Syntax Notation One (*ASN.1*). The encoding rules which are applicable to the defined abstract syntax are the Basic Encoding Rules (*BER*).

3.8. Integration of TMN and IN

IN is a generic, service-oriented architecture, intended to be used for all kinds of services (real-time or management) on top of call-control type services. TMN is a generic, management-oriented architecture, intended to be used for all kinds of management services. Obviously, the IN and TMN architectures overlap. For instance, one TMN application such as billing and one IN application such as Freephone must be tightly related because Freephone billing should be handled in a consistent way with TMN billing. This shows that, unless both IN and TMN architectures are made more consistent, the interconnection of IN and TMN applications would be difficult. It is not possible to support two independent architectures while applications on both architectures must interoperate. Also, IN is just one part of the whole network, and as such should be managed with TMN. The integration of TMN and IN can be considered as an evolution path to TINA [Appel93].

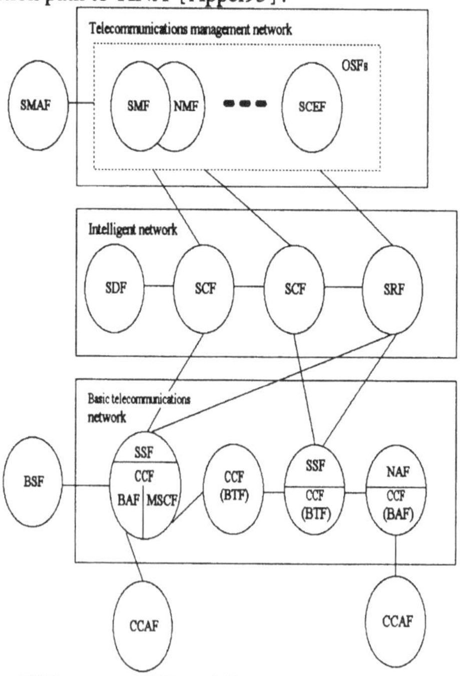

Figure 15. The TMN and IN concept. [Wyatt91]

Figure 15 shows network-related functions required for IN architecture: the Basic telecommunications network, Intelligent Network, and the Telecommunications Management Network.

The Basic telecommunications network is commonly known as the Public-Switched Telephone Network (*PSTN*), this network controls basic telecommunications services (for example, local and transit/toll switching, voice and data calls) offered to a user. It detects whether control of a call should be transferred to the IN. The Intelligent Network manages intelligent telecommunications services offered to a user. It includes specialized telecommunications functions, such as customized announcements, voice regognition, encryption, and network reource assignments. At present, TMN controls telecommunications support for basic telecommunications network and IN functions. In the future, TMN will include functions such as service creation, service provisioning, service deployment, and service management.

Both in TMN and IN, the challenge is to ensure a global consistency of all interconnected applications, while allowing for evolution of some applications. This shows that while IN and TMN architecture are to be integrated, they both must evolve towards a unified target architecture to be more flexible. [Appel93]

3.8.1. Comparison of IN planes to TMN planes

The IN Conceptual Model represents different points of view to the users, customers and operators. The TMN planes describe, however, different management-related aspects. The correspondence of these architectures is shown in this section.

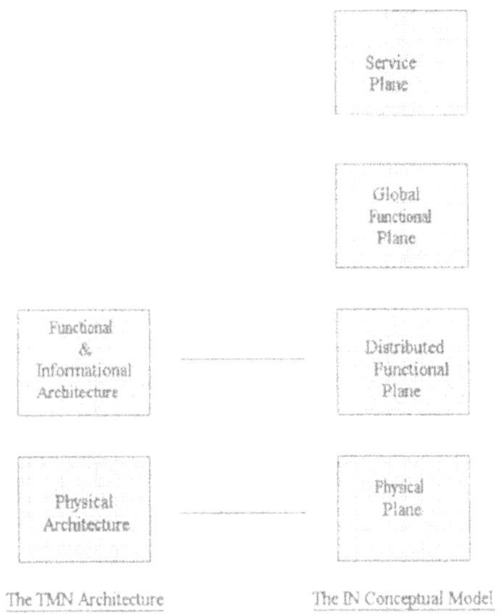

Figure 16. Correspondence of IN planes and TMN architecture planes.

The Service Plane represents the service from the user's point of view. The TMN architecture does not directly provide with this kind of aspects. The Global Functional Plane represents with the service designer's point of view of the services. The TMN architecture does not directly provide with aspects of Global Functional Plane. Distributed Functional Plane represents the fucntional parts of the IN architecture and the relations between them. This is quite the same as the TMN architectures Functional Architecture. The relations between DFP parts corresponds to the TMN Informational Architecture. The lowest layer of IN architecture corresponds straight to the Physical Plane architecture of INCM (Figure 16).

In order to avoid multiple definitions of management it is possible that IN will be managed through TMN concept. This is very well stated, because TMN has been widely accepted as a telecommunications management concept.

3.9. Future IN Capability Sets

The main CS1 capabilities support flexible routing, flexible charging and flexible user interaction [Q1211]. Only limited mid-call interruption facilities are supported. It is not expected that significant capability will be provided within CS1 for services occurring during the active phase of call, for multiparty or multimedia services, for services requiring the direct manipulation of call topology such as mobility or conference calling or non call associated signalling as needed in mobility. Such capabilities, as well as standards for SMF and SCEF capabilities, are expected to be provided in CSs beyond CS1, starting with CS2, on which work began in 1992. Refinements of CS1 will continue during 1994. The CS2 with non-call associated signalling, SDF and management interfaces will be available in 1995. CS3 providing terminal mobility is to be completed in 1997.

Thus, the work beyond CS1 will provide support of mobility, multimedia calling; support of services affecting a call in the active phases where several subscribers may be affected (Type B Services); standards for feature interaction mechanisms; standards for creation, deployment, and management of service logic; and support for complex call topology management. However, it seems to be so that CS2 will continue to address only Type A services.

REFERENCES

[Abram92] Abramowski, St., et al., A Service Creation Environment for Intelligent Networks, Philips Research Laboratiories Aachen, 1992

[Appel93] Appeldorn, Menso, Kung, Roberto, et al., TMN + IN = TINA, IEEE Communications Magazine, April, 1993

[Garra93] Garrahan, James, Russo, Peter, et al., Intelligent Network Overview, IEEE Communications Magazine, March, 1993

[Homa92] Homa, Jonathan, Intelligent Network Requirements for Personal Communications Services, IEEE Communications Magazine, Vol 1., 2, 1992

[Lauta93] Lautanala, Kari, Veijalainen, Kari, Intelligent Network Architecture and Services with the DX200 Switching System, Workshop proceedings: Workshop on Intelligent Networks, Lappeenranta University of Technology, 1993

[Lehti93] Lehtinen, Pekka, Performance and Overload Modelling of SCP and SSPS of an IN, Workshop proceedings: Workshop on Intelligent Networks, Lappeenranta University of Technology, 1993

[Modar90] Modaressi, Abdi, Skoog, Ronald, Signalling System No: 7: A Tutorial, IEEE Communications Magazine, July, 1990

[Nerys91] Nerys, C., Operations Systems For Intelligent Networks, AT&T Technology, vol. 6, no. 2, 1991

[Q1201] CCITT Recommendation I.312/Q.1201, Principles of Intelligent Network Architecture, CCITT, 1992

[Q1203] CCITT Recommendation Q.1203: Intelligent Network Global Functional Plane architecture, October, 1992

[Q1204] CCITT Recommendation Q.1204: Intelligent Network Distributed Functional Plane architecture, COM XI-R 208-E, April, 1992

[Q1211] CCITT Recommendation Q.1211: Introduction to Intelligent Network Capability Set 1, COM XI-R 210-E, April, 1992

[Q1214] CCITT Recommendation Q.1214 &5: Distributed Functional Plane for Intelligent Network CS1, COM XI-R 213-E, April, 1992

[Q1218] CCITT Recommendation Q.1218: Interface Recommendation for Intelligent Network CS1, COM XI-R 217-E, April, 1992

[Raat93] Raatikainen, Kimmo, A Framework for Evaluating the Performance of IN Services, Workshop proceedings: Workshop on Intelligent Networks, Lappeenranta University of Technology, 1993

[Roger90] Rogerson, David, The Intelligent Network: Market Strategies, Ovum Ltd., 1990

[Wyatt91] Wyatt, George, Barshefsky, Alvin, et al., The Evolution of Global Intelligent Network Architecture, AT&T Techical journal, Summer 1991

2

Intelligent networks for personal communications

Hans Bisseling, Ericsson Telecommunicatie BV
etmjobi@crosby.ericsson.se ,
Jos den Hartog, Ericsson Telecommunicatie BV
etm.etmjdha@memo.ericsson.se ,
Ericsson Telecommunicatie BV
IN Application Laboratory
PO Box 8, 5120 AA Rijen, The Netherlands
Tel: +31 1612 29911, Fax: +31 1612 29699

Abstract

Service mobility and the personalisation of services are main concerns of a modern telecommunication environment. In addition, co-operation among services and a simple presentation form to the end-user are of extraordinary importance as well. To fulfil these requirements, a user defined environment called Personal Services Communication Space (PSCS) is under development by the RACE II project Mobilise (R2003). This concept has to take into account all involved players and their specific requirements. The main features of PSCS are: personal mobility based on UPT, personalisation to have personal working environments for end-users, and interoperability to have effective interworking between different services offered on heterogeneous networks. The PSCS Conceptual Framework is primarily based on the Intelligent Network Conceptual Model (INCM) with extensions taken from Open Distributed Processing (ODP).

During the development process of PSCS, difficulties were encountered in composing PSCS services and services features in an implementation independent way by Service Independent Building Blocks (SIBs) of Capability Set 1 (CS-1). As a result important enhancements are proposed:

The introduction of recursion and parallelism into the concept of SIBs to allow IN service engineering to be more object-oriented.

To break down service features into a chain of SIBs that are executed sequentially is difficult, and will become much more difficult for future complex features. To handle these difficulties more effectively, recursion is introduced which enables grouping of SIBs into bigger SIBs, i.e.,

SIBs can be nested. In addition, SIBs are allowed to trigger new service logic which can be executed in parallel.

The introduction of domains in the GFP.
The IN network being regarded as a single entity neglects the different stakeholders involved in IN services. Different network operators and service providers are already visible in for instance mobility services such as UPT. originating, terminating, and home domains can be related to different stakeholders. Services may also cover more than one domain in a sense that a management service is partly implemented in an IN network, and partly in a TMN network.

Even end-users and subscribers will have their own domain in the future. This idea is reflected by the notion of the PSCS Flexible Service Profile (FSP), in which end-users have their own personalised service logic.

1. INTRODUCTION

The RACE II project Mobilise (R2003) is a four year project (1992 - 1995) and its objective is to define a concept for personal communication [Mobilise D4]. Service mobility and the personalisation of service conditions are one of the main concerns of a modern telecommunication environment. In addition, smooth co-operation among diverse services and a simple presentation form to the end-user are of extraordinary importance as well. To fulfil these main requirements in new telecommunication systems, a user defined environment called Personal Services Communication Space (PSCS), is envisaged to be developed. The main features of the PSCS are:

Personal mobility
Personal mobility means that an end-user can use any network access point and any terminal while being identified through the same number (identity) and charged to the end-user's personal account. PSCS is considered to be an extension of UPT [ETSI NA7], which offers a personal mobility service.

Personalisation
The PSCS concept for personalisation is that end-users have personal working environments, that can easily be managed by subscribers and configured by end-users. Subscribers can control the service delivery to their end-users, and define limitations on the service usage, depending on the situation the end-user is in at a certain moment. End-users are then allowed to configure their personal environments within these limits.

Interoperability
Interoperability describes the capability of the system to support effective interworking between different services, offered on heterogeneous networks.

The PSCS Conceptual Framework [Mobilise D4] is primarily based on the Intelligent Network Conceptual Model (INCM) [ITU-T Q.1200] with extensions taken from Open

Distributed Processing (ODP) [ISO ODP]. This paper discusses enhancements related to the IN's Global Functional Plane (GFP). [ITU-T Q.1203]

2. ACTORS AND ROLES IN PERSONAL COMMUNICATIONS

The PSCS conceptual design starts at an *enterprise modelling* stage which has similarity to ODP's enterprise view. Enterprise modelling is needed for PSCS in order to put the service features into the right context, to identify on-line and off-line (contractual) relationships and to relate the services to the domain interfaces.

A key feature of Mobilise is to approach the PSCS from a user's perspective rather than from a technological viewpoint, to ensure that final results will be readily acceptable. However, it is recognised that any set of requirements must consider technological constraints. While the conceptual work on PSCS is user-driven, consideration is also given to how it can be implemented.

Requirements are placed on a framework for personal communication from all the different stakeholders involved in the deployment and operation of PSCS: End-User, Subscriber, PSCS Service Provider, Application Service Provider, Network Provider and Access Provider. These PSCS stakeholders are characterised by their roles and the mutual relationships between them, see Figure 1.

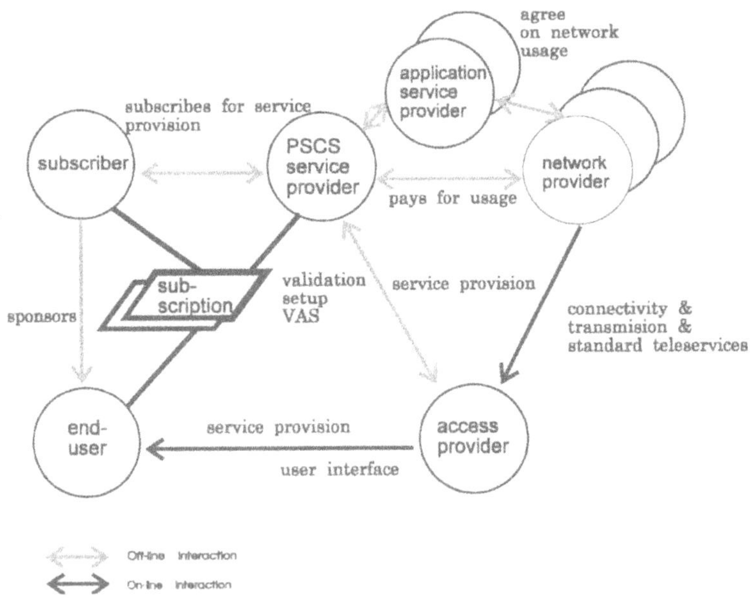

Figure 1. The PSCS Enterprise Model.

It is important to understand that this model relates entities in their business roles and not corporations. The same corporation (or parts thereof) might act in different roles in this model, simultaneously or at different times. This distinction is often referred to as a distinction between *actors and roles*, whereby a certain role can be played by one or several actors; and whereby a certain actor might play one or several roles simultaneously. Therefore, Figure 1 shows the mutual relationships between the different roles that are relevant for PSCS.

Every entity in this model is linked via a contractual relationship to each other. A central role plays the PSCS subscription which links the PSCS service provider with both the subscriber and the end-user. Some of the relationships materialise during actual service use into physical interconnections. The increasing demand for flexibility pushes increasing parts of the relationships into the on-line state. Therefore business relationships for service provisioning have to be taken into account carefully in service design.

End-Users define their personal service environment using different roles and services, adapted to their personal needs. To be allowed to use services, the end-users have to agree with the subscriber on the particular rights which will be associated with them. These are given in a form of a *PSCS subscription.*

The *Subscriber* subscribes to and pays for services offered by the service provider, or offered by different service providers. The subscriber is generally an organisation with an interest in maintaining a balance between quality and economy of service on behalf of the real end-users of the services, e.g. its employees.

It is possible that the end-user and the subscriber are the same person; this would characterise most of the current residential market for PSCS. There the need has been identified to relate the cost of communication more clearly to the usage of the particular person as a family in fact is a small organisation.

On the other hand, the end-user could be an employee of the subscriber, which would be a corporate organisation. The middle ground is characterised by a many-to-many relationship between end-user and subscriber, such as a consultant working for several clients, possibly even within the same corporation. This relationship is modelled by the provision of various subscriptions linked to the same end-user.

The *Network Provider* provides the infrastructure for the interconnection of several sites with specified basic delivery conditions. He is providing standardised services, so-called 'network services', such as basic telephony and data communication.

The *Service Provider* is generally an organisation that commercially manages services offered to subscribers. He is responsible for agreements with network providers on network usage, and with access providers on usage of access facilities.

In PSCS, two classes of service providers are identified:

- The *PSCS service provider* offering a PSCS framework service which is a platform for other services enhancing and unifying network services.

- The *application service provider* offering one or many (tele-) communication services on top of the PSCS service platform, e.g. personalised information services, message service, directory service, etc.

The *Access Providers* ensures that the end-users can get access, through the provision of appropriate terminals or access networks, to the public network in order to be able to use the services.

3. ENHANCED SERVICE ENGINEERING WITH INTELLIGENT NETWORKS

To break down IN services and service features into a set of service components is difficult to do; therefore, the concepts to identify IN service components need to be powerful to ease the process of *service engineering*. Especially for PSCS type of services this service engineering is very difficult when it is based on Global Functional Plane (GFP) concepts for Capability Set 1 (CS-1) [ITU-T Q.1213]. Regarding CS-1 the following problems were identified:

1. The CS1 SIB definition is not recursive and CS1 SIBs are too low-level.

   The (de-) compositioning process to implement services and service features based on CS1 SIBs is difficult due to the one step mapping of services features to SIBs. It is not possible to decompose services and their features into high level functions or processes and refine these later on.

2. The execution of a chain of SIBs is only sequential, whereas parallel execution is needed as well.

   At a very low level one can regard the execution of a service as a sequence of consecutive actions, but on higher levels there is a need to distinguish from these fine-grained sequential activities and rather talk about parallel interworking activities or processes.

3. The introduction of non-call related service execution implies the need for a well defined process definition.

   In CS1, IN services are only call related in which the BCP triggers GSL. In CS2 non-call related service execution is introduced for mobility services, for instance for location updating. To be able to model this in the GFP a well defined process definition is needed.

4. No domains exists in the GFP: the IN network is regarded as a single entity.

   The IN network being regarded as a single entity neglects the different stakeholders involved in IN services. Different network operators and service providers are already visible in for instance the DFP regarding mobility services such as UPT: originating, terminating, and home domains can be related to different stakeholders. And even end-users and subscribers will have their own domain in future, in which they have their own unique service logic which is adapted to individual needs.

To be able to map effectively IN service features onto the Global Functional Plane, composition and decomposition techniques, sequential execution but also parallelism are necessary. To be able to perform composition (bottom-up) and decomposition (top-down), high level SIBs are introduced. And service processes are introduced to allow for parallel execution. But within each service process, the service logic is executed sequentially. To have a clear understanding of these concepts, the notions of *service processes*, *high level SIBs (HLSIBs)* and of *SIBs* were identified. SIBs are considered as the smallest service component, they are not further refined in the GFP. HLSIBs, however, are SIBs which are composed out of other (HL)SIBs. And service processes, that can be executed in parallel, encapsulate a chain of (HL)SIBs that are executed sequentially. Furthermore, by using service processes, HLSIBs and SIBs for GFP modelling gives also good opportunities to further decompose/refine SIBs and service processes at a later stage.

These solutions identified by Mobilise have resulted in concrete proposals to ITU-T SG11 (Melbourne, 1 -10 March 1994) and ETSI NA6 (Vienna, 25-29 April 1994), who have adopted these concepts to enhance the GFP concepts for CS-2 [ITU-T Q.1203, ITU-T Q.1290].

3.1. Modularity

One of the basics of object orientation is modularity [Meyer, OO]. Whereby the level of modularity depends on criteria such as modular decomposability, composability, understandability, continuity and protection. In addition, principles such as explicit interfaces and information hiding are to be observed to ensure proper modularity. In IN the basic notion of modularity is the Service Independent Building Block (SIB). This section shows that this modularity can be applied recursively by introducing the notion of High Level SIB (HLSIB).

HLSIBs, that are as normal SIBs executed sequentially, support abstraction mechanisms as composition and decomposition. By composition, SIBs can be defined out of smaller SIBs, forming a HLSIB. On the other hand, the decomposition technique provides for a top-down refinement of a SIB (i.e. a HLSIB), allowing to partition the granularity of a HLSIB in smaller building blocks that can be reused, see Figure 2. A HLSIB stands for abstraction by hiding service logic and parts of the Service Support Data (SSD) that is considered to be local to the HLSIB. Call Instance Data (CID) is considered to have a more global nature. HLSIBs have the following additional characteristics:

- HLSIBs can be composed out of other HLSIBs and SIBs only.
- The lowest level of HLSIBs contains SIBs only, i.e., no further detail is visible on the GFP.
- One of the (HL)SIBs within a HLSIB is the first to be executed; therefore, HLSIBs have only one entry point (logical start), the same as with normal SIBs. But, as with normal SIBs as well, HLSIBs can have one or more exit points (logical ends).

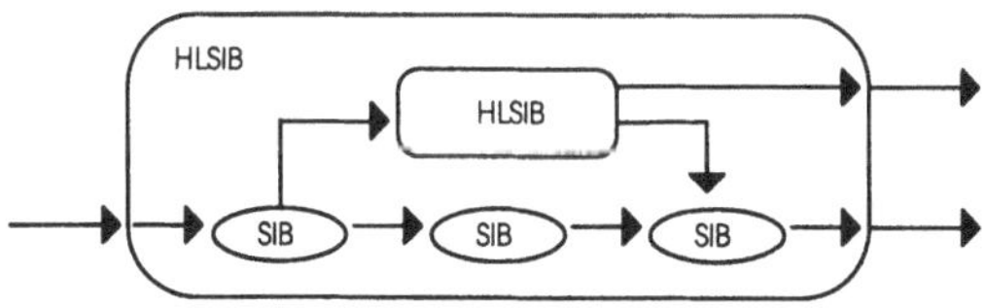

Figure 2. Recursive Concept of High Level SIBs.

So, by using HLSIBs the initially one-step functional decomposition of services into SIBs, interconnected via Global Service Logic (GSL), becomes much more flexible and more object-oriented.

3.2. Communicating Sequential Service Processes

For many current and especially future IN services parallelism is needed. It is even inevitable when parallelism is inherent to the service. And it is often wanted to improve Quality of Service (QoS). To allow parallel processing of activities in a service, a new type of service component is needed which is called service process. This means that a service may now comprise of more than one service processes (each of them containing a chain of SIBs) that are executed in parallel at a given time. This parallel execution is illustrated in Figure 3.

The spawning of a new service process is achieved by a Point Of Initiation (POI) and synchronisation between parallel service processes can be achieved via Points Of Synchronisation (POSs). Therefore, interprocess communication capabilities are needed: a Spawn SIB to spawn new service processes and Send and Wait SIBs for synchronisation purposes and data exchange of CID.

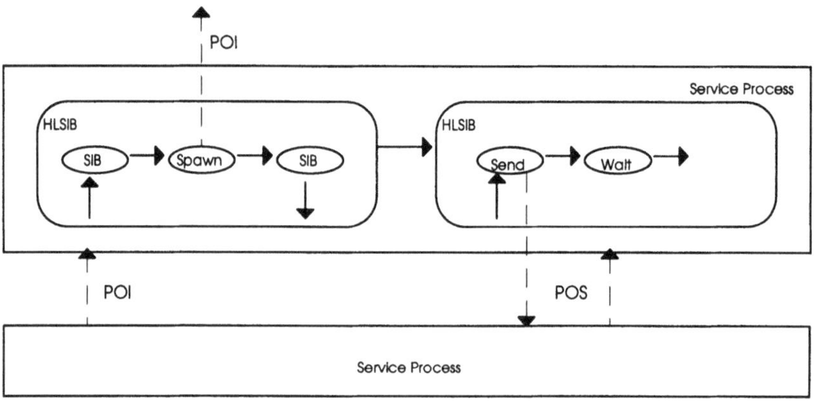

Figure 3. Communicating Sequential Service Processes.

A service process represents a chain (service logic) of SIBs or HLSIBs which are executed sequentially, but in parallel with other service processes. Service processes have the following characteristics:

- Service processes can be composed out of HLSIBs and SIBs, that are executed sequentially.
- One of the (HL)SIBs within a service process is the first to be executed; therefore, service processes have only one entry point initiated via a POI.
- Synchronisation between service processes can be performed by POSs.
- Service processes need mechanisms to send, receive and process POIs and POSs and to use the attached data.
- Call Instance Data (CID) is considered to be local to a service process, but global within that service process. Data exchange is performed explicitly via POIs and POSs.

- The BCP can be regarded as a specialised service process.

Synchronisation between two service processes can be achieved via Points Of Synchronisation (POSs). A POS is a functional interface between service logic of two service processes over which asynchronous communication is initiated. This means that a particular SIB in the sending service process has capabilities to send a POS to another service process that is executed in parallel. After the POS has been send this sending service process can continue its execution. The receiving service process, however, has to wait until the POS has arrived. This means that the execution of a particular SIB in the receiving service process has to be suspended until the POS has been received. If the receiving service process is not yet suspended at the time the POS has been received, the service process must buffer the POS. Full synchronous communication can be achieved by both SIBs of the two service processes that are performing a handshake with two POSs.

Service processes consists of the actual service logic between Start and End SIBs, see Figure 4. But they also include a set of HLSIBs used within the service processes. In Figure 4 for example, HLSIBs x,y,z are included in service process A. And within a service process these HLSIBs can be reused several times, for instance HLSIB y is used within the service logic of the service process itself, within HLSIB x and z.

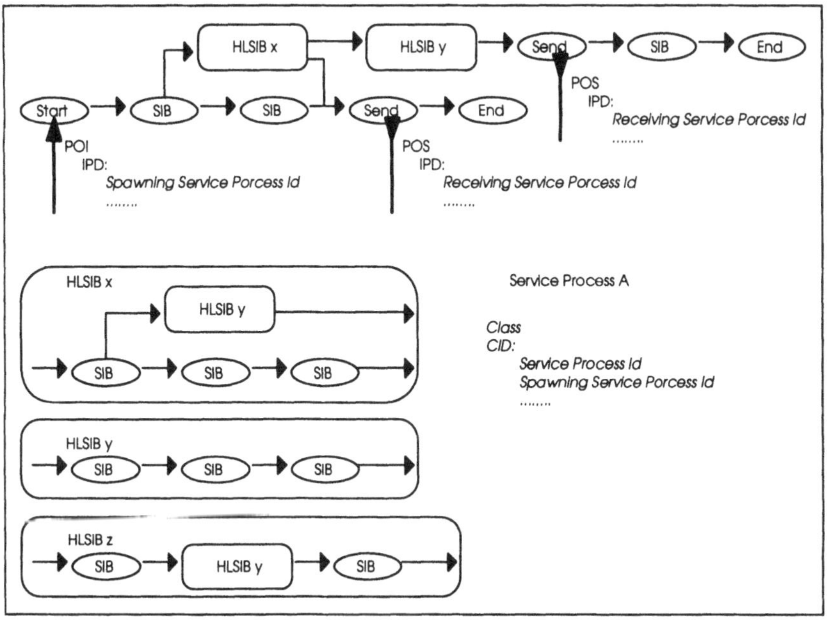

Figure 4. Modelling Service Processes.

Besides the asynchronous way of communication between service processes, it might be useful to have synchronous communication as well. Think of Remote Procedure Calls (RPCs) in which a remote procedure (different service process, or even different domain) is called and the calling side is waiting for the result. This could be modelled by HLSIBs. In Figure 4 for instance, HLSIB z is not used by the service process itself, but it offers other service processes to invoke service logic of that service process via an RPC.

The service process of Figure 4 contains two Send SIBs that might be able to send two different POSs to another service process. For instance, this service process might send a POS with certain data attached in normal cases or a specific POS with error information in abnormal cases. But the receiving side does not know beforehand which synchronisation event to expect. Therefore it should be able to expect both. A way to handle this problem is by allowing for multiple threads. The next figure shows an example of *multiple threads* in which two different Wait SIB are interconnected via the continue outlet. Both Wait SIBs will wait for a certain POS at the same time, but only one outlet (one thread) will be executing when the expected POS arrives.

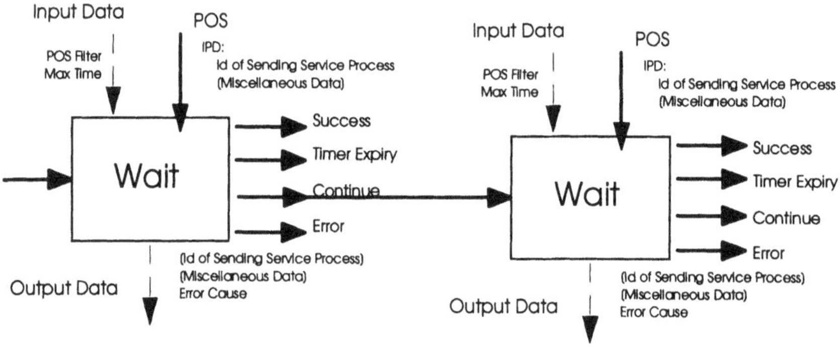

Figure 5. Multiple Threads within a Service Process.

3.3. Domains

Domains, which encapsulate predefined roles in the IN structured network are modelled by service processes that communicate over domain boundaries. Therefore, the boundaries between different domains identify logical interfaces. Within such a domain, the IN structured network is regarded as a single entity.

The domains are visible on the GFP and this visibility is illustrated in Figure 6. This figure shows as an example how a service process of the service provider (e.g. a UPT service provider) can be put on top of the service processes of the network provider (e.g. offering basic IN to the UPT service provider). POIs and POSs are used to communicate over domain boundaries and are also used explicitly to exchange data between service processes and therefore between domains. So, the set of POIs and POSs between two domains define exactly their logical interface. Furthermore, to have full control within a certain domain, service processes may not exceed domain boundaries.

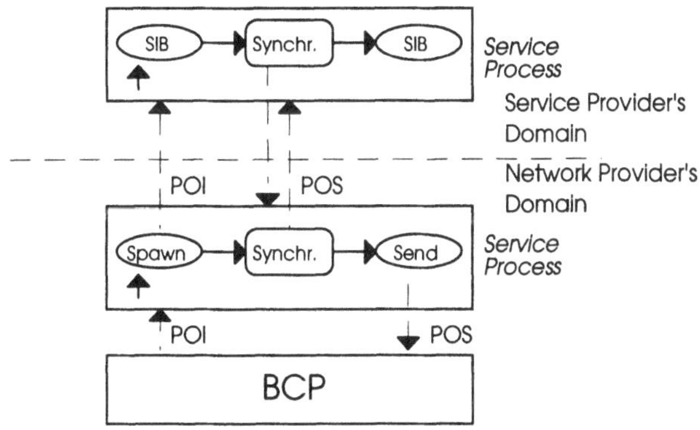

Figure 6. Domains on GFP.

What we have obtained now is not new, in fact we have modelled logical interfaces between domains which are called computational interfaces in ODP.

4. APPLICABILITY TO PERSONAL COMMUNICATIONS

In PSCS each stakeholder as shown in Figure 1 has his own domain in the GFP. Which means that also End-Users and Subscribers have their own domains. This is also recognised in services proposed for IN CS-2, such as Customized Call Routing (CCR) in which the subscriber's domain is queried during an incoming call to get further instructions (call processing and routing information) before attempting to complete the call. In PSCS such a service is called routing schemes [Mobilise D12].

In this section we will show how such a service can be implemented by using the presented concepts for modularity, communicating sequential processes, and domains. To do so we will use a scenario in which the subscriber's domain and end-user's domain is queried for further instructions. The result is shown in Figure 7.

For each of the involved stakeholders requirements concerning the processing of incoming PSCS calls are to be dealt with:

The PSCS service provider

The PSCS service provider handles all incoming PSCS calls but queries the subscriber's domain for further instructions. It provides the subscriber's domain with information such as A-number, B-number. As result a C-number to forward the call to is expected. If no response is received in time, the query fails and a notification message has to be send.

The subscriber

The subscriber's domain distributes incoming calls according to the time the call has arrived. Incoming calls will be forwarded to a mailbox, a secretary or to the end-user himself. When the end-user's domain is queried, it is provided with the A-number, B-number. As result a C-number to forward the call to is expected, this destination is checked by the subscriber as the subscriber has to pay for the service. If no response is received in time, the query fails and the call is forwarded to the help desk.

The end-user

The end-user checks whether the caller is actually on his VIP list. If so, the call will be forwarded to his current location. Otherwise the call might be rerouted (time dependent distribution) to a mailbox.

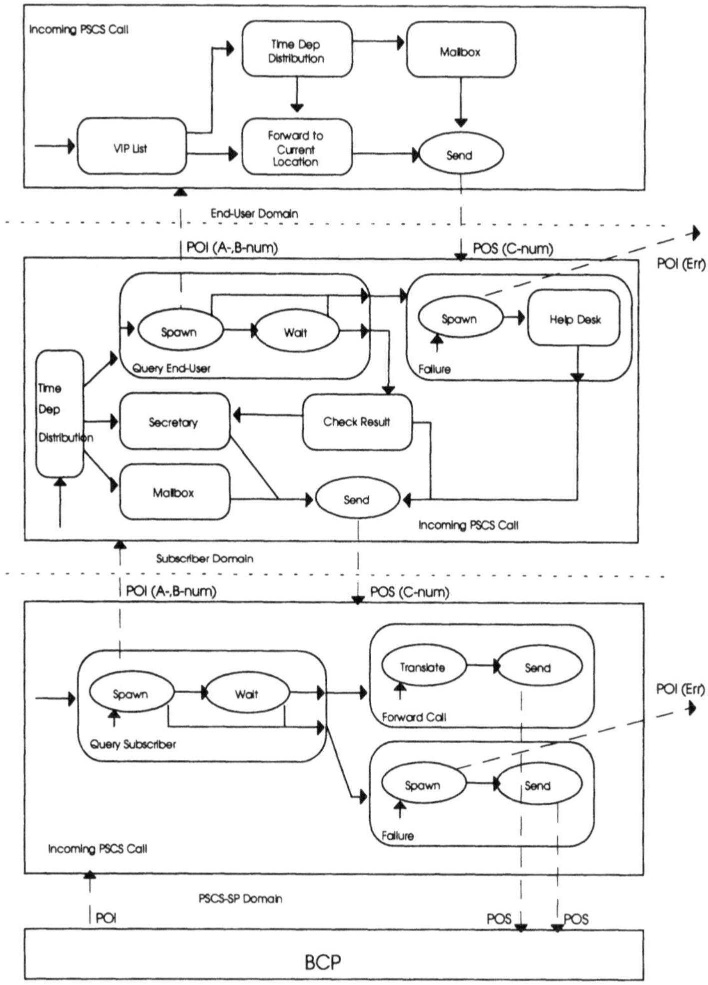

Figure 7. End-users and subscribers in control of their own domains.

This scenario is realised with IN service components such as service processes, high level SIBs and SIBs regarding the PSCS service provider's domain, the subscriber's domain, and the end-user's domain. There are several ways in which this scenario can be mapped onto the DFP. For example, all domains could be mapped onto one SCF executing all the service logic. Or, the domains could be distributed over several SCFs and even a private SCF owned by a subscriber. And, as the logical interfaces between service processes and therefore between domains are well defined, parts of the service can easily be supported by non-IN technology such as intelligent terminals (e.g. PDAs). Such a device can simply be regarding as a black box, it needs certain input (e.g. A- and B-number) and generates certain output (e.g. a C-number).

5. CONCLUSION

This paper presents enhancements of the CS-1 concepts for Global Functional Plane to realise personalised service for subscribers and end-users. The two main enhancements are:

- The introduction of concepts for recursion (HLSIBs) and parallelism (service processes) to allow IN service engineering to be more object-oriented.
- The introduction of domains in the GFP.

SIBs are considered as the smallest service component, they are not further refined in the GFP. HLSIBs, however, are SIBs which are composed out of other (HL)SIBs. And service processes, executed in parallel, encapsulate a chain of (HL)SIBs that are executed sequentially. Furthermore, by using service processes, HLSIBs and SIBs for GFP modelling gives also good opportunities to further decompose/refine SIBs and service processes at a later stage.

The notion of domains is needed because in PSCS each stakeholder has in fact his own domain, including end-users and subscribers. This will enable personal communications adapted to personal needs in which IN and non-IN architectures work together.

6. REFERENCES

[ITU-T Q.1200] International Telecommunication Union, Standardization Sector, "Q.1200: Q-Series Intelligent Network Recommendation", ITU-T, Study Group XI, March 1993.

[ITU-T Q.1203] International Telecommunication Union, Standardization Sector, "Q.1203: Intelligent Network - Global Functional Plane Architecture" Q-Series Intelligent Network Recommendation", ITU-T, Study Group XI, October 1993.

[ITU-T Q.1213] International Telecommunication Union, Standardization Sector, "Q.1213: Global Functional Plane for Intelligent Networks CS-1", ITU-T, Study Group XI, March 1993.

[ITU-T Q.1290] International Telecommunication Union, Standardization Sector, "Q.1290: Glossary of Terms Used in the Definition of Intelligent Networks", ITU-T, Study Group XI, March 1993.

[ETSI NA7] ETSI NA7, Technical Report ETR NA-70201, "Universal Personal Telecommunication: General Service Description", July 1992.

[ISO ODP] ISO/IEC JTC 1/SC 21/N 7053, "Working Draft - Basic Reference Model of Open Distributed Processing - Part 1: Overview and Guide to Use".

[Meyer OO] Bertrand Meyer, "Object-oriented Software Construction", Prentice Hall International, 1988.

[Mobilise D4] RACE Mobilise (R2003), Fourth Deliverable of the Mobilise Consortium, "PSCS Concept: Definition and CFS, Draft Version", April 1993.

[Mobilise D12] RACE Mobilise (R2003), Twelfth Deliverable of the Mobilise Consortium, "PSCS Specification and CFS: Architectural Framework - Draft Version", December 1993.

3

Service creation from IN to mobile and broadband

Caroline Knight
Hewlett-Packard Laboratories
Filton Road
Stoke Gifford
Bristol BS12 6QZ
UK
TEL:+44-272-228040
FAX:+44-272-228972
email:cdfk@hpl.hp.com

1. SERVICE CREATION NOW AND FUTURE

This paper describes the functionality of the research version of HP's Service Creation Environment (SCE). Some of the design decisions and reasons behind them are given.

Section 2 describes the current HP lab's SCE looking particularly at supporting the call model, providing user customisation, aiding programming abstraction and enabling the programming of multiple network elements.

Section 3 extends the model to include mobile services and section 4 discusses broadband services.

Section 5 goes beyond the programming of services in two areas; the user's interface to services and at a different type of service based on multi-media.

Section 6 concludes the paper and the references and glossary are in sections 7 and 8.

This paper does not go into the underlying transport technology although there is an assumption of an IN-like programmability, nor does it address billing or feature interaction. The architecture we are using is based on Bellcore's AIN model, not CS.1.

2. SCE DEVELOPMENT

We have placed service creation within a lifecycle model. It is important to see service creation in context, it is not an end in itself but comes after a phase of definition, which might include marketing and major customers inputs, and is followed by the deployment, monitoring and use of the service. Throughout the lifetime of the service there will be modifications and then finally phasing out, possibly with something else as a replacement.

Our approach has been to build various tools to support services throughout the lifecycle to edit, test, monitor and improve services.

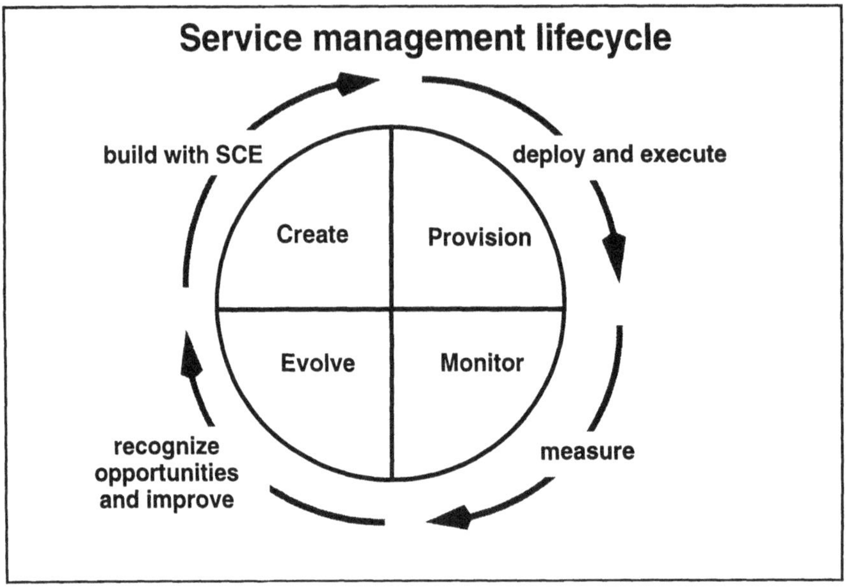

Figure 1.

It is useful to note where the major constraints to service creation come from, namely:

- Interfaces with switches and other network elements
- Customer interfaces (telephones)
- Interfaces with billing
- Security restrictions
- Legal and regulatory restrictions

This paper concentrates on the first two of these, though one aspect of providing a restricted, more secure, interface for end-user customisation is addressed.

The work we have done has been based on a standard AIN architecture [Bellcore `90]. During the initial design and testing of a service, the switch is simulated and a cut-down version of the service management system (SMS) is used to provision the service on the service control point (SCP), software telephones (xphones on the diagram) provide number pads, on and off-hook etc. but there is no real voice path. This all runs on a single workstation. Once the service is ready a real switch can be connected and the service tried out and tested further. When we extended this architecture to include a voice-based intelligent peripheral (IP) we used real voice paths from the start as it is much more convincing to test such a service live, though again the logic of the program can be tested separately first.

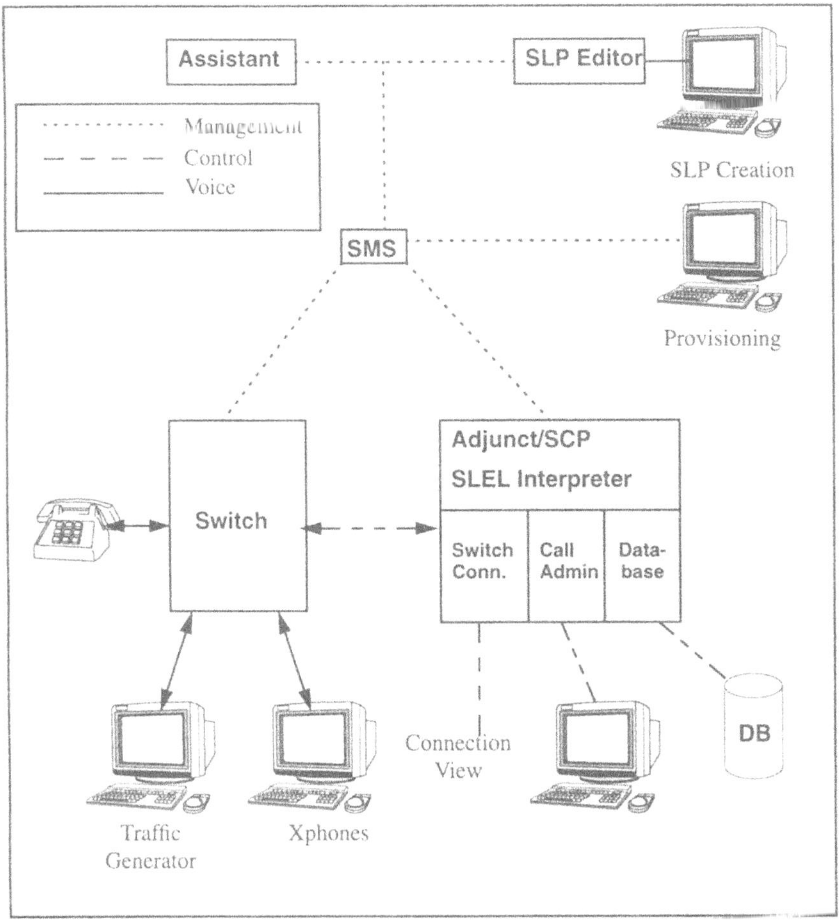

Figure 2.

These are the components:

- The Assistant is described in the section on supporting the call model.
- The SLP editor provides three different levels of programming abstraction: textual programming language, a graphical SDL-based view and service independent building blocks (SIBs). See Wray 94.
- The SMS is the service management system, services are down loaded onto the SCP via this and the appropriate initial trigger checkpoints (TCPs) set on the switch enabling the service for a specified line.

- The switch can be any switch for which a call model is definable. Phones, either real or simulated are connected to the switch.

The SCP executes the service logic language (SLEL), it also provides a connection view of the network which is used when testing a service to reflect the current state of the connection segments and legs.

2.1 Supporting the call model

The key to IN is the call model. We have an explicit representation of the call model in a tool called the assistant. Using this tool the service logic programmer has a guide to which events and trigger checkpoints are available and which must be handled. The assistant can also translate a service between two different call models (if the new call model is rich enough to support the service). Even if the programmer does not use the assistant to guide or translate, it is useful as an active picture of the call model.

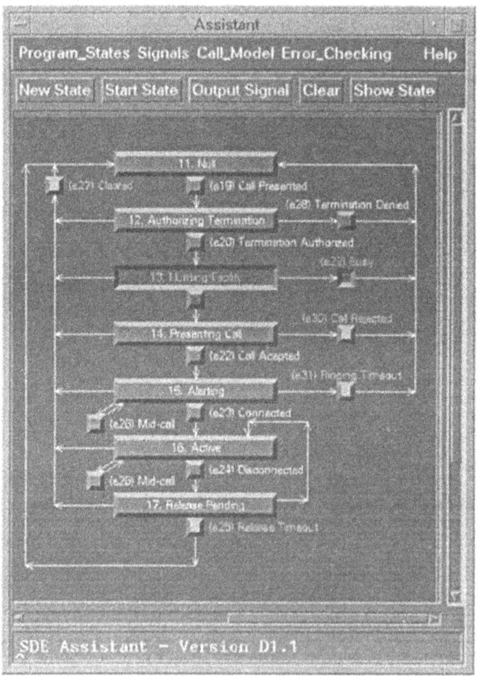

Figure 3.

When the user clicks on an event of interest, the appropriate trigger checkpoints are highlighted.

2.2 Providing user customisation

An important concept in service creation is that of on-site, real time customisation, using customer premises equipment (CPE). This is something which can make a service more valuable to the user, but carrier and service providers will be keen to ensure that customisation is a secure activity. Therefore to simplify the task and to ensure the restrictions necessary we have a tool which allows simple default call processing rules to be created. In user studies these were found to be a more effective representation than the more common decision tree when modelling policy information. The rules are in fact compiled into decision trees but they provide a neatly condensed view of the information needed:

		Attempt	Call fail reason	Day	Time	Try...
✓	Busy 1	1	Busy	Mon-Fri	08.00-17.00	1111
✓	NA 1	1	NoAnswer	Mon-Fri	08.00-17.00	2222
✓	2nd try	2	-	Mon-Fri	08.00-17.00	3333
✓	Default	-	-	-	-	4444

Figure 4

This example shows a hunting policy where during office hours the first attempt at redirection depends on the reason for failure so far:

- If the line is busy then the call is forwarded to 1111
- If on the other hand the call hasn't been answered after a specified number of rings then it is forwarded to 2222.
- No matter which of these were used, if they fail then the call is forwarded to 3333.
- Out of hours and at weekends the last rule is used and the call forwarded to the office answer phone on 4444. This is also the last resort if 3333 fails to answer.
- Notes that the "-" is a wildcard and therefore the least specific value that can be given. The rules are applied so that the most specific, relevant rule is applied first.

It is simple for the user to update days, times, target numbers as well as adding new rules whilst all these choices are restricted to only these variables and to ranges of acceptable values for these variables. The user states the general rule then enumerates the exceptions separately.

2.3 Aiding abstraction - SIBs

The graphical layout of SDL provides an extremely good way of describing and showing the control flow of a service. However, it has some major limitations:

- A real service is hard to see all at once on the screen. Good zooming in and out reduces this problem, but at some level the sheer size needs to be reduced to aid understanding.

- The only way to re-use bits of code is by copying and pasting. This can lead to errors and makes updating complicated; it also fails to provide a clear indication to the viewer that the same actions are being used in more than one place.
- Another consequence is that one is less likely to make generalisations that help in the understanding of the structure. Such generalisations require some form of input variables and some way of returning results.

The textual solution to this is a subroutine or function, the graphical solution is a service independent building block (SIB).

Therefore the SIB is an important aid to writing and reading services. Our SIBs take arguments, and return results.

Figure 5.

This SIB is "Gather Digits". It takes two arguments: a prompt which is played, e.g. "Please type in your personal identification number now" and number which indicates the number of digits required. The prompt is optional and the programmer may remove it.

One difference from a standard programming subroutine is that when programming graphically it is natural to have multiple exit points.

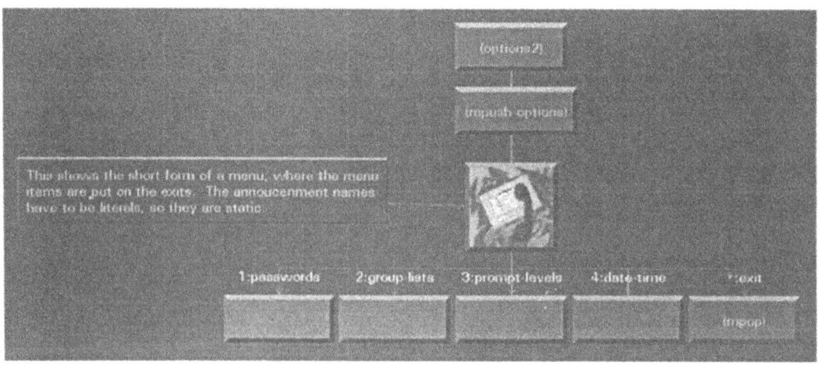

Figure 6.

This menu SIB has five normal exit points which correspond to the choices available to the caller with the associated results.

2.4 Multiple network elements

We have extended the IN architecture to add in a voice-based intelligent peripheral (IP). This is an example of having multiple programmable elements in the network which must be orchestrated to provide services. The IP is added to the architecture thus:

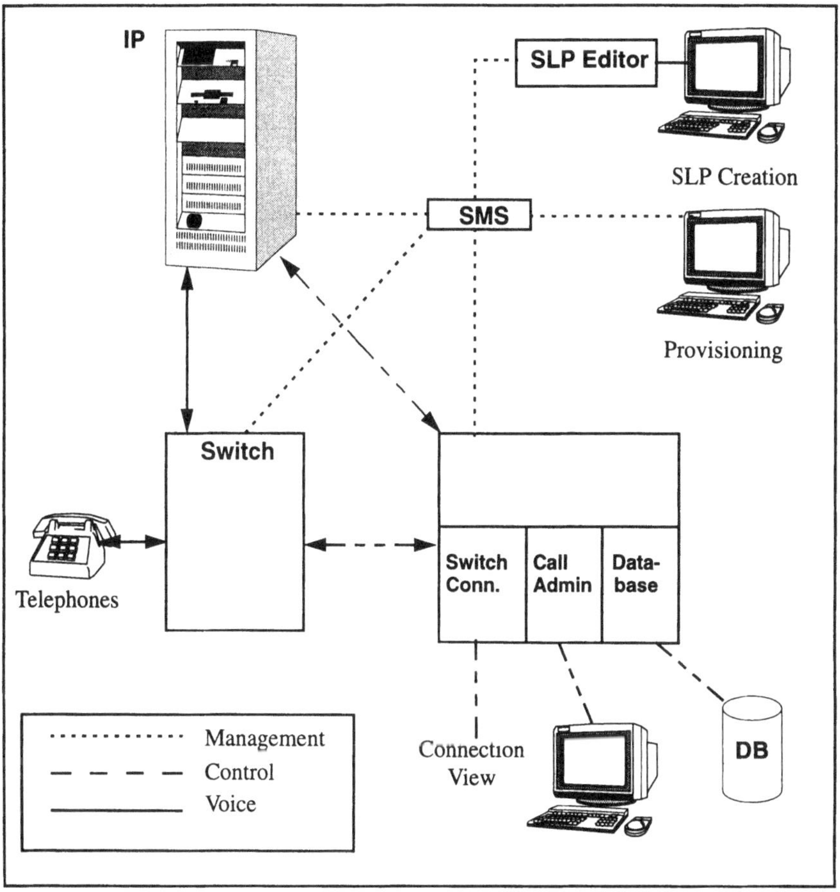

Figure 7.

The SMS provides a management path to all components and the SCP has a control path to both the switch and the IP. The switch treats the IP as any other element connected via voice paths and it may be physically remote via trunk lines.

Using SIBs we packaged up the interface to the IP. Here are the SIBs for "Play Announcement", "Gather Digits", "Record voice message" and "Menu".

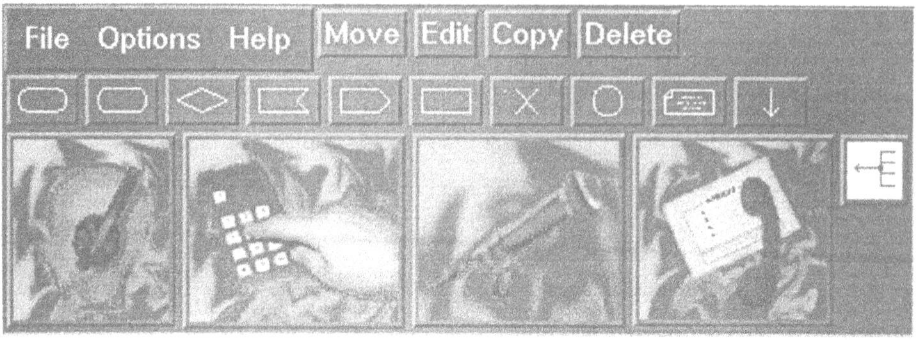

Figure 8.

Using these the service writer can make use of the voice-based IP without deeper knowledge of its functioning. One can use these SIBs to define further SIBs such as "Check authorisation" the IP part of which might look like:

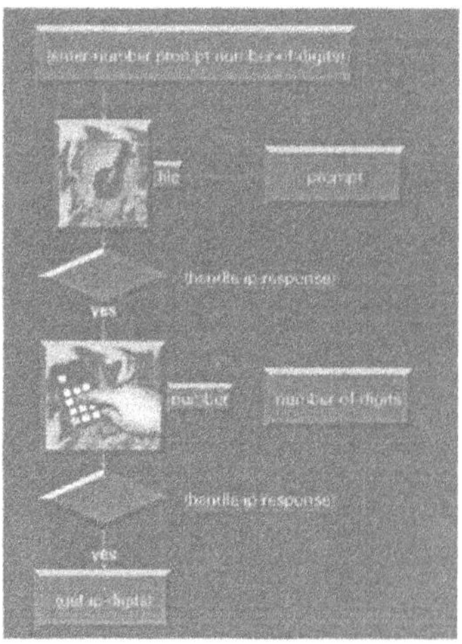

Figure 9.

To complete the SIB one would need to check the number with a secure database of PINs and might put in a loop to allow the user a couple of tries before failing. This and other enhancements such as producing an alert after a specified number of failed attempts from the same user could all be packaged inside the "Check Authentication" SIB. When further enhancements are made these can be automatically made available to all other services using this SIB.

3. EXTENDING TO MOBILE

In mobile the whereabouts of the subscriber is of key importance. Therefore location tracking and authentication for secure communication are needed. For example GSM uses Visitor Location Registers (VLRs), Home Location Registers (HLRs) and Authentication Centres (AuCs) to keep track of the dynamic user.

Consider a few potential mobile services:

- Roadworks information based on location
- Location of stolen car
- Distribution of tasks to mobile workers, e.g. taxis, roadside assistance
- Broadcast alerts relevant to area e.g. flood warning

To enable these services a call model is needed to package up the interactions with the mobile-specific components of the network thus enabling the service writer to concentrate on the logic of the service rather than the specifics of particular databases or protocols. For instance a key event is the change of location which could trigger various services. Other triggers which are common amongst mobile phones are on/off, subscribe/unsubscribe. Strictly speaking these are not elements of a call model but more of a general mobile service model, where calls are only one element of the service. Some parts will more naturally be packaged as SIBs, especially those which require inputs or return results.

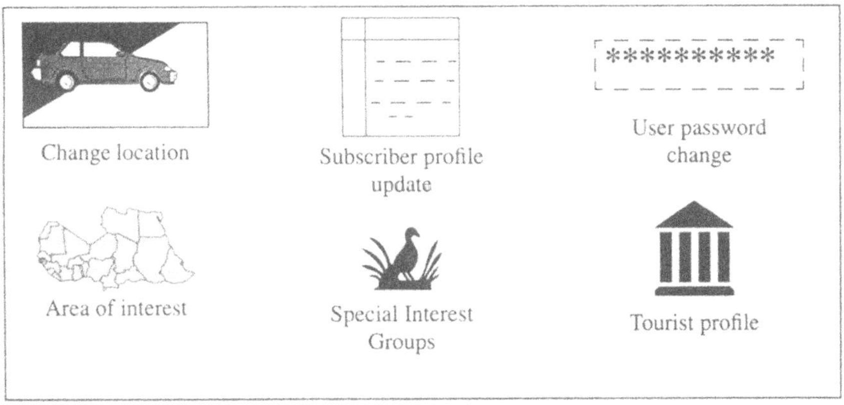

Figure 10.

4. BROADENING THE BAND

Another aspect of telephony that is changing the way people use and think about services is the greater bandwidth becoming available. This leads to potential video-based services. The standard ones which are quoted are:

- Tele-shopping,
- Tele-banking,
- Video on demand and
- Video phones.

Adding in video is a major change. Mobile is voice based, land-based telephony is voice based, the IP previously mentioned is voice based. Video requires changes in the user's equipment to display images. However from the service providers' point of view incorporating video requires many of the same elements as other new network elements; some form of call model defining the course of an interaction and the potential interrupt points and triggers plus the ability to package up reusable elements as SIBs.

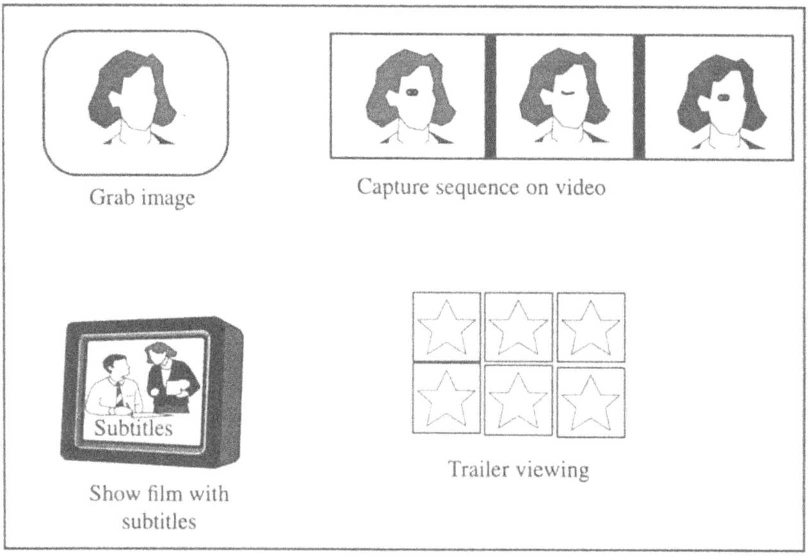

Figure 11.

The user's interface to the service becomes both more flexible and more apparent. Many SIBs could package up frequently used user interactions such as:

- Trailer viewing - this is a way of quickly selecting a particular video, or position in a video. The video might be a film, a magazine, or a music video for instance. The trailer viewer shows stills from key points and when selected shows a burst of video or an edited trailer, the length of time will vary with the purpose.

- Recording and editing the welcome message on your video answer phone - or more generally recording and editing a short video. This needs to be an easy function to perform and not require the complexity of a full editing suite.
- Navigation menus, agents or services to aid the user in finding their way to fun, games, information and friendships.

Most of the current emphasis is on person-to-machine interaction rather than the traditional role of telephony which is person-to-person interaction. Both need to be addressed and they need to be combined to allow, for instance:

- Shopping with friends
- Family banking
- Film clubs
- Education, for instance a group tour of historic places where one also interacts with teachers and other students
- Distributed music making
- Multi-user games

Starting with this goal is important as there are consequences on the hardware which users will need; not only must the device be able to receive video, data and high quality sound but it must also be able transmit all three as well.

5. BEYOND PROGRAMMING

There is more to a service than the writing, debugging and provisioning on switches and other network elements. One of the main reasons for the lack of use of services currently available, especially on business site PBXs, is the user's interface to those services - the phone. Some phones have been created with buttons to support specific services but this is not a viable solution in a world of ever increasing new services.

5.1 Tomorrow's telephone

Smart phones of various descriptions are available, however there is a tendency towards unusable complexity. The difference between today's portable computers and tomorrow's telephones is small; it will be important to draw the line correctly between flexibility and the ease of use of a purpose-built solution. The user and designer will both be faced with choices. The shape of phones now and soon to be:

- A traditional handset, wired into a single location and supporting voice with only the barest level of support for services via the number pad and hook. An LCD number display if you are lucky.
- Next generation handsets with built in softkey screens, typically dependent on a particular switch or PBX e.g. ADSI
- A mobile phone, with limited battery power but flexible location. Supporting voice but likely to have better support for a small set of services through specific buttons. A small LCD screen.

- A settop which connects your television to a video server. It may be smart enough for some shopping services to be implemented. It can use the TV screen for a much more flexible interface. This is not portable.
- A video phone to talk to those with whom voice contact is not enough. With a well designed screen you get the advantage of images and flexible control of services. Maybe it will even be touch sensitive.
- A palmtop computer phone that plugs into an ISDN line and provides access to new services.

One of the benefits of incorporating a screen for even a purely voice service is that many ordinary services become more attractive. Here is an example of a conference call:

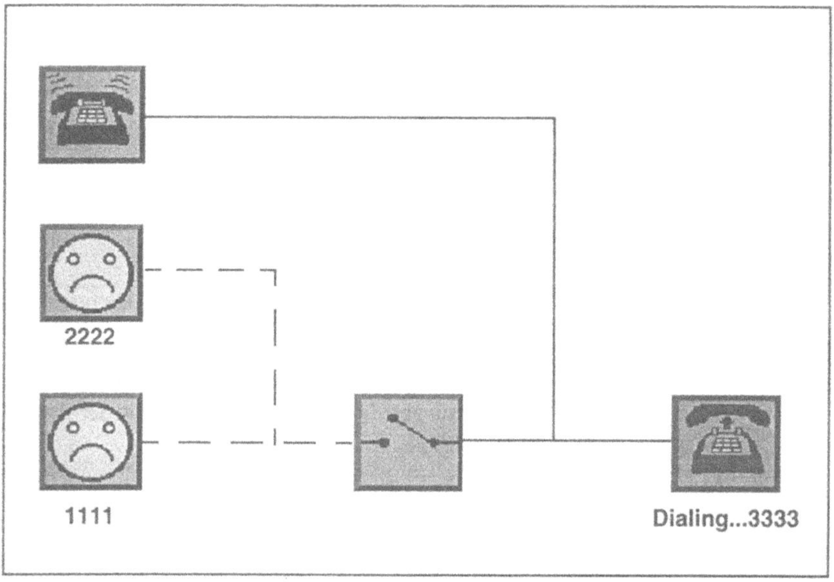

Figure 12.

The initiator has already connected with 1111 and 2222 who are now on hold whilst 3333 is being dialled. The current call models make it difficult to provide all participants with the same view but ideally everyone wants to be able to see who is in the conference and to be kept up-to-date as people leave or join.

5.2 Service development support for multi-media

There are several tools already on the market which support the creation of animation, video, with sound-tracks, interactive stories etc. The creator is more of a director or writer than a programmer though there are many elements in them which are akin to the programming of services. At the high-level where there are only a few well packaged interactions with the mechanics of telephony such tools could be adapted to service writing. To enable this sort of service creation we must look at what aspects of telephony will need to be supplied.

A call model that combines all the possibilities is required. Each network element will have its own subset of the call model. The potential states and triggers need to be defined. However this degree of complexity will need to be packaged up into worthwhile chunks that are embodied in SIBs. To clarify this let us take an example:

A Story At Bedtime

Imagine:

- Grandma writing a picture story for all her grandchildren
- She is effectively customising the service provided to her
- The service providers have a means of taking her graphics (stills, animation and video) and her sound (voice, special effects and music). They put this on a form of IP which her grandchildren are given the password to.
- The children telephone the IP, give the password and then play with Grandma's active story whilst she, several countries away is already asleep. They are themselves in two different towns and discuss the story whilst playing it together.

Maybe the simplest version of this is where the user, Grandma, simply customises a supplied story by including the names and pictures of the grandchildren and their pets. But suppose Grandma has drawn some cartoon characters and wants to dictate the plot and types of choices available. Now the tools need to include animators which take a few graphics and manipulate them to provide interesting continuous cartoons, plot editors which allow a flow of control to be mapped onto these characters actions and various types of events where choices can be made, the characters will also need voices and sound effects, all these need to be either under the control of Grandma or taken care of in an acceptable manner. Some of the aspects which require direct support during use are:

- Video display
- Synchronised soundtrack
- Additional voice connection between players
- Control of interaction
- Multiple views - players make different choices from each other so their views are all different - if the choices are mutually exclusive multiple worlds are needed
- Shared space - players might alter the environment as experienced by the other players within the same story world, other spaces might always be seen as new
- Representations of individual players

6. CONCLUSION

The scope of services is increasing rapidly; the requirements on service creation are growing equally. To enable networks to incorporate more than just switches the service writer needs either detailed models of each component or neatly packaged SIBs. An important element in service uptake will be the user's interfaces to those services - they must be flexible yet simple.

7. REFERENCES

[Bellcore 90] "Advanced Intelligent Network Release 1 Network and Operations Plan", June 1990

[Wray 94] Mike Wray and Mark Syrett "Service Creation Using the Hewlett-Packard Service Creation Environment" Intelligent Network `94 Workshop, Heidelberg, Germany May 1994

8. GLOSSARY OF TERMS

These are mostly three letter acronyms:

ADSI:	Analog Display Services Interface
AIN:	Advanced Intelligent Network
AuC:	Authentication Centre
Conn view:	The connection view, synonymous with c-view
CPE:	Customer Premises Equipment
DB:	database
GSM:	Global system for mobile communications
HLR:	Home Location Register, see also VLR
HP:	Hewlett-Packard
HPLabs:	Hewlett-Packard Laboratories
IN:	Intelligent Network
IP:	Intelligent Peripheral, generally voice-based
ISDN:	Integrated Services Digital Network
PBX:	Private Branch Exchange
SCE:	Service Creation Environment
SCP:	Service Control Point
SDL:	Service Description Language
SIB:	Service Independent Building block, synonymous with SIBB
SLP:	Service Logic Program
SMS:	Service Management System
TCP:	Trigger Check Point
VLR:	Visitor Location Register

4

Nokia's IN solution for fixed and cellular networks

Pekka Lehtinen
Markus Warsta
Nokia Telecommunications
P.O. Box 33
02601 Espoo, Finland

1. INTRODUCTION

The development of Intelligent Networks (IN) was started by Bellcore in USA more than ten years ago in order to support the Regional Bell Operating Companies in the new deregulated telecommunications environment. The original goal was to enable the network operators to effectively introduce and manage new services by means of a centralized database in a Service Control Point (SCP).

In the past years, several IN specifications have been developed, including IN 1, AIN Rel.0 and AIN Rel.1. The Advanced Intelligent Network Release 1 (AIN Rel.1) of Bellcore contains a service-independent standard interface for the queries sent from the Service Switching Points (SSPs) to the SCP.

The International Telecommunications Union (ITU) has worked out its Intelligent Network Application Part (INAP) specifications for the SSP-SCP interface, the so-called Capability Set 1 (CS.1). The European Telecommunications Standards Institute (ETSI) has completed its corresponding ETSI CS.1 Core specifications in 1993. It is expected that the CS.1 recommendations will be adopted world-wide. This would offer open interfaces between the SSP and SCP resulting in vendor independence.

Signalling System No.7 (SS7), fault-tolerant computing, on-line transaction processing, and today´s database technology have enabled the development of efficient IN elements. This paper presents Nokia´s IN solution.

2. FUNCTIONAL ARCHITECTURE OF INTELLIGENT NETWORKS

The main elements of the IN architecture of the ITU and ETSI are:

- Service Switching Point (SSP),
- Service Control Point (SCP),
- Service Management System (SMS),
- Service Creation Environment (SCE), and
- Intelligent Peripheral (IP).

The SSPs and SCPs are interconnected by the SS7 network. The SMS and SCP are connected by data communication protocols.

SSP provides the end-users an access to the IN services. It contains trigger points for detecting service access codes and sending service requests to the SCP. It also performs the required network-related operations based on the information received from the SCP.

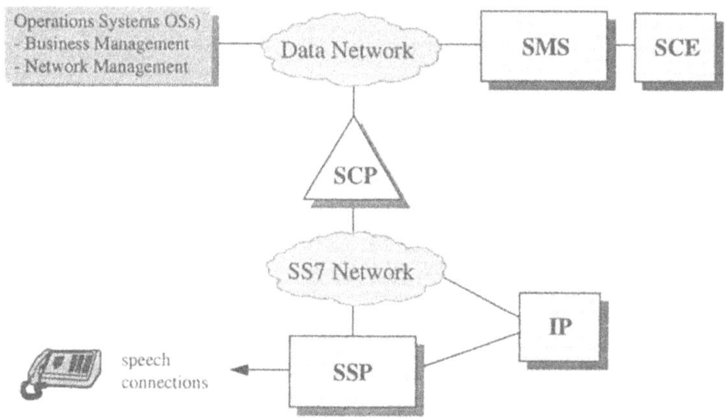

Figure 1. Functional Architecture of the IN.

In the SCP, Service Logic Programs (SLPs) contain the control logic of the service functionality, and the database contains the required information.

SMS includes IN management applications and capabilities to control the service providers. It is used

- to download the service database to the SCPs,
- to provide interfaces with service subscribers and service providers for the management of service parameters, and
- to gather measurement data.

SCE contains tools for the development of SLPs. The software development can be carried out e.g. in UNIX workstations running the SCE.

3. NOKIA´S IN APPROACH

3.1 Objectives

Service complexity, signalling capabilities, network coverage and revenue sharing are the variable factors, that determine the requirements of each customer. Nokia offers solutions for different customer environments. A good solution meets the customer's current economic and technical needs and allows an evolution towards the future network architecture.

Network operators are facing new challenges due to new trends of customer expectations, regulatory environment and information technology. However, the limitations of the current networks must be taken into account. The migration to a new network environment and new services must be smooth in order to ensure the protection of investments.

The objective of Nokia´s IN approach is to offer solutions, that meet the following market drivers and network operators´ requirements:

- reaction to the customers´ needs and new market opportunities,
- competition in the deregulated environment,
- introduction of open systems environment,
- generation of revenue from the existing investment,
- low initial investment,
- low investment in reusable equipment and a long-term plan of new equipment,
- end-users´ awareness of grade-of-service and availability.

A variety of signalling systems are in use, including decadic and multi-frequency signalling systems. Some networks are widely using the SS7, while others are in the middle of a transition towards the SS7. The network operators´ main concern is often the creation of an IN service offering environment within the confines of the network signalling capability.

In co-operation with Hewlett-Packard, Nokia is implementing a combination of open computer and telecommunications technology for providing Intelligent Network capabilities to fixed and mobile network operators. This IN architecture enables a unified solution to fixed and mobile networks in accordance with the new ETSI CS.1 Core specification.

3.2 Centralized IN Architecture

Nokia has developed a general service-independent SSP-SCP interface specification based on the CS.1 and the Transaction Capabilities Application Part (TCAP).

The main elements of Nokia ´s solution for a centralized IN architecture are:

- DX 200 SSP (including the IP functionality),
- Unix SCP,
- Unix SMS.

Nokia´s DX 200 switching system is highly modular. It is based on a distributed call control architecture. These characteristics facilitate a flexible introduction of the IN features and other new capabilities.

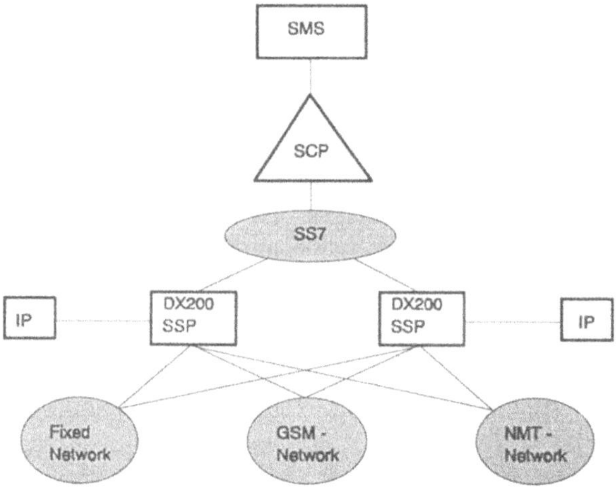

Fig. 3 Centralized IN Architecture

DX 200 SSP

DX 200 SSP can be configured for the IN services only, but it can also contain the functionality of a normal exchange. The capacity can be easily upgraded to a very high level by adding processing units.

DX 200 SSP can perform Global Title translation. Thus, it can send queries to multiple SCPs. This is useful, if the SCCP function is not generally available in the network. An important aspect is the ability to add new triggers or change the parameters of the triggers in the SSP by using Man-Machine Language (MML) commands.

DX 200 SSP has a built-in IP functionality for sending announcements and collecting Dual-Tone-Multi-Frequency (DTMF) digits as requested by the SCP. In addition, an external voice announcement machine can be installed.

The first release of the DX 200 SSP was based on the Bellcore AIN Rel.1 specifications supporting the functionality needed for the freephone, premium rate, personal number, calling card and VPN services. Subsequent releases of the SSP also meet the ETSI CS.1 Core requirements.

Unix SCP

The Unix-based SCPs compliment a centralized SCP by decentralizing the control intelligence of local-type services and mass calling applications at lower levels in the network. It is based on the CS.1 specifications.

The Unix SCP is based on the standard Unix platform of Hewlett Packard. The use of standard high-volume Unix computers based on the RISC technology ensures a declining trend of the price-performance ratio. Fault tolerancy is achieved by a duplicated processor. The processing capacity is 400 transactions per second. It can be increased by adding one or more mated SCPs which have an identical database. The maximum size of the service database is 500 MBytes of Random Access Memory.

Unix SMS

The Unix SMS contains the master database of the underlying SCPs. Database changes are automatically propagated from the SMS to the SCPs. If end-users are allowed to make changes in the service database, e.g. C-number updates, these changes are also propagated to the SMS and further on to the mated SCPs.

The SMS includes applications for service management, configuration management, fault management, and performance management. SMS is used to gather service-traffic-related measurement data from the SCPs.

The user interface of the SMS is based on a local area network (LAN). Unix-based workstations connected to the LAN include a variety of tools for the operating personnel to perform operations and maintenance tasks. They also provide a Service Creation Environment (SCE) to the network operator for the development of new services and the modication of the existing ones.

Unix SCE

In the service creation process, several high-level steps can be identified, such as service definition, specification, development, verification, deployment, and version control.

SCE can be used during the full service creation process. It is implemented by HP and is running on Unix workstations, which can be connected to the SMS via a LAN. It contains the following components:

- a graphical interface for rapid service creation (editing, validation),
- a simulated SMS for initialising the database with structural information of the subscribers' data records,
- a software-simulated SCP,
- a software-simulated SSP, with the ability to simulate the call traffic.

SSP-SCP Interface

The INAP between the SSP and SCP is based on the ETSI CS.1 specifications. The application messages are carried in TCAP messages. The SCCP and the Message Transfer Part (MTP) constitute the lower protocol levels.

INAP supports several advanced service features, including

- call-origin-based blocking or filtering (based on area code, country code, or A-subscriber´s number),
- routing, depending on additional dialled digits,
- routing, depending on call characteristics (e.g. requirement for a digital connection),
- follow-me diversion,
- C-number update,
- rerouting in the busy and no-answer cases,
- follow-on calls in the case of the Calling Card service (i.e. sequence call, extension of the INAP),
- charging options (e.g. alternate billing, billing code, or tariff, extension of the INAP).

3.3 Decentralized IN Architecture

In many telecommunications networks, the SCPs are centrally located and the SSPs are distributed at the transit exchange level. This solution has a number of advantages:

- The implementation and deployment of new services is fast.
- The IN service management is centralized in a few network nodes.
- The IN architecture is well suited for nationwide services, such as network-wide routing and call-distribution services.

On the other hand, some services (e.g. televoting) generate heavy traffic. This may cause overload problems, if the traffic is concentrated on a few SSPs and a centralized SCP.

A number of services are not suited for implementation in a centralized IN architecture. Some services are strictly local by nature (e.g. incoming call screening), or the response time delay of a centralized SCP is unacceptable, or a centralized SCP causes capacity limitations. In addition, some services require more information about the caller than what can be provided by the SSP at the transit level. Due to these requirements, it is obvious that the SSP functionality must be migrated down to the local exchange level as far as SS7 is available. From the viewpoint of the SCP, this evolution trend has the following advantages:

- short response time to end-users,
- possibility to increase the processing capacity according to customer requirements,
- capability to customize services for local use.

Several SCPs may be needed in order to offer a sufficient coverage of the network. The SCPs can be allocated to networks, regions, or services depending on the network, service characteristics, and customer requirements.

A great benefit of the above mentioned IN architecture is the low cost of implementing a true transitional IN architecture, including the reuse of service applications. This offers the network operator an opportunity to take advantage of the SS7 network without neglecting other parts of the network.

Depending on the signalling system, it may be possible to transfer the A-subscriber's identity from a local exchange to the SCP. The A-subscriber's validation in the SCP enables the creation of a charging record for each identified subscriber. In particular, Premium Rate service with various charging functions can be provided. A number of other services can be provided to the subscribers who are connected to the SCP.

3.4 Customer Implementations

Nokia has been in the forefront of IN development, providing advanced services based on a pre-IN architecture and subsequent true IN solutions based on open service-independent interfaces.

An AIN Rel.1-based IN solution has been delivered to Telecom Finland. The SSP exchanges are based on the DX 200 switching platform. In the first phase, Freephone, Premium Rate, Personal Number, and Calling Card Validation services were implemented.

Nokia has pioneered the development of the service-independent general SSP-SCP interface. Based on the AIN Rel.1 specifications, the INAP protocol, SS7 and TCAP have been implemented. In spite of its many advantages, the AIN Rel.1 specifications are incomplete. Adaptations and additions were also necessary due to the differences between the telephone networks in USA and Finland in the area of numbering schemes, charging procedures, and signalling capabilities.

Nokia has delivered a DX 200-based Service Switching and Control Point (SSCP) to Tele 2, a Swedish network operator. In this case, the AIN Rel.1 specification is used in the interface between the Service Switching Function (SSF) and the Service Control Function (SCF) integrated into the same exchange. The Application Service Elements (ASEs) are transferred via a high-speed message bus, instead of using the SS7. A good example of the capabilities of the DX 200 switching platform is that the same DX 200 SSP is also used as an international switch.

The DX 200 SSCP of Tele 2 provides access validation services that enable the end-users to make international calls via the network of Tele 2, even if the terminal is not directly connected to this network. The SSCP contains a subscriber database, which is used to validate the identity of the calling station and its rights to make the call.

The first contract to deliver an open IN solution based on the Nokia-HP IN architecture has been signed with Tele 2. This will enable Tele 2 to widen the service offering.

3.5 Further IN Development

It is evident, that the IN standardization activities will continue. Many issues are subject to further standardization, including global IN services, service portability, service feature interworking, and the interfaces between the SCP, SMS and SCE .

It is foreseen, that there is a growing market for services, that provide the mobility of persons and services in addition to terminal mobility. This implies a need to receive and make calls by using a personal number across multiple fixed and cellular networks. In addition, the same personalized service features should be available to all the end-users, irrespective of their terminal. There are also many cellular-network-specific requirements, such as a need to know the location of the calling mobile terminal in the case of an IN-based emergency call.

The functional integration of IN mobility services arises several questions. How to interchange customer profile and call history information between the service nodes? What method can be used to access the location registers? Nokia's strategy is to have a common IN platform for the fixed and cellular network elements in order to ensure future-proof solutions to the evolving mobility requirements.

It is obvious, that SSF will become an optional part of the DX 200 switching platform, including exchanges in fixed and cellular networks. This capability will offer the network operator a great flexibility regarding the network configuration and the customer-specific service options.

Integration of the SSF into a mobile services switching system will provide an environment, where the combined capabilities of a GSM/DCS network and IN can be fully utilized. The importance of openness is emphasized by the operators. It can be achieved by designing the SSP-SCP interface and the call model in accordance with the ETSI CS.1 Core specification.

5

Service creation environment as a software development platform

Andy Bihain
GTE
PO BOX 152092
Irving, TX 75038
USA
214-718-6670
arb1@Gte.com

James White
Tandem Telecommunications Inc.
1255 W.15th Street
Suite 900
Plano, TX 75075
USA
214-423-5383

1. INTRODUCTION

This paper documents the process view of the Service Creation Environment(SCE) and how this view can be physically implemented to support the end-to-end process of the Intelligent Network(IN) as viewed by GTE. GTE has defined the base sets of capabilities within IN as the Products and Services Infrastructure. This allows any new developments to draw upon the base set of capabilities without having to rewrite services that are needed by the application.

1.1 Vision

The vision of the SCE is to build software that can be compiled, tested, provisioned and run in multiple target platforms in an automated fashion.

The following goals should allow GTE to achieve the vision show above:

- Vendor Independence
- Service Independence
- Separation of "What" from "How" of application
- Separation of application from the data
- Separation of "customer" view and "network" view of application
- Reuse of software components

Vendor Independence--By the use of open system concepts and standards GTE seeks to achieve vendor independence at the applications level.

Service Independence--The infrastructure must be kept independent of the services offered so it will not have to be alter for eve service that is developed like today's services.

Separation of "What" from "How" of application--This goal allows GTE to build reusable components without worrying about the underlying platform structure from the application.

Separation of application from data--This goal allows GTE to build applications without regard for embedding data in the application which would require rewriting the application each time data fields changed.

Separation of "customer" and "network" view of the application--This goal allows GTE to write applications so that the customer has a view of the applications that he can relate to and the network has it's view of the application for execution.

As you can seen by these goals GTE is building a very standards based open systems environment that will allow us to accomplish our vision so we can break the main frame life cycle paradigm for applications.

2. PROCESS VIEW OF SERVICE CREATION ENVIRONMENT

Service Creation Environment Function(SCEF)-Provides a work-bench to create AIN services that utilize resources within the GTE environment to produce Service Independent Element(SIE) (Note: difference between SIE and SIB's is that SIE's contain all functions not just call processing)

Service Description(SD)-Ideas for a new service are received from any source and a description is drafted from a customer perspective with the call processing aspects of service behavior. An approach to service pricing and customer billing is proposed and an estimate made of potential market size. Other service descriptions are compared to see if would be a service enhancement or a new service needs to be built. Potential feature interaction screening on a static level will take place here.

Service Specification(SS)-The SS addresses all aspect of service behavior from real-time call processing logic to support issues, service provisioning, maintenance and billing. This service is then viewed from different user perspectives(customer, sales/marketing, service assurance, etc.) which is used to generate a consistent service profile. The service profile is used to specify all views and requirements as part of the SS.

Service Analysis(SA)-A Service Specification/Service Profile is analyzed in detail from a number of different perspectives. The purpose of the analysis is multifold:

- Ensure completeness of SS/SP
- Consider impact of proposed new service on existing, deployed services
- Estimate cost of deploying service
- appreciate impact of Service Deployment upon GTE support structure.

Service Approval(SAP)-At this point in the process flow, a detailed SS/SP has been defined, Service Combination profiles have been generated for the new service executing in combination with existing services, detailed analysis of these profiles are then put forward for management approval.

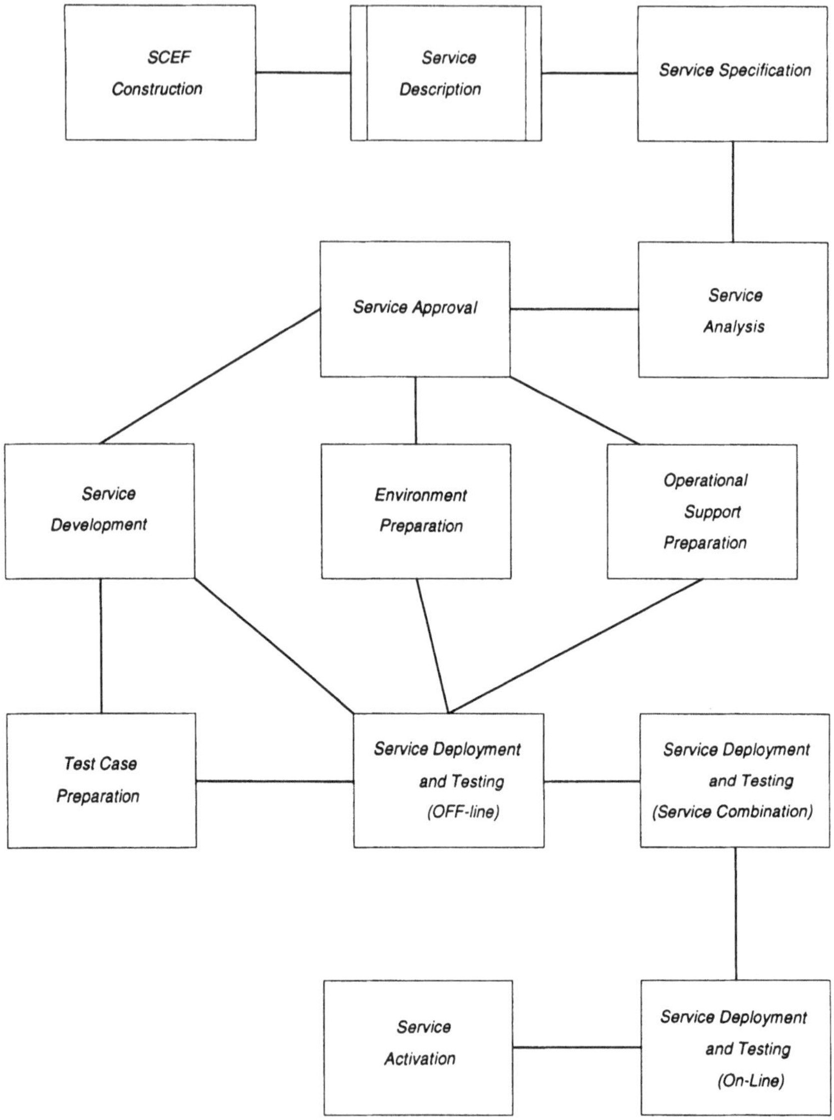

Figure 1. SCE process

Service Development(SD)-After approval, an implementation program has been put into place and now Service Development can take place. Service Development is defined as the process of creating all elements that will run across each of the target platform environments in support of all processes. Part of the SD process includes building test cases for the testing

Environment Preparation(EP)-Prior to deployment of the new service package to the GTE environment several preparation activities must be completed:

- Local Resources are deployed in target environment
- Both the off-line and on-line test environments are prepared for service testing.

Operational Support Preparation(OSP)-The support environment is prepared ahead of time for the introduction of the new AIN service and all outstanding non-technical issues are resolved to ensure that nothing will prevent the timely deployment of the service.

Service Deployment and Testing-Off-Line(SDTOFF)-There are two distinct phases of off-line service testing:

- Service testing in an off-line test environment
- Service Combination testing in an off-line environment

The off-line test environment is a mirror of the physical network elements but isolated in an off-line environment for testing.

Service Combination Deployment and Testing-Off-Line(SCDTOFF)-After testing the service in an stand alone mode it is necessary to test it in combination with other deployed Services for the following reasons:

- Magnitude of feature interactions considered during SS/SA were correctly noted.
- Uncover feature interactions at functional level not discovered during analysis.
- Discover feature interactions resulting directly from structural design of new service.
- Verify correctness of combination packages developed for new services.

Service Deployment and Testing On-Line(SDTON)-The processes involved in the deployment of a service to an on-line environment are essentially the same as for the deployment of that service to an off-line environment, however, the set of test to be performed in the on-line environment represents a small subset of the tests performed in the off-line environment.

Service Activation-Upon successful completion of service testing, the service is formally activated, if it is a revision of a previous service then all the service data, subscription data, and pending updates will be converted. Configuration data will be different for each version of the service and should not be converted.

3.0 TOOL SELECTION PROCESS

The complexity of this environment and what we are trying to build with reuse concepts dictates this development must be an Object Oriented(OO) development. The OO paradigm allows GTE to accommodate a very large development and design effort across multiple target platforms and systems while achieving a high degree of code module reuse.

3.1 OO Analysis and Design Tool

Rumbaugh's Object Modeling Technique(OMT) was chosen as the Object-Oriented analysis(OOA) technique. GTE selected VisualWorks from Martin-Marietta as the initial design tool running on a PC. After initial analysis, the model is being input into Soft Ware through Pictures(SWP) by IDE. SWP allows us to generate C++ header files that are imported into an Object-Oriented Design Code(OODC) toolset. Rumbaugh's methodology was chosen because most of our designers are familiar with it and our OODC environment has built a direct import interface into SWP. We have also developed extensions to OMT to offset some of the weaknesses in design of the OMT methodology so we can use recursive design with class and class structure charts using state transition diagrams.

3.2 OO Model Code Development

We are taking the C++ header files generated by SWP and then inserting them into a toolset called Teknekron Communications Systems BaseWorx Object Services Package(OSP). OPS features include generating graphical user interfaces, object servers, object communication and distribution, external communications, GUI and managed objects code generators.

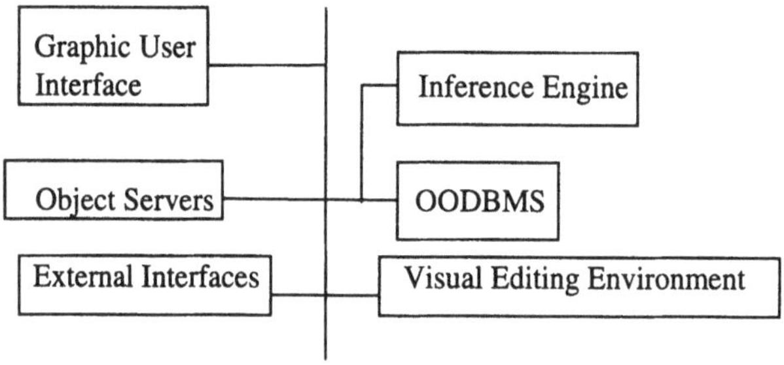

Figure 2. Software Backplane Services

The above listed diagram is a visual representation of the TCSI's OSP core and the elements that the software backplane concept is supporting. The graphical user interface is for creating code that is used by the toolsets not applications. The applications management operations, administration, and management(OAM) is supported in the Object Server Module. For the applications this includes configuration, fault, performance and security management. This server is designed to support both on-line transaction processing systems(OLTP) and network management based systems(CMIP). This approach allows us to build for any type of environment and legacy system that we must support in our applications migration.

The Object Server Module also provides for the interface into the OODBMS which is the Verstant OODBMS.

The communications module provides for message based communications using the communications core platform(CCP), a transaction manager(TM) and rapid transit(RT) paradigms. The object based communications is defined in the applications entity communications(AEC) service which provides access to object within a single process and across process boundaries. The AEC service provides a connection-oriented, asynchronous client-server protocol for intra-process, inter-process and inter-computer communications. Manager/Agent communications is provided via the CCP or CMIP protocol stack and the OSI tools provided by the CMIP stack vendor.

The OO middleware vendor provides the connectivity into the software backplane where my other toolsets are also communicating with each other.

3.3 Object Oriented DataBase(OODB)

In our application we had the following needs that had to be met by our database.

1. The network applications model has complex data structures and relationships that could be captured either via graph structure or object structures.
2. The need of extremely high performance with fine-grained concurrence control.
3. Because this application is mission critical applications must ensure data integrity, be highly available, and support very large databases.
4. The applications operate in geographically distributed environments and require support for distributed databases with the ability to move data across various nodes in the network.
5. The applications require that the DBMS maintain versioning control.
6. Because these applications are event driven, they require the ability to deliver and receive notification of network-wide events.

These were the reasons for choosing an OODB. Versant was selected because they are based around the concept of logical object identifiers(LOIDS). Because an object's LOID in Versant never changes and is never reused, the semantic integrity of database is assured. For transaction integrity, proven techniques of locking, and both physical and logical logging. Versant provides support of 281 Tera-Objects per database and 65,000 databases per network.Versant is also providing the tools to support a 365x24 operation without taking a database of line for backups and keeping the database integrity within the network.

3.4 Inference Engine(IE)

Due to the complex nature of the applications that the SCE is writing software for the ability of having applications make decisions based upon business or process rules is an absolute requirement. Also the ability of feeding back information from the applications will allow us to change business requirements only once and they will be propagated through out the systems as required.

Because of these requirements the tool ART-IM from Inference Corporation was chosen to provide these features as the software is generated and then distributed across the run-time environment. During the OO Design process business rules will be identified and process rules will be identified and then distributed across the run time environment.

3.5 Visual Programming Environment(VPE)

The use of a VPE for the SCE is a mandatory requirement. The ability of using this environment for programmers that are not fully trained is a principal that has been established by the business. Under current evaluation is the Tandem Service Creation Environment (SCE) 2000 to fulfill this role.

The SCE2000 consists of a programmer's workbench, the service designer's visual service editor, and a third layer a service provisioning environment. The SCE2000 workbench provides a structured means of managing the creation of service primitives(discrete operational components). Currently written in GNU Perl it's now being ported to GNU C++. The workbench is capable of producing API's, state machines, switch triggers, or downloads to a batch environment. The SCE2000 relies on the GNU rcs product to manage source files and also provides for distributed management of primitives by allowing for the local and remote generation through the use of makefiles.

The SCE2000 workbench employs revision control to provide a layered philosophy of subsystems, programs, and modules. The subsystem would contain a simple service comprised of programs. From a control standpoint root would be created to own the subsystem. For each new subsystem root privilege would be required. Subsystem/Programs naming follow standard UNIX conventions. Programs represent lower-level services required to create a subsystem and programs are comprised of modules.

Modules(primitives are equivalent to Service Independent Building Blocks(SIBB), and constitute the lowest-level of functionality possible in the system. Primitives are implemented using C++ classes are wrappers, their functionality comes from C/C++ source code compiled and made available in pseudo-library format. Programmers are responsible for primitives, however, check-out occurs at the program level in order to access and/ or modify primitives.

Primitives provide the lowest level of functionality in the SCE, with a web of networked primitives constituting a service. The workbench environment provides a pre-programmed logic that eases the integration of primitives. A menu option will create a primitive template file(PTF).

Primitive Template Files are basically interface definitions for a primitive. This is used both when the primitive is generated and when the primitive object is used in the service design stage. When the primitive is defined a secondary file containing the primitive's icon representation in bitmap format is also created. This bitmap file contains a default bitmap and can be modified by service programmer.

PTF files define the number and types of the primitive's input and output parameters, as well as the names of the primitive's successors. Parameters can be defined using pre-defined data types, or user-defined data type can be created for the primitive(structures or blobs).

Definition and usage of the user-defined data types requires direct modification of source code files. Primitive successors are essentially valid pathways that can be traversed once the primitive has completed processing. These pathways are specified simply by providing the names of other primitives.

WorkBench also creates Service template files, which define the connections between the primitives. These files allow changes to be made to the Service without requiring a recompile of the entire system. Such changes are limited, however, to those involving the logical connections between primitives.

Wrapper code is generated from the SCE2000 workbench and consists of two C++ source code files, one header file and one C source code file. Tandem will be changing the entire system to C++ to facilitate inheritance.

Within the service design environment primitives are plugged-in to create a service. Tandem is providing a level of integration so that you will be able to modify a primitive and then dynamically refresh the design environment. Next is the Visual Service Environment(VSE).

The VSE is an object-based service definition system. Using the primitive components developed under the SCE2000 workbench, services can be created by dragging and dropping primitives from one or more palettes, and then connecting these new service components to define both logic flow and data pathways. Palettes are where primitives can be grouped together and one or more palettes being available for selection for each service created. This is modeled after the standard UNIX library convention. Primitive palette windows work cooperatively with the service creation window to provide drag-and-drop capabilities. You cannot drag-and-drop between palettes yet. Blocks are a complexity management tool and exist at the same level as services, and can house any combination of Service, Primitive or blocks. Each block can contain up to 10,000 components and the entire system supports up to 10,000 level of nesting.

As you can now see why the tool sets were chosen and how GTE is looking to integrate them to provide a comprehensive environment for our programming staff. Now let's look at the physical implementation of the SCE process and how GTE is going to automate the Service Specification Tool.

4. SCE LOGICAL IMPLEMENTATION

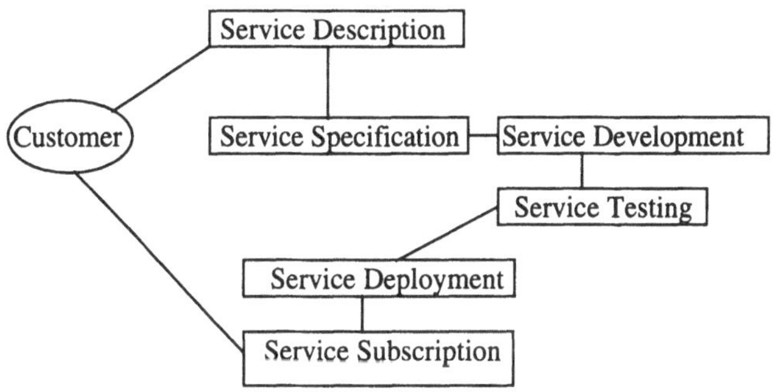

Figure 3. Logical SCE Diagram

Customer-This customer gives a service idea to a GTE representative as a set of capabilities.

Service Description-This is the first pass at capturing the idea of the customer and putting into graphical/textual description from a customer viewpoint.

Service Specification-This is the second pass at the service description putting it into the network view point to build.

Service Development-Coding the service out of the Service Independent Elements.

Service Testing-Testing the service against the customer and network criteria.

Service Deployment-When a service is deployed within the network ready for subscription.

Service Subscription-The service is network ready for the customer to subscribe , use and be billed for it.

4.1 Service Specification Tool

The service specification tool need to be built so that it can extract information from our corporate data bases and build a customer profile from the information present in the GTE databases. Then the profile can be run against an AI matrix which contains the capabilities sets of the Intelligent Network. Then a list of capabilities questions can be generated for the SS tool. The sales engineer can then take the tool and work with the communications manager to test the capability concepts for his network. After the questions are answered then an analysis is done against the capabilities and service suggestions would be made.

The Sales Engineer(SE) would then be able to visually present service concepts to the customer and refine them there. Upon agreement, the SE would bring the model back to feed into the SCE process. All views of the service could then be built with the customer profile as the baseline information for the SCE. With a generated model looking at all of the views necessary for the SCE and rules applied from the inference engine then most of the code could be generated automatically.

5. CONCLUSIONS

Because we have looked at the Intelligent Network as an end-to-end process, GTE is developing the integration of the SCE/SMS necessary so that we can overcome the deficiencies of our legacy systems. We are migrating to a new systems environment because of Process Re-engineering. This has given us the capabilities of using corporate data as an asset by removing corporate data from legacy applications. We are building an object view which looks at each process from it's unique view point and are building capabilities and rules to accomplish this. The end-to-end process view allows GTE to build an infrastructure that allows to build our products and services in a very rapid manner with substantially less cost then our previous environment.

6

IN service specification using the KANNEL language

Kari Granö[a], Jarmo Harju[b], Tapani Järvinen[a], Tapani Karttunen[b], Tapani Larikka[b], Jukka Paakki[a]

[a]Department of Computer Science and Information Systems, University of Jyväskylä P.O.Box 35, FIN-40351 Jyväskylä, Finland

[b]Department of Information Technology, Lappeenranta University of Technology P.O.Box 20, FIN-53851 Lappeenranta, Finland

Abstract

KANNEL is an application-oriented language for protocol engineering, with integrated support for all the main aspects of implementing typical communications software. Intelligent network services with distributed and interacting components share a number of central features with communications protocols, thus bringing in the possibility of applying protocol engineering techniques within the area of intelligent networking. To verify this view, the KANNEL language is applied to the specification of an intelligent calling card service. The specification includes both visual and textual notations of KANNEL.

1. INTRODUCTION

Recent active research on application-oriented languages promotes the view that application areas with specialized nature, such as protocol engineering, should be supported with dedicated tools founded on high-level concepts that match with the central characteristics of the applications. This view is in sharp contrast with the simplistic approach of engineering every application using general-purpose languages, such as C.

KANNEL [GHP94] is an application-oriented language designed especially for protocol engineering. While the design principle behind most of the conventional languages in this area, such as Estelle (Extended State-Transition Language) [BuD87], LOTOS (Language of Temporal Ordering Specification) [BoB87], SDL (Specification and Description Language) [BeH89], ASN.1 (Abstract Syntax Notation One) [Ste90], and TTCN (Tree and Tabular Combined Notation) [ISO89], has been on providing support for solving only some narrow problems within protocol engineering, KANNEL aims at covering the whole discipline in a uniform manner. In this way the need for using several different languages and several unrelated software tools to implement a protocol disappears since the whole task can now be managed under a single language and its integrated environment. The most notable features of KANNEL to support integrated protocol engineering are hierarchical finite state machines, distribution, interfaced layering, and encoding/decoding facilities.

While the main application area of KANNEL is protocol engineering, it can well be used as the implementation language of general (distributed) software systems. This more versatile side of KANNEL calls for more general software engineering capabilities which include object-orientation, visual design facilities, and an integrated tool set with a graphical user interface.

Intelligent network (IN) services are by their nature specialized distributed applications with the service specific logic residing in SCP, the service control point. The functionality of the service is realized in close co-operation with SSP, the service switching point, which is an integral part of a modern telephone exchange. In their interactions, SCP and SSP follow the standardized protocols described e.g. in ETSI's CS-1 standard [ETS94]. Thus, there are similar elements in engineering of IN services and in typical protocol engineering problems, which brings in the possibility to apply KANNEL in the specification of IN services.

In this paper we apply the KANNEL language to specifying a calling card service that allows telephone users to use any telephone for outgoing calls, charged to their own account [Kar94]. Authentication is normally based on a personal access number together with a personal identification number (PIN code) but a set of trusted A-subscriber numbers can be provided, authorized without PIN code. Moreover, the user can change his/her PIN code when using the service. The service also provides for specially priced short numbers, screening of unauthorized numbers, and special numbers with configurable destination. We describe the KANNEL specification of the calling card service, concentrating on the central SCP component.

Intelligent network services are tightly coupled with the existence of modern telephone exchanges (even though a well defined IN service is independent of the implementation of the underlying switching network). Since KANNEL does not directly support low-level physical aspects of telecommunications, it can be argued that KANNEL is not a full-fledged implementation language for IN applications. To overcome such obvious problems, we do not completely specify the SSP component of the calling card service but instead incorporate it into the KANNEL specification as an external entity, assumed to be implemented by some suitable means. This also holds for the database module of the service (SDP), due to the lack of decent database facilities in KANNEL. However, the communication protocol between all the main components of the service (SSP, SCP, SDP) is fully included in the specification.

We proceed as follows. The KANNEL language and its programming environment are summarized in Section 2. The personal call service and its specification in KANNEL are presented in Section 3. Finally, conclusions are drawn in Section 4. The service specification in textual KANNEL is included as an Appendix.

2. THE KANNEL LANGUAGE AND ENVIRONMENT

Starting from the development of packet networks and telephone exchanges with stored control, the role of software has become more and more important in communication systems. This is largely due to the increased intelligence built into distributed systems to facilitate the supply of versatile services through communication networks. Protocols are the key elements in providing this intelligence, and during the last twenty five years we have seen a tremendous rise in the scope and complexity of protocols on various levels of communication architecture.

Protocol engineering may well be the most complex field of software engineering where (semi-)automatic implementation based on formal techniques has been set as a realistic goal. There are five well-known languages that have been designed to formally specify distributed systems, and to generate communication software from specifications.

While there has been extensive research and development efforts on protocol design, the proposed languages usually have a restricted focus of application. For instance, Estelle and SDL can be used for defining the internal control behavior of a protocol entity (as an extended finite state machine) with a very abstract notion of data, and ASN.1 is intended solely for describing the conveyed data units without specifying any operational characteristics of the protocols. LOTOS, in turn, is too abstract for being implemented efficiently with current technology.

A survey of protocol tools used by Finnish industry is given in [AHG93]. The conducted experiments clearly show that the low quality and limited scope of the applied languages and the inconsistency of the software tools based on them make practical protocol work overly laborous. The essential conclusion about the current state-of-the-art is that although all aspects of protocol engineering (or, more generally, distributed communicating systems) can be handled using current languages and tools, the task is rather hard to carry out systematically and consistently.

The KANNEL language is an evolving attempt to overcome the shortcomings mentioned above. In protocol engineering, the usual approach has been to separate the design of a protocol from its implementation, and to consider functional logic and concrete data units as orthogonal issues. We believe that a unified scheme which would provide tools for all phases of protocol engineering is a desirable goal. To reach this, KANNEL integrates special protocol engineering facilities with general software engineering principles.

The key aspects of KANNEL that are related to software engineering in general are briefly described below.

object-orientation:

KANNEL supports the class and object concepts of object-oriented languages. There are two fundamental object categories in KANNEL: value objects are always represented by value, while reference objects are represented by reference. This division is a compromise between efficiency and orthogonality - representing all objects via references would be inefficient, and it is desirable to have the primitive types, such as integer, part of the class system.

In addition, objects may be local or distributed; this distinction affects the way the object can be communicated with.[1] Local objects are flexible to use, but do not fulfill strong distribution semantics.

KANNEL has adopted its notion of classes most notably from the Sather language [Omo93], in particular the facility of incremental superclassing and the explicit separation of (multiple) inheritance from code reuse.

---

[1]KANNEL does not offer a completely transparent distribution in the manner of e.g. Emerald [Jul88]. The reasons for this lie in efficiency and application area.

interfaces:
An interface class is used to define the abstract service interfaces of sets of classes. An interface class has no implementation (in this sense it resembles the deferred class concept of Eiffel [Mey88]). Subtyping in KANNEL is based on interface classes.

code reuse:
KANNEL allows a class to reuse the implementation of another class. Code reuse has no effect on type compatibility; it can be simply regarded as textual inclusion. This reflects the opinion that subtyping and code reuse are two distinct things that should be kept separate [Ame89].

visuality:
The visual components of KANNEL support the design of behavioral components of a protocol. It has been argued that a visual notation can greatly enhance the understandability of complex constructs [Har88]. In particular, this seems to be true for state machines.

integration:
Because KANNEL can be used to describe all the relevant aspects of a protocol, the supporting tools can be integrated. This implies that we do not need to generate code by several distinct tools and then integrate the code fragments with hand-written code. Instead the code generation process is fully automated in the sense that the resulting code need not be complemented.

maintenance:
The KANNEL environment supports multiple specifications that represent the system on different levels of abstraction. The integrity of these specifications is preserverd by the environment and changes to one specification are automatically propagated throughout the system.

Beside these general software engineering aspects, KANNEL provides for the following dedicated facilities of protocol engineering:

distribution:
In KANNEL, the service interfaces provided by a distributed class are separated from their implementation into a channel. This decision has the following advantages:

- Separate compilation is facilitated. The monolithic class mechanism employed in e.g. C++ implies that even changing the private part of a class requires recompilation for all clients. By separating these concerns (and following the tradition of Ada, Turbo Pascal, Modula-2, etc.) more flexibility is gained.
- The unidirectional nature of messages is supported better. By giving different views into the service, related messages going in different directions may be syntactically attached together. This leads to a more concise representation of the mutual relationship between a client and its server.
- When channel is made an explicit language construct, we may better grasp the notion of transfer syntax.
- The view concept provides static security and more readable specifications.

A channel is a variant of an interface class that specifies a multiway relationship for a set of distributed classes. Thus, the channel specifies the abstract service employed between a set of (typically two) peers.[2] The service is split into a set of disjoint views [KMR92], each grouping a set of messages that travel in the same direction. A channel may also specify the transfer syntax used to represent the concrete content of the messages.

communication:
In KANNEL, a transfer syntax is a class definable by the protocol writer. Each transfer syntax specifies a method of encoding/decoding all values of both basic and compound types. Functions for encoding and decoding data types are automatically created and applied when needed.

The idea of transfer syntax in KANNEL is similar to that in ASN.1. However, the realization of this idea is more high-level and advanced as a language.

concurrency:
KANNEL supports concurrency in terms of active objects. At most one thread of execution is active within each active object at a time. This approach, although excluding e.g. concurrent read access, provides a simple and safe solution to mutual exclusion problems.

functional logic:
KANNEL has adopted the statechart formalism [Har87] with some modifications as the means to describe the concurrent semantics of an active object. A statechart specifies the exact set of messages the object will respond to in a given state and thus maintains its integrity --- this can be viewed as stating as a precondition those messages that are valid for a given state. The communication model is asynchronous so as to offer greater efficiency and flexibility in communication operations.

synchronization:
As pointed out above, only one thread can be executing within an active object at a time. This implies that we need not consider synchronization within an active object.

layers:
Communication protocols are conventionally organized as layers. A layer (when both parties are concerned) consists of peer processes and the interconnecting channel. KANNEL employs static groups to create a layer by interconnecting the peer processes with a channel. To support the common practice of layered communication, the layers can be refined into more concrete ones (that is, a mapping from a layer into its lower neighbour can be systematically engineered).

reactivity:
Protocol engines are reactive in the sense that their operation is directed by incoming messages that typically make some request. Depending on the message, the engine may respond to it by, for example, sending some other message to another engine. The reactivity properties of active

[2]This is a more general concept than "client-server", as it recognizes the fact that these roles are usually view-dependent, i.e. that a server is also a client (for someone else).

objects are defined by statecharts. They define the set of incoming and outgoing messages, and the way each incoming message is responded to.

While the intended application area of KANNEL is protocol engineering, its key facilities are quite usable in other branches of software engineering as well, for instance in the implementation of distributed systems or intelligent networks. This more general side of KANNEL will be demonstrated in the next section where the language is applied to specifying an intelligent calling card service. The specification also serves as an example of a KANNEL program. Note, however, that while KANNEL as a complete programming language can be used to implement any (computable) system, it does not necessarily provide the proper level of abstraction in special areas other than protocol engineering.

We are currently implementing a protocol engineering environment that supports the KANNEL language [JGH94]. Our environment aims at integrating all the central tools needed in protocol engineering, thus making it possible to specify a protocol and to generate the corresponding implementation with a single tool set.

The main component of the environment is a visual KANNEL editor which is the default tool for writing a protocol specification. The specifications written with the visual editor are translated into the purely textual form of KANNEL which can also be used directly as the specification language. A compiler will translate the (user-written or internally produced) textual form into some general purpose programming language. Our intention is also to implement transformators between KANNEL and current protocol languages, so as to support interfacing with existing protocol standards and specifications. A long-term plan is to provide graphical analyzers and simulators directly on the visual KANNEL level. The tool set will be running under a common graphical user interface.

3. PERSONAL CALL SERVICE

3.1 Service description

The personal call service consists of two functional entities. These entities are access/authentication and placing outcall/PIN change.These two entities can be further divided into eight functional parts.

A user accesses and identifies him/herself to the system by entering a user specific access number. There are two ways of authentication. The first one is to ask the user to enter his/her personal identification number (PIN). The number of retries in case of a false combination of PIN and access number is limited. In the case of a false combination, the user is prompted to enter again both the access number and the PIN. When the maximum number of retries is exceeded, the user is notified and the call is terminated. This also blocks the service from the user. When the user has entered the required information correctly, he/she is asked to make an outgoing call or to select the PIN change procedure. The PIN can be either of fixed or of variable length. The other way of authentication is to check for an A-subscriber number. A user can have a limited number of A-subscriber numbers for which a PIN is not required.

The outgoing calls are checked for authentication of the particular user. If the entered number is allowed for this user, the call is completed; otherwise the call is abandoned. The verification is made in terms of global and user specific screening lists. There are two different types of outgoing numbers: short numbers of one or two digits, and special service and network numbers of at least three digits. When the user enters a number, the system checks from a database that a physical correspondence exists. If a short number or a special service number has been entered, the system gets the actual physical number from the database and connects the call; otherwise the call is simply connected to the entered number (if allowed).

After invoking the PIN change procedure, the user is asked to enter his/her old PIN. If the entered PIN matches the current PIN, the user is asked to enter a new PIN. The new PIN is checked for validity (e.g., that it is not 1111). If the new PIN is invalid, the operation is abandoned. If the entered new PIN is valid, the system asks the user to re-enter it. If the re-entered PIN matches the previously entered one, the database is updated; otherwise the old PIN remains valid.

After any completed operation the user is prompted for another operation. This is the so called global follow-on feature.

At any state of the whole process, the user can close the personal call service by hanging up the phone.

3.2 Service specification in KANNEL

In this section we specify the personal call service described above with the KANNEL language. The presentation is mostly graphical but fragments of textual code are given when a more elaborate presentation is appropriate. A complete textual presentation can be found in the Appendix.

3.2.1 General architecture

The system is composed of three main components (see Figure 1):

ssp (service switching point) is in essence the user interface to the system.

scp (service control point) is where the intelligence resides and, as such, the main subject of our interest. scp interacts both with ssp and sdp.

sdp (service data point) is the database which holds the information about user identification and authentication, short numbers, number screening, etc.

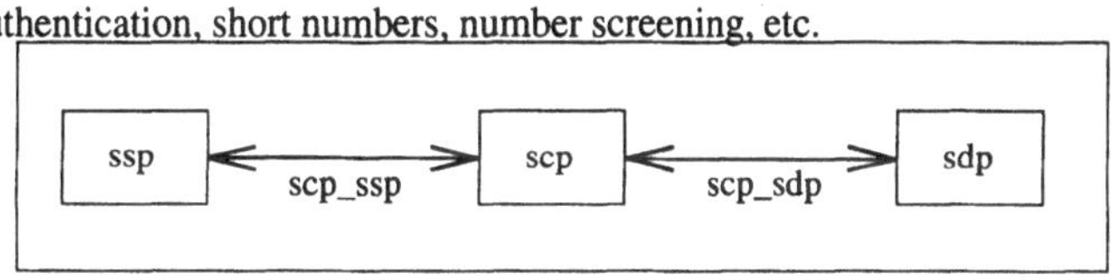

Figure 1. General architecture

Of these, scp is our main target, whereas ssp and sdp are merely supporting entities for which we provide only an abstract description. In Figure 1 the outermost block corresponds to the group concept of KANNEL. The inner blocks correspond to classes, and the bidirectional arrows denote channels.

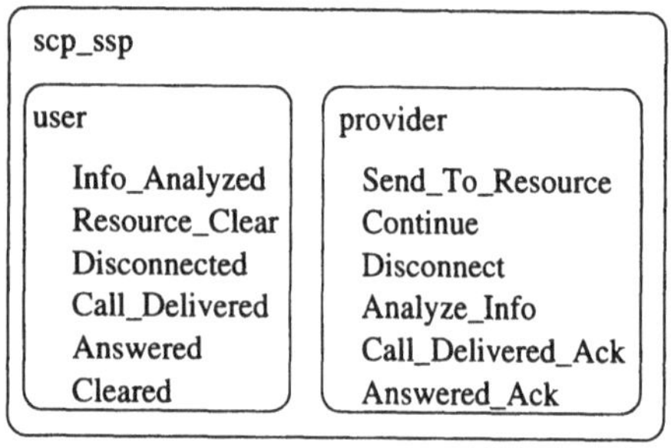

Figure 2. Structure of the scp_ssp channel.

To specify how scp interacts with ssp and sdp, we fix the structure of the channels scp_ssp and scp_sdp. As shown in Figure 2, the channel scp_ssp provides two views, namely user and provider. The user view conveys messages (Info_Analyzed, etc.) from ssp to scp. The provider view, in turn, conveys messages (Send_To_Resource, etc.) from scp to ssp. Similarly, Figure 3 illustrates the structure of the channel scp_sdp.

For illustrating how a message is specified in KANNEL, let us have a look at Query_Result for the user view of the scp_sdp channel:

```
class Query_Result key 14 is
        pin_needed: boolean;
        pin: integer;
        pin_ok: boolean;
end
```

This message has three parameters, the boolean-valued pin_needed and pin_ok, and the integer-valued pin. The message has a key value 14 which is used to encode it uniquely over a transfer syntax. The message system of KANNEL resembles the approach taken in the data description language ASN.1 [Ste90] with a facility of automatic encoding and decoding under a complete transfer syntax. We omit a further elaboration in this case, and assume that the underlying transfer syntax is one of the standard ones, akin e.g. to the basic encoding rules (BER) of ASN.1.

Having the general framework defined, we can now move on to refine the internal structure of the service control point, scp.

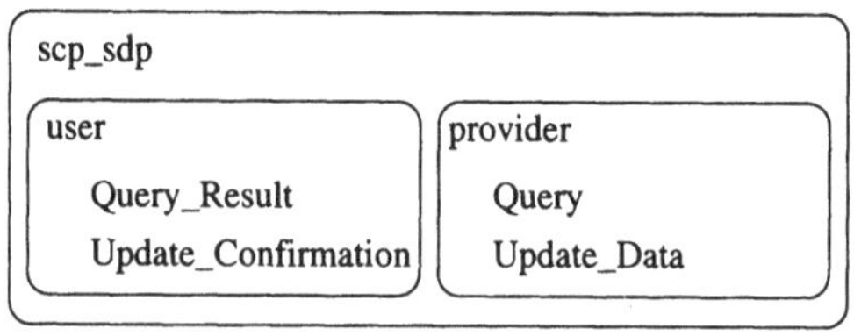

Figure 3. Structure of the scp_sdp channel.

3.2.2 Service control point

The service control point mainly takes care of controlling and informing ssp, by using the services offered by sdp. The duty of scp is largely to receive messages from ssp and sdp, and to send appropriate replies back to them.

Figure 4 sketches the complete visual specification of the personal call service. It has the same basic elements as Figure 1. However, now the specification of scp is much more detailed. The state machine of scp consists of two main states, *login* and *outcall_PIN_change*. User identification and authentication are handled in state login, while an outgoing call and a PIN code change are handled in state outcall_PIN_change.

State *login* is composed of five substates of which *idle* is the default, i.e. the state entered whenever *login* is entered. State *idle* has one transition labelled Info_Analyzed/Query. The interpretation of this label is to fire the transition whenever the signal (or message) Info_Analyzed is received. Upon reception of the signal, the message Query is emitted. Finally, control moves to state *query*.

State query has two transitions, labelled Query_Result(PIN not needed) / e and Query_Result(PIN needed) / Send_To_Resource("Provide PIN") (here e stands for an empty action). Considering the signal names only, this may seem to make the machine nondeterministic. However, taking a look at the corresponding textual code fragment for state query, we see how this is solved:

```
state query arcs
        Query_Result -> {
                if Query_Result.pin_needed then
                        real_pin := Query_Result.pin;
                        ssp ! Send_To_Resource("Provide PIN");
                        auth
                else
                        outcall_PIN_change
                end
        }
end
```

Here Query_Result is the triggering message, and the associated action is enclosed in curly braces. The destination state is either *auth* or *outcall_PIN_change* depending on the Boolean value of Query_Result.pin_needed.

In other words, on a single incoming message we can make further decisions about the following actions according to e.g. parameters of that message. In this case we check the

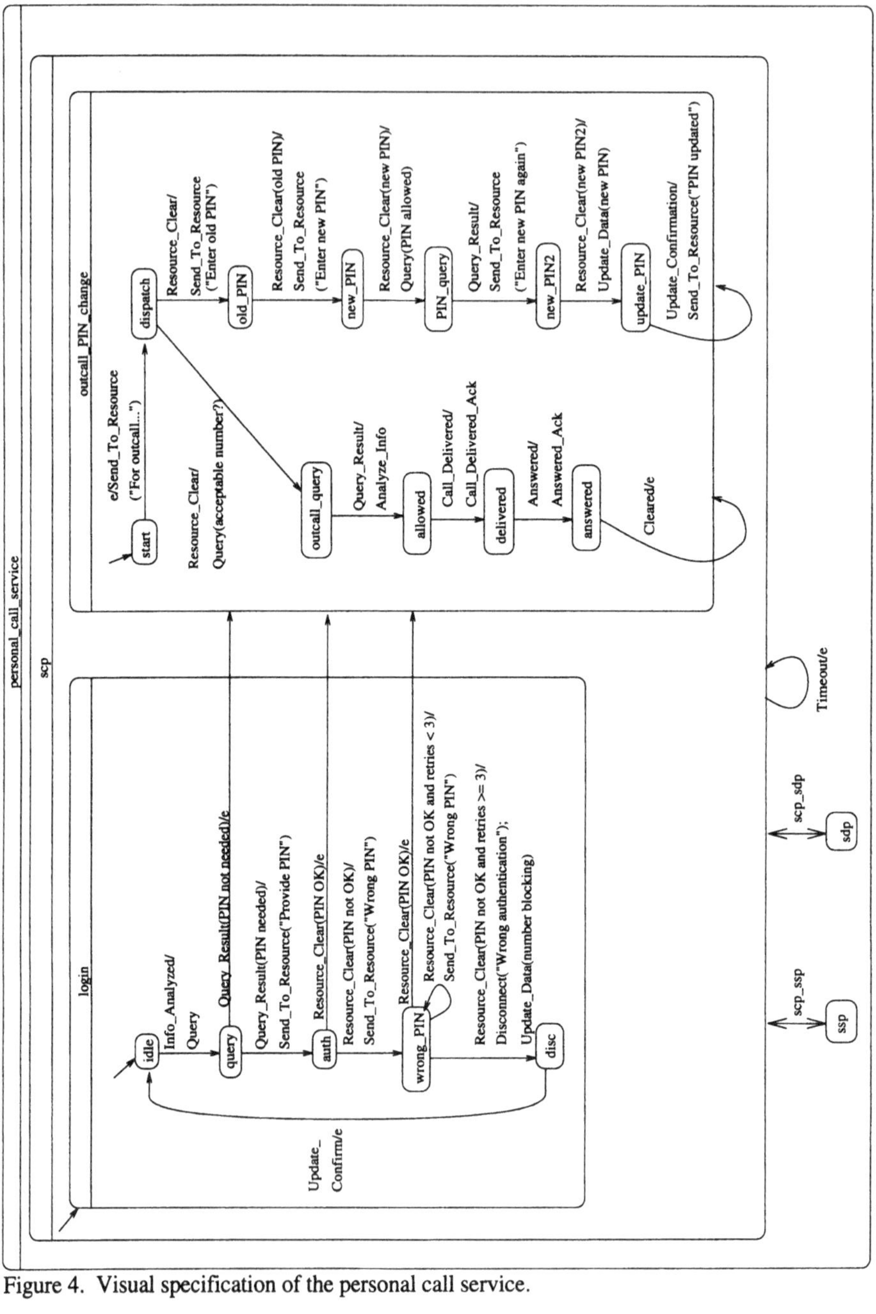

Figure 4. Visual specification of the personal call service.

pin_needed parameter and decide whether or not the user has to supply a PIN.

If a personal call service user calls from a known telephone machine, he/she does not need to supply a PIN code. Otherwise a PIN is requested from the user who may retry for three times. If the user fails to provide a correct PIN code three times in a row, his/her personal call access number is blocked.

Another substate of scp is *outcall_PIN_change*. This state deals with outcalls and the PIN change procedure. It has 11 substates, of which *start* is the default. The user selects either one of the two services by entering a destination number or the #-symbol. In substate *dispatch* we choose the correct path to accomplish the selected service.

If the user selects the outcall service, the machine moves to state *outcall_query*. Here scp makes a query to sdp to check whether the user is allowed to place an outcall on the indicated number, and whether it is a special number or a short number.

In case the user selects the PIN change procedure, he/she is prompted for the old PIN. If he/she provides the correct PIN, the new PIN shall be entered. If the new PIN is acceptable (i.e., it is a combination of digits which is not too easy to guess), the user is prompted to provide the new PIN again. If the entered PINs match, the database at sdp is updated by the new PIN.

3.2.3 Service switching point and service data point

From the viewpoint of the service control point (scp), the service switching point (ssp) and the service data point (sdp) are abstract, supporting entities which can be seen as black boxes. For this reason we provide only a high-level specification for them, demonstrating the abstraction capabilities of KANNEL. In Figure 4, ssp and sdp are just plain boxes, connected to scp with channels. In the textual form of KANNEL this is specified by prefixing the class definition by the keyword external:

```
external class ssp
ports
        scp: scp_ssp(in provider, out user)
end

external class sdp
ports
        scp: scp_sdp(in provider, out user)
end
```

As these classes are external, we do not specify their functional behavior. All we need to know about them is the interface, which is specified by providing the ports clause. As usual, the external entities (here: ssp and sdp) are assumed to have a complete definition or implementation whose details, however, are irrelevant for the main subject (here: scp).

3.2.4 The complete system

We conclude the building of the personal call service by collecting together all the components presented above. The visually specified parts of the complete system are shown in Figure 4 except for the channel specifications which are given in Figures 2 and 3. The group combining ssp, scp, sdp, and the channels is textually defined as follows:

```
group personal_call_service
ports
local
        r: ssp;
        s: scp;
        t: sdp;
is
        s.ssp.attach(r.scp);
        s.sdp.attach(t.scp);
end personal_call_service
```

The textual form makes use of port names that correspond to the endpoints of channels in Figure 4. The channel structure and the distributed object structure of the system are static, that is, this part of the system configuration cannot be adjusted while running the system.

While the presented KANNEL solution to the personal call service can be regarded appropriate for a specification, it falls somewhat short for a running implementation. In particular, a complete implementation should pay attention to the possibility of abnormal cases, e.g. reception of unknown (or damaged) messages by the scp, or reception of nothing while scp is actively executing its state machine. Also the user interaction with the ssp should be checked for correctness. For timeout situations (see Figure 4) there would be a special timer system (a KANNEL class), and for the uninterpretable communication situations the implementation would be prepared for "any" input matching none of the accepted messages. For reasons of space, these abnormal cases are not handled in our specification.

CONCLUSIONS

We have applied the protocol engineering language KANNEL to the specification of an intelligent calling card service. Due to the special nature of KANNEL, we have concentrated on those aspects of the service that are most protocol oriented: the interaction between the main components of the system, the service switching point (ssp), the service control point (scp), and the service data point (sdp).

The main functionality of the application resides at the scp. Hence, our specification is also founded on the description of scp's functional logic. In our opinion, the protocol engineering paradigm pursued by the KANNEL language proves to be quite suitable at least for this aspect of the IN concept. We have described the secondary components of the system, ssp and sdp, as external entities whose details are not given in KANNEL. Note, however, that it would be quite possible to completely specify ssp and sdp in KANNEL as well, even though the abstraction level needed by their special characteristics would probably not be sufficiently covered by KANNEL. This observation can be tuned to a view of future development of the language: To more properly support the implementation of IN applications, KANNEL should provide dedicated facilities for describing databases (in this case: sdp) as well as (abstract) user interfaces (in this case: ssp).

The PROFILE project (PROtocol Facilities in an Integrated Language and Environment) on designing and implementing the KANNEL language has been running since the end of 1992. The prestudy phase and the language design phase of the project have been finished, and we are currently in the implementation phase, working especially on the visual KANNEL editor.

An initial implementation will be finished during 1995, after which a final industrialization phase has been planned to take place.

ACKNOWLEDGEMENTS

The PROFILE project is mainly funded by the Technology Development Centre of Finland (TEKES). The project is supported by an industrial steering group consisting of Ari Ahtiainen (Nokia Research Center), Tom Leskinen (Nokia Cellular Systems), Olli Martikainen (Telecom Finland), Heikki Oukka (X-Net), Jari Rautiainen (Nokia Telecommunications), Jari Simolin (Nokia Telecommunications), and Asko Vilavaara (Technical Research Center of Finland). The ideas and suggestions of the steering group are gratefully acknowledged. The working environment has been provided by Nokia Research Center.

REFERENCES

[AHG93] Arvonen K., Harju J., Granö K., Paakki J.: A Survey of Software Tools in Protocol Engineering. Report 1, Department of Information Technology, Lappeenranta University of Technology, 1993.

[Ame89] America P.: Issues in the Design of a Parallel Object-Oriented Language. Formal Aspects of Computing 1, 1989, 366-411.

[BeH89] Belina F., Hogrefe D.: The CCITT Specification and Description Language SDL. Computer Networks and ISDN Systems 16, 4, 1989, 311-341.

[BoB87] Bolognesi T., Brinksma E.: Introduction to the ISO Specification Language LOTOS. Computer Networks and ISDN Systems 14, 1, 1987, 25-59.

[BuD87] Budkowski S., Dembinski P.: An Introduction to ESTELLE: A Specification Language for Distributed Systems. Computer Networks and ISDN Systems 14, 1, 1987, 3-23.

[ETS94] ETSI (European Telecommunications Standards Institute): Intelligent Network (IN); Intelligent Network Capability Set 1 (CS 1), Core Intelligent Network Application Protocol (INAP): Part 1: Protocol specification. Draft PrETS 300 374-1, 1994.

[GHP94] Granö K., Harju J., Paakki J., Järvinen T.: Proposal for a Protocol Engineering Language. Technical Reports TR-6, Department of Computer Science and Information Systems, University of Jyväskylä, 1994.

[Har87] Harel D.: Statecharts: A Visual Approach to Complex Systems. Science of Computer Programming 8, 1987, 231-274.

[Har88] Harel D.: On Visual Formalisms. Communications of the ACM 31, 5, 1988, 514-530.

[ISO89] ISO (International Organization for Standardization): The Tree and Tabular Combined Notation (TTCN). ISO/IEC DIS 9646-3, Information Technology - OSI Conformance Testing Methodology and Framework, Part 3, 1989.

[JGH94] Järvinen T., Granö K., Harju J., Paakki J.: An Integrated Environment for Protocol Engineering. In: Proc. Nordic Workshop on Programming Environment Research (NWPER'94) (B.Magnusson, G.Hedin, S.Minör, eds.), Lund, 1994. Report LU-CS-TR: 94-127, Department of Computer Science, Lund University, 1994, 177-193.

[Jul88] Jul E. et al.: Fine-grained Mobility in the Emerald System. ACM Transactions on Computer Systems 1, 6, 1988, 109-133.

[Kar94] Karttunen T.: Personal Call Service Description. Draft, Telecom Finland, 1994.

[KMR92] Koivisto J., Malka J., Reilly J.: Object-Oriented Implementation of Telecommunication Software Using the OTSO Enviroment. In: Proc. IFIP TC6 4th Conf. on Information Networks and Data Communication (INDC-92), Espoo, Finland, 1992, 293-304.

[Mey88] Meyer B.: Object-Oriented Software Construction. Prentice-Hall, 1988.

[Omo93] Omohundro S.: The Sather 1.0 Specification. Unpublished; available electronically as file manual-1.0v5.ps.Z in directory /pub/sather via anonymous ftp to ftp.icsi.berkeley.edu.

[Ste90] Steedman D.: ASN.1 - Tutorial \& Reference. Technology Appraisals Ltd., 1990.

APPENDIX TO ARTICLE "IN SERVICE SPECIFICATION USING THE KANNEL LANGUAGE":

TEXTUAL KANNEL SPECIFICATION OF THE PERSONAL CALL SERVICE

```
-- KANNEL specification of the personal call service
-- Messages SSP -> SCP

class Info_Analyzed key 0 is
    aid: integer;
    xyz: integer;
end

class Resource_Clear key 1 is
    pin: integer;
    change_pin: boolean;
    telephone_number: integer;
end

class Disconnected key 2 is
end

class Call_Delivered key 3 is
end

class Answered key 4 is
end

class Cleared key 5 is
end

-- Messages SCP -> SSP

class Send_To_Resource key 6 is
    msg: string;
end

class Disconnect key 7 is
    msg: string;
end

class Continue key 8 is
end

class Analyze_Info key 9 is
end

class Call_Delivered_Ack key 10 is
end

class Answered_Ack key 11 is
end

-- Messages SCP -> SDP

class Query key 12 is
    aid: integer;
    xyz: integer;
end
```

```
class Update_Data key 13 is
    pin: integer;
end

-- Messages SDP -> SCP

class Query_Result key 14 is
    pin_needed: boolean;
    pin: integer;
    pin_ok: boolean;
end

class Update_Confirmation key 15 is
end

-- Other messages

class Timeout key 16 is
end

class User_Release key 17 is
end

-- Channels

channel scp_ssp is
    view user     -- ssp -> scp
        Info_analyzed;
        Resource_Clear;
        Disconnected;
        Call_Delivered;
        Answered;
        Cleared

    view provider    -- scp -> ssp
        Send_To_Resource;
        Disconnect;
        Continue;
        Analyze_Info;
        Call_Delivered_Ack;
        Answered_Ack
end

channel scp_sdp is
    view user    -- sdp -> scp
        Query_Result;
        Update_Confirmation
    view provider    -- scp -> sdp
        Query;
        Update_Data
end

external class ssp
ports
    scp: scp_ssp(in provider, out user)
end

external class sdp
ports
    scp: scp_sdp(in provider, out user)
end
```

```
class scp
ports
    ssp: scp_ssp(in user, out provider);
    sdp: scp_sdp(in user, out provider);

local
    real_pin: integer;          -- pin returned by sdp
    user_pin: integer;          -- pin given by the user
    const max_retry: integer;         max number of retries in
authentication
    retry: integer;              -- retries so far
    init () is
        max_retry := 3;
        retry := 0;
    end


state (login, outcall_PIN_change) is

    state login (idle, query, auth, wrong_pin, disc) is

        state idle arcs
            Info_Analyzed -> {
                sdp ! Query(Info_Analyzed.aid, Info_Analyzed.xyz);
                query
                }
        end

        state query arcs
            Query_Result -> {
                if Query_Result.pin_needed then
                    real_pin := Query_Result.pin;
                    ssp ! Send_To_Resource("Provide PIN");
                    auth
                else
                    outcall_PIN_change
                end
            }
        end

        state auth arcs
            Resource_Clear -> {
                if Resource_Clear.pin = real_pin then
                    retry := 0;
                    outcall_pin_change
                else
                    retry := retry + 1;
                    Send_To_Resource("Wrong PIN...");
                    wrong_pin
                end
            }
        end

        state wrong_PIN arcs
            Resource_Clear -> {
                if Resource_Clear.pin = real_pin then
                    retry := 0;
                    outcall_PIN_change
                else
                    if retry < max_retry then
                        retry := retry + 1;
```

```
                    Send_To_Resource("Wrong PIN...");
                else
                    ssp ! Disconnect("Wrong authentication...");
                    sdp ! Update_Data(upt number blocking);
                    disc
                end
            end
        }
    end

    state disc arcs
        Update_Confirm -> { idle }
    end

end login

state outcall_PIN_change
    (start, dispatch, outcall_query, allowed, delivered, answered,
     old_PIN, new_PIN, PIN_query, new_PIN2, update_PIN) is

    state start arcs
        nil -> { ssp ! Send_To_Resource("For outcall..."); dispatch}
    end

    state dispatch arcs
        Resource_Clear -> {
            if Resource_Clear.change_pin then
                ssp ! Send_To_Resource("Enter old PIN");
                old_pin
            else
                sdp ! Query(Resource_Clear.telephone_number);
                outcall_query
            end
        }
    end

    state outcall_query arcs
        Query_Result -> {
            if Query_Result.allowed_number then
                ssp ! Analyze_Info(AMA);
                allowed
            else
                ssp ! Send_To_Resource("Number is not allowed");
                start
            end
        }

    state allowed arcs
        Call_Delivered -> {
            ssp ! Call_Delivered_Ack;
            delivered
        }
    end

    state delivered arcs
        Answered -> {
            ssp ! Answered_Ack;
            answered
        }
    end
```

```
        state answered arcs
            Cleared -> { outcall_PIN_change }
        end

        state old_PIN arcs
            Resource_Clear -> {
                if Resource_Clear.pin = real_pin then
                    ssp ! Send_To_Resource("Enter new PIN");
                    new_PIN
                else
                    ssp ! Send_To_Resource("Old PIN not valid");
                    outcall_PIN_change
                end
            }
        end

        state new_PIN arcs
            Resource_Clear -> {
                user_pin := Resource_Clear.pin;
                sdp ! Query(Resource_Clear.pin);
                PIN_query
            }
        end

        state PIN_query arcs
            Query_Result -> {
                if Query_Result.pin_ok then
                    ssp ! Send_To_Resource("Enter new PIN again");
                    new_PIN2
                else
                    ssp ! Send_To_Resource("Selected PIN not acceptable");
                    outcall_PIN_Change
                end
            }
        end

        state new_PIN2 arcs
            Resource_Clear -> {
                if user_pin = Resource_Clear.pin then
                    sdp ! Update_Data(Resource_Clear.pin);
                    update_PIN
                else
                    ssp ! Send_To_Resource("Entered PINs do not match");
                    outcall_PIN_change
                end
            }
        end

        state update_PIN arcs
            Update_Confirmation -> {
                Send_To_Resource("PIN updated");
                outcall_PIN_change
            }
        end

    end outcall_PIN_change

arcs
      Timeout -> {
            scp
      }
```

```
      User_Release -> {
            Basic_Call
      }
end scp

group personal_call_service
ports
local
    r: ssp;
    s: scp;
    t: sdp;

is
    s.ssp.attach(r.scp);
    s.sdp.attach(t.scp);
end personal_call_service
```

7

Performance analyzer for Intelligent Network

Jorma Jormakka
Communications Laboratory
Helsinki University of Technology
Otakaari 5, 02150 Espoo, Finland
E-mail: Jorma.Jormakka@hut.fi
tel. 358-0-4512360, fax. 358-0-4512345

Abstract

This paper describes a prototype of a tool which can be used in dimensioning Intelligent Networks (IN). A EURESCOM pre-study of currently used dimensioning methods for IN showed that conventional methods and tools for dimensioning telephone networks do not as such suit to IN and that the commercially available simulation tools and environments are not easy to use as dimensioning tools. EURESCOM P308 project and it's continuation project aims to developing methods and later prototyping a suitable tool. The prototype tool described here is not the tool to be developed by the EURESCOM project though the author has been working on the project E-P308. This tool can be used as input to the EURESCOM prototype which is expected to address a larger part of dimensioning problems. The tool described here is only limited to evaluating some Grade of Service parameters and to analyzing the effects of load control on the net.

MAIN PARTS OF THE TOOL

The tool consists of two editors by which the network structure and IN protocols are modeled, a simulator, an approximate solution generator and a plotter.

Network editor

Figure 1 shows the network editor. It is a graphical editor and will be used for studying televoting in a network of one Finnish telephone operator. There is a strong interest for this service for instance audience voting in television programs and the basic question with this service for an operator is what amount of televoting traffic can be allowed. Televoting as a source of revenue for an operator may not be very important but this service can disturb the network and the grade of service for this IN service should be on acceptable level, that is, the level expected from this service. The network is modeled by the editor and the model is run with a simulator which tracks the amount of generated traffic, blocking probabilities, interarrival time distributions, service time distributions and other relevant features. Currently the televoting is not using SCP and the implemented version of this network is basically a circuit-switched net but the model will be upgraded when the underlying network structure is changed. Interesting problems here are a possible unfairness of televoting blocking probability and the effects of different congestion control solutions. The telephone images in the editor are traffic sources where the rate, distribution, time-dependancy of the distribution and user impatience statistics are given. Other elements are channels and switches which are expected to model the existing equipment in sufficient detail from traffic point of view.

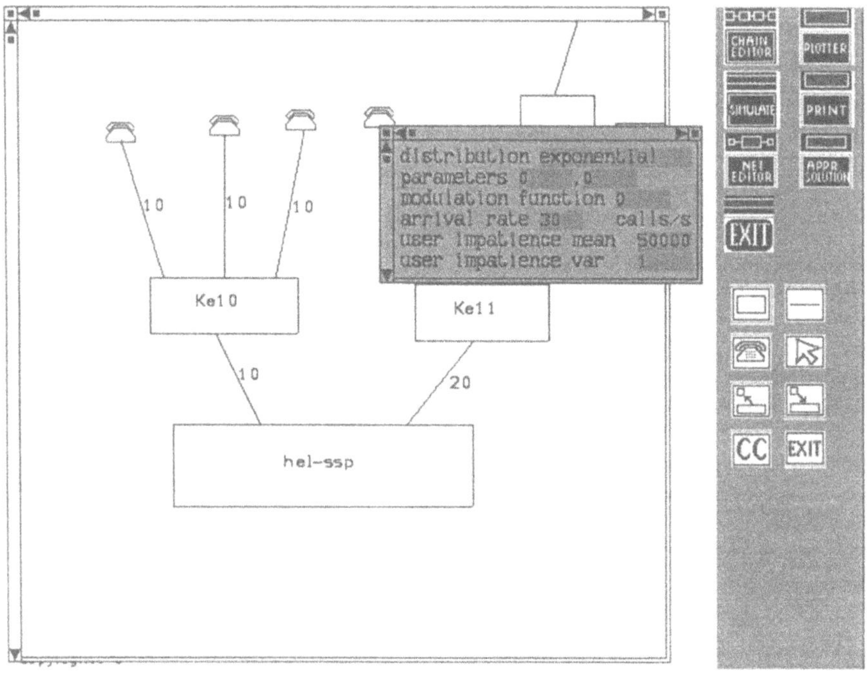

Figure 1.

Work-load-chart editor

Another editor is used to describe IN protocols. The editor is seen in figure 2. In this editor an IN service protocol is described as a work load chart by which here means a message sequence chart where each message is assigned a work load. Each work load are given the mean and the distribution. In a communication channel the work load is the transfer time and in a network node the work load is the processing time. Channels and nodes are both described as elements in the editor and the editor has for each element a set of parameters including number of servers, waiting room size, processor sharing mode in a server, selection rule for a free server and load control filter method. The main usage of this editor is to model signaling messages for a given IN service and then use either the simulator or another part of the tool, the approximate solution which will be described later, to obtain some grade of service parameters such as connection establishment and termination delays and rejection probability for the signaling part. This description of the network by work loads is phenomenological: the protocol is known and will be standard but the actual equipment can be very different and of it's performance little may be known. This solution avoids the problems of modeling the actual equipment and models only the protocol on some layer, say TCAP-layer. The equipment appears then only in the work load of the messages and in the few selected parameters for the elements.

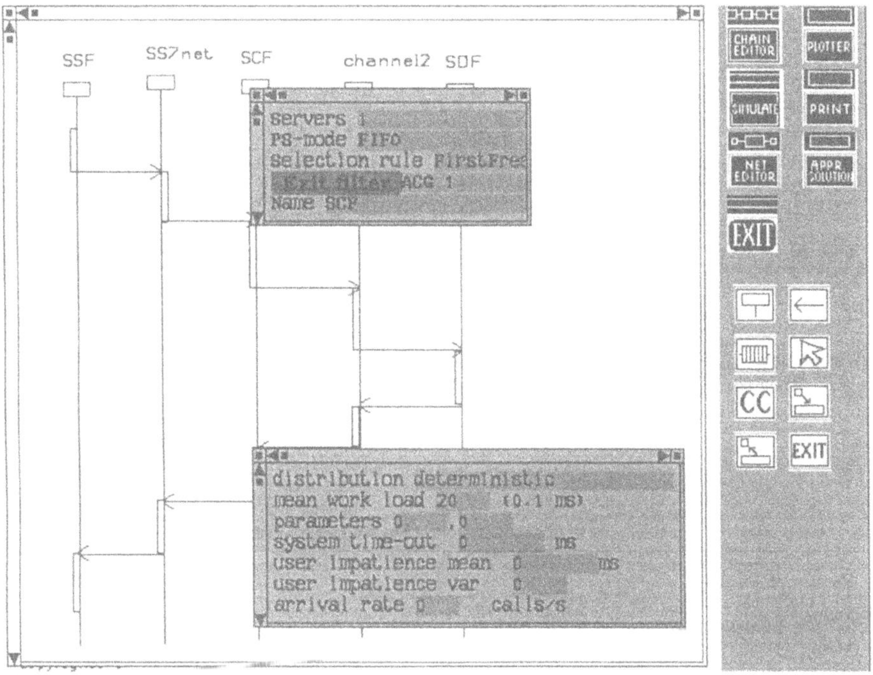

Figure 2.

Simulator

The simulator is seen in figure 3. It is not a discrete event simulator but of type which occasionally is called a scan-next-event simulator. The simulator keeps in each task the amount of received work and remaining work for each job and selects from all active jobs next time quanta which will be processing up to the next event which is the time when the number of active jobs changes. In a discrete event simulator (DES), which is the solution used by most simulators, the ending time of each job is called an event and all these events are put in time order to an event list which a scheduler is reading. The difference between these simulator types is that the amount of received service for each job is not available in a DES-simulator which makes difficult for instance implementation of such processor sharing algorithms where the selection of which job gets next time quanta depends on the amount of received or remaining service. Simulation of different processor sharing algorithms was seen essential to IN since SDF is likely to be a database and this was the main reason that the simulator was implemented as a scan-next-event simulator. A DES simulator is in general faster but the implemented simulator is sufficiently fast for the intended applications. Mostly the simulator is needed for determining grade of service parameters and for selecting a good load control algorithms. These applications involve only a relatively small network and the simulator performance is not so important as the ability to correctly model different features.

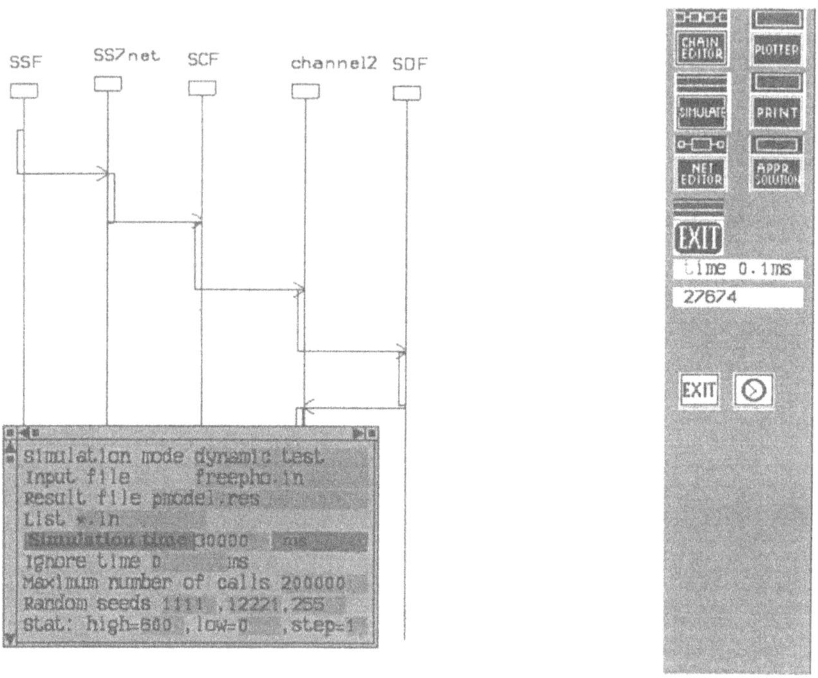

Figure 3.

Approximate solution finder

This dimensioning tool is not basically a simulator though a simulator is a necessary part for investigating the real behavior of the simulation model. Simulation can, however, be slow and it gives results as curves and not as expressions from which follows that generality of the results is often not clear. The approximate solution finder has another approach: Also here the editors are needed to model the network and protocols. But the models are not simulated, instead using the information in the models it is in many cases possible to solve some traffic related parameters using other mathematical and algorithmic methods than direct simulation. Applying the methods described in EURESCOM P308 project [1] this part of the tool creates a very simple model for the network as seen from SSP: it approximates the whole network as a one-server queue and a load dependent delay matching the maximum throughput to the queue and matching end-to-end delay to the load dependent additional delay for two different loads. This model may seem a bit too simple but if we ignore call rejection, then the whole network could be modeled as one queue with some unknown service time distribution, so our model is actually even unnecessarily complicated.

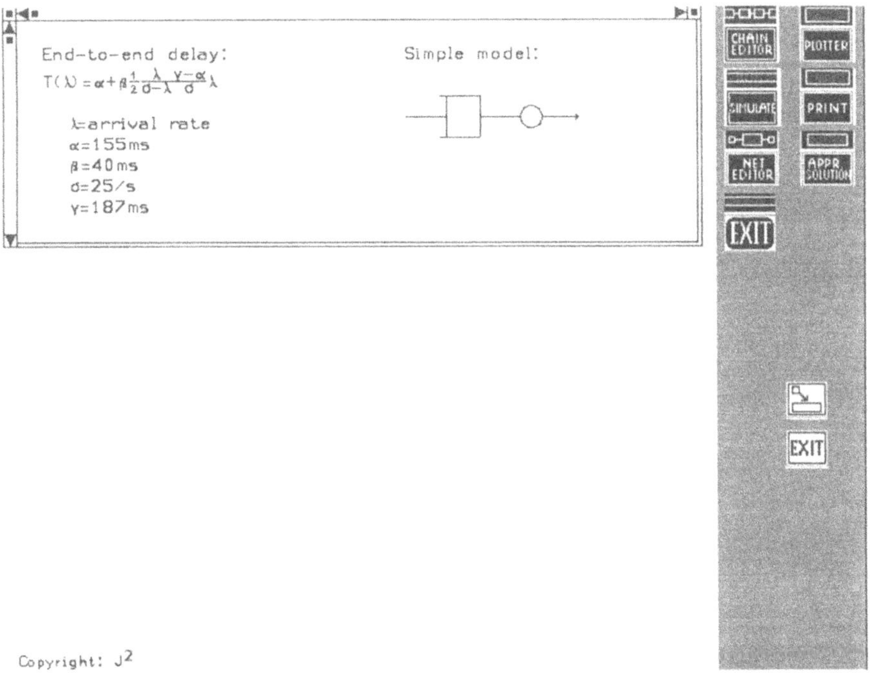

Figure 4.

The idea behind this model is that since, under mild conditions, all queues create in high load exponential delays, we separate the exponential part to a 1-server queue with similar high load behavior and the remaining part turns out to grow almost in a linear manner in many cases. The real growth of the remaining part is much more complicated but never the less, this simple model is useful as a rough model for calculating end-to-end delays. The tool uses an algorithm [2] for determining the queue service time and the additional delay parameters. In this model most of the more complicated features of the network are omitted but [2] describes some ways to take them into account.

The proposed way to include load control into this simple model is to have a finite waiting room for the queue and to modify the arrival process from a typical Poisson process, which may be a good approximation for some IN services such as FreePhone, to a process which is leaving a call gapping algorithm when Poisson process is entering the algorithm. It is not difficult to calculate the filtered process for some call gapping algorithms, see [2] for formulas and [1] for derivations. Knowing that real life SSPs and SCPs use several methods for load control one may wonder if the suggested ways are sufficient. It is likely that transfer from the load control parameters in this model to the actual load control parameters will not be one-to-one.

One way to include time-outs in steady state is to modify the service time in the one-server queue from the balance equations as in [2]. In [2] are suggested ways to include other features into the model such as rejection caused by automatic call gapping, however work of EURESCOM P308 on these formulas is not finished and so the actual methods may be quite different.

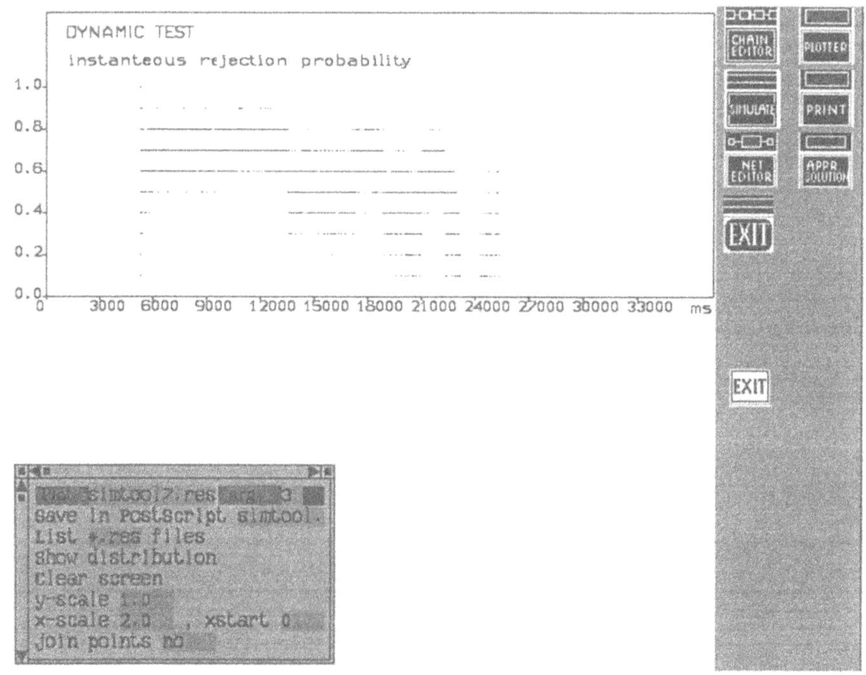

Figure 5.

The principle of the approximate solution finder is seen in figure 4. Whatever the actual methods for finding formulas for Grade of Service (GOS) parameters will be from P308 project, the tool calculates what it can calculate from the models defined with the editors and produces formulas for some GOS-parameters, such as connection establishment delay and rejection probability. These formulas can be used in another dimensioning tool which optimizes the network capacities. It is necessary to stress here that this presented tool is not the only dimensioning tool that would be needed, this tool helps to find some GOS-parameters, study the effects of load control algorithms and maybe something else. There are other needs, such as calculation of traffic matrices, optimization, planning which this tool does not at all address.

Plotter

Plotter (figure 5) is used for viewing the results. It can create a PostScript file for printing. A feature which is lacking from many commercial simulators is plotting a given curve against the results. In this plotter this task is easy.

The tool contains also a print-button by which the figures in this paper are created. It dumps the screen as a compressed color image to a file from which the image can be read to a text processor.

EXAMPLE OF DIMENSIONING PROBLEM, LOAD CONTROL ALGORITHM

As an application of the simulator in IN dimensioning figures 6-13. There the problem is to determine a good way to control overload. In IN load control is made in several ways and different call gapping algorithms have a central place. In [3] an adaptive load control algorithm for call gapping in circuit-switched network is presented. It uses occupancy of a node and tries to keep the occupancy at a target level, say 0.85. This way of dynamic control is also suggested in P308 [1] for the circuit-switched part.

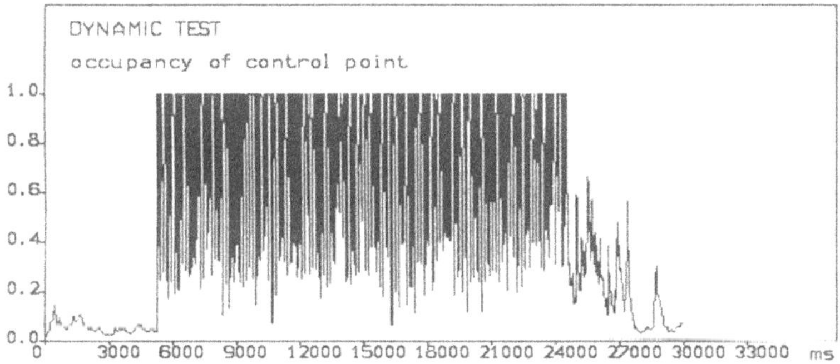

Figure 6. Occupancy using the algorithm in [3] for the simulated signalling network. The control does not manage to keep the occupancy in a given value.

For the signaling network this performance measure is worse than some other performance measures such as queue length in the control point or relative wasted time defined as received service/total waiting time for a message for two reasons: The occupancy is well-defined only in a steady state, under changing traffic conditions occupancy must be averaged and then arises a question how to calculate effectively this measure, in the simulations occupancy was averaged using a circular buffer over 100 last changes weighted by the length of the periods between changes, such calculation is a relatively heavy operation. Another problem with occupancy is that it is sensitive to overload only if the target value is much below 1. Since the signaling network is basically a waiting system, maximum performance as measured by number of serviced calls as suggested in [3] is obtained by having all servers busy if there is work to be done. Even if additional work to be done by the network nodes is taken into account, it is probably better to have some small queue in SDF and SDF occupancy close to 1 in case overload control is used. Then a suitable measure of load is queue length or relative wasted time which is closely related to the queue length. A suitable measure of goodness is then not the number of calls serviced because this measure will be, in all cases, almost the same as the number of calls serviced without any overload control, since it is limited by the service speed if servers are mostly busy. Better measure of goodness is the average waiting time or as in the following simulations, the queue length or the relative wasted time. All of these measures of goodness are related and give similar results. Control of the waiting times can be made by controlling either queue length or relative wasted time.

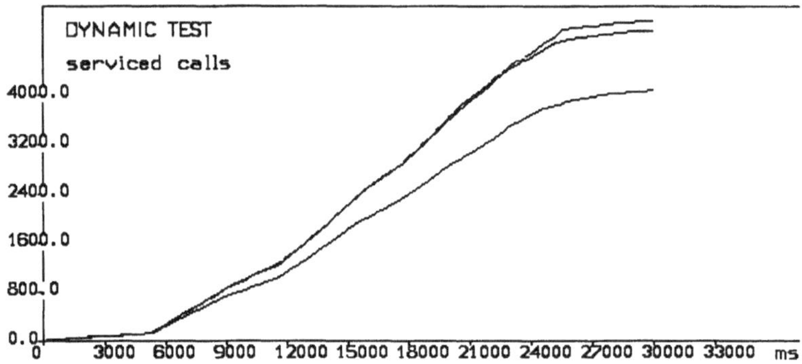

Figure 7. Number of serviced calls is smaller for the algorithm using occupancy (lowest curve) than for SQLA (simple queue length alg.) for the obvious reason that more calls are being served if the servers are all the time busy. The highest curve is SQLA and the middle curve uses constant gap size set to nominal maximum allowed gap size, then occasionally servers are not busy when the offered rate is low resulting to a bit lower performance than SQLA.

The tool contains a dynamic test where for 5 s (seconds) comes Poisson distributed traffic with a rate that can be serviced (in the example 25 calls/s), then from 5 s to 10 s the rate is raised to a high level (1000 calls/s) and then lowered again to the acceptable level (25 calls/s). The call gapping algorithm in all simulations was the basic call gapping algorithm where time is divided into time slots (gaps) of fixed length and the first call in each gap was accepted. Other variants of call gapping are also implemented in the tool.

A simple queue length control algorithm (SQLA) where two gap lengths were used changing them when the queue length in the control point (SDF) reached given upper and lewer bound levels results to some level of oscillation. The first oscillation is higher than the upper level (figure 8). As analyzed in [1] the first high oscillation is caused by the calls which have already arrived to the system before the overload is noticed in the control point. These calls create a queue which will not have time to resolve. Similar phenomenon occurs also using other measures of load. A simple improvement is to first stop all traffic for a short time by putting gap length to a high value and then lower the gap to the maximum level that can be serviced. This rejecting all arrivals for a short time allows the queue to resolve and waiting times are reduced (figure 11).

The simple queue length control algorithm has another problem when the offered call rate goes fast down, the algorithm changes the gap size many times (figure 12). In the simulations there comes another peak in queue length, this is caused by retrials since in the simulations rejected calls are retried 2 times with retrial time normally distributed with mean 5 s and variance 1. The control algorithm behaves rather poorly in the falling call rate part. One easy improvement is to change the gap length to a smaller value gradually. In figure 13 the gap length is decreased with a linear function of queue size instead of putting it abruptly to a small value as in SQLA. The second peak in queue length is reduced and in general the behaviour is better.

These simulations as such do not strongly recommend any dynamic load control algorithm since the model of the actual net has major importance. In these simulations good performance was obtained by an algorithm which, when gap should be increased puts it first to a high value for a short number of arrivals to the control point and then sets it to the maximum level that can be serviced. In the opposite way, decreasing rejection probability by decreasing gap length is made gradually.

It is also possible to simulate with the tool several sources, i.e., several SSPs using same SCF and SDF. It is just these kind of situations where dynamic load control is needed, if only one SSP is generating traffic to SCP, then call gapping could be all the time on with cap size put to the maximum level of traffic which can be serviced. With several SSPs, one way is to use two levels: a garanteed arrival rate for SSP and a maximum allowed arrival rate from one SSP. Dynamic control of gap size is used to change the traffic coming from SSPs between these levels depending on the activity of other SSPs. SCP must keep track of the SSP where call originates. One computionally effective load measure is queue length obtained as a difference of calls entering the control point and calls leaving the control point, such counter is kept for each SSP controlled by the SCP.

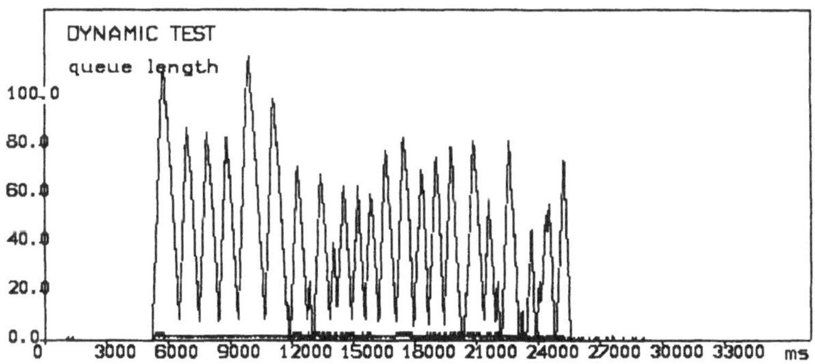

Figure. 8 SQLA can result to oscillating queue length (upper curve) but the oscillating can be made smaller by selecting the upper and lower value more suitably. Notice that SQLA results to one high oscillation when the rate changes (5 s and 10 s). The lower curve is queue length when a constant gap of maximum allowed rate is used.

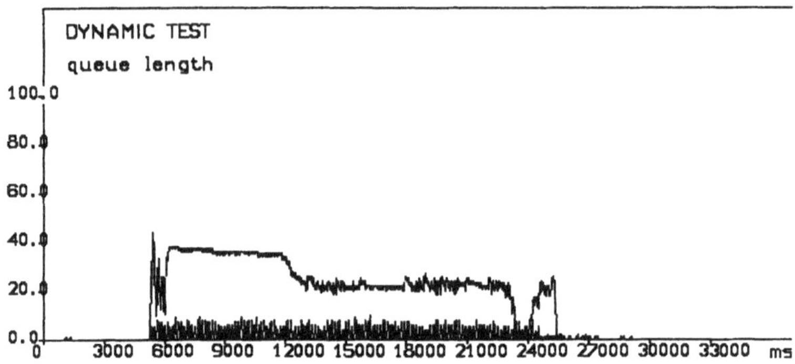

Figure 9. Queue length in the control point using wasted time (upper curve) and occupancy (lower curve).

The relation (received service time/total waiting time) in the controlling element is here called wasted time, as a load measure it is similar to the queue length and an algorithm similar to SQLA can be made using this measure with upper and lower bounds for load and two values for gap size. This algorithm gives similar results to those of SQLA since the load measures are closely dependent. The upper curves in figures 9 and 10 show respectively the queue length and the serviced calls if wasted time is measure and bounds 0.1 and 0.5. In the lower curves the algorithm is as in [3] (in [3] it the algorithm is not intended for signalling network).

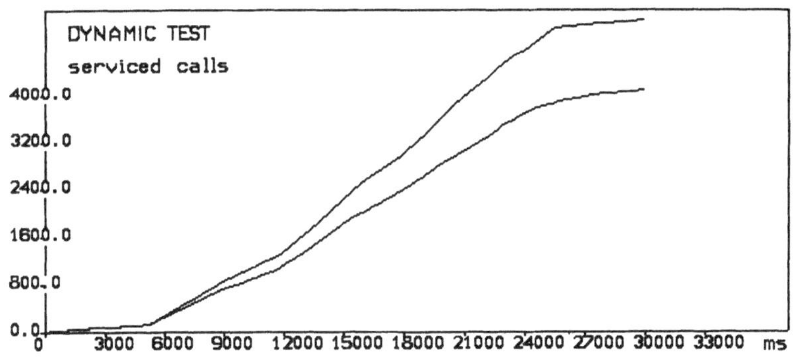

Figure 10. Number pf serviced calls using as load measure wasted time (upper curve) or occupancy (lower curve).

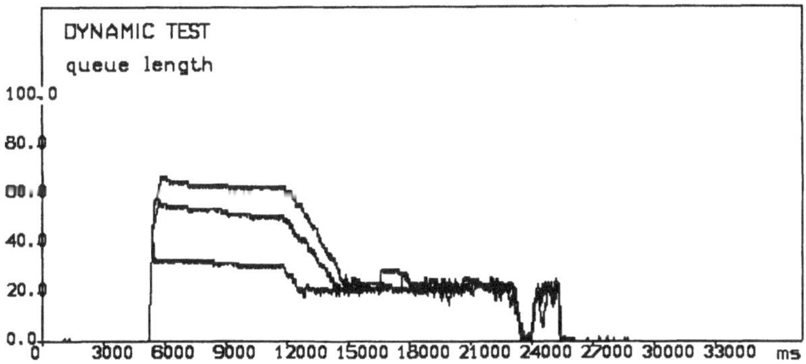

Figure 11. By stopping arrivals (putting gap size to a high value) for a small time, the waiting time of a call can be reduced. The waiting time depends on the queue length. In the lower curves the arrivals are stopped, highest is only SQLA.

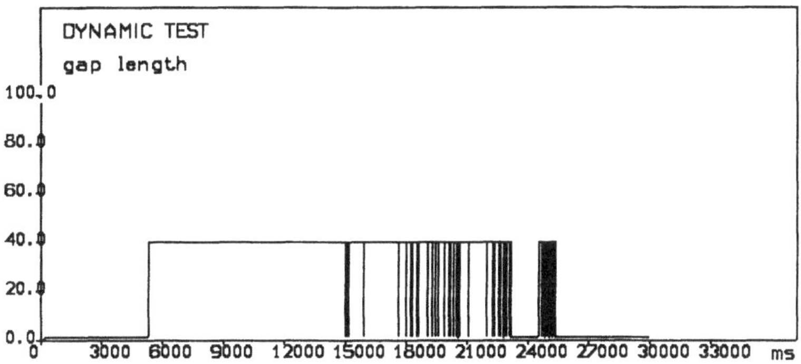

Figure 12. When the arrival rate goes down, SQLA has difficulties with stabilizing, in the figure the gap size changes many times. This results also to longer waiting times for calls than what is necessary.

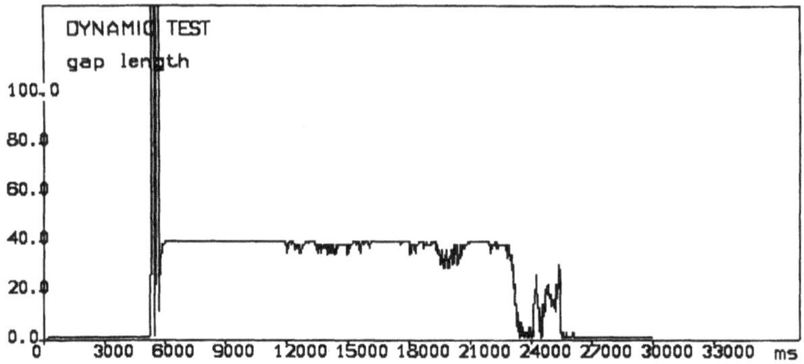

Figure 13. By reducing the gap size gradually, as in the algorithm above, the improved behaviour gives shorter queue lengths and consequently shorter waiting times.

ADDING NEW FEATURES TO THE TOOL

A user would probably sooner or later want to modify and add new features to the dimensioning tool. It is probably not possible to create a tool which has all of the necessary features implemented and parametrized so that a user can only change parameters. Reasons why this is not likely is that IN field is changing rapidly, there are non-standard solutions in existing networks and different solutions will always exist for achieving good performance, so a tool cannot model performance of any IN network using a small number of options. Many commercial simulation tools have solved this problem by creating a special simulation language by which any system can be described, the tool itself is often licensed software product of which sources are not given. In this tool a user makes changes by writing new C-code and all source code of this tool is available and can be modified at will. The advantage of this is that there should not be unclarity of what the tool does. The penalty is that changing somebody else's code is not one of the pleasures of the world. There are currently two major parts in this tool which should be understood by a modifier, the simulator and the graphical interface.

The simulator is a short piece of code, in each element queues as linked lists are kept for each server and a common queue to all servers. A scheduler is selecting next time quanta by asking from each element the time to the next instance of finishing service and selecting the minimum of these proposed times for each element and the time to next arriving call. All servers are advanced by this time quanta. The principle in the simulator is that a call is a chain of messages and the messages are queued in the server queues, this is not quite the same as an ordinary queueing network. When a call is generated, it is assigned a time-out which is the minimum of system time-out and user impatience time. Time-outs are checked between every time quanta from a special time-out list which is made effective by special pointers. In general, understanding the simulator code should be easy.

The tool makes use of a user interface development tool GRAPH created by the author. GRAPH is a part of a medium size software kit (about 40000 lines of C-code) which contains in addition to GRAPH a database, a special C-language compiler and interface to transport service and interface routines to some operating system calls. This set of tools forms a collection of local interfaces so that protocol software written in sligthly restricted C can be ported to another system implementing these interfaces (user interface, database interface, transport interface, system call interface) and also moved through net and compiled with the special C-language compiler. This enables fast creation of high level services such as multi-player games or other graphical programs which should work in the net but will not be standardized as application layer protocols, a feature which might be interesting in IN service creation once graphical services appear in large scale. Other parts of the kit than GRAPH will not be described here.

There exists several good tools to develop user interface such as MS-Windows, X Window System, Visual Basic so it may feel unnecessary to have a new tool for that. GRAPH is probably easier to use than most user interface development tools, especially for creating graphical editors. However, the simplicity is largely a result of lacking many advanced options. GRAPH has ready the following features:

- press buttons, icons, cards, basic graphical routines
- menu, text and vector graphics window with operations: resize, move, change to icon, scroll vertical and horizontal, pop, pull, rotate)
- vector graphics, that is movable and resizable objects made out of line segments and hot points where can be menus for implementation of a graphical editors
- fonts, fixed and proportionally spaced, created with a font editor
- two-dimensional editor for making bit map pictures, for instance for icons
- three-dimensional editor for making pictures which can be rotated in three dimensions
- animation editor for making animation objects
- creation of PostScript and compressed images (*.pcx)

GRAPH is object based, an objects has three functions: makeobject, object, freeobject. A display manager is managing the graphical objects. GRAPH is written for 286-486 DOS PCs but porting GRAPH is relatively easy since the dependency of operating system is only in extended memory handling, display of files in a directory and in mouse handling. Dependency of graphics display and processor is only in basic drawing routines (read and display a rectangle, bit, byte in a screen). In total, the non-portable code in GRAPH is a very small and isolated part of the code most of which is portable C-language code.

Main advantage in implementing the dimensioning tool prototype with GRAPH is that this interface tool contains several ready window types which in many other interface tools would be made with lower level coding. The connection to the simulator is also implemented and easily modified as is the plotter for viewing results from ASCII-tables produced by the simulator.

Using GRAPH for creating a new editor to the dimensioning tool would involve creating a new press button to the main menu. The press button image is made with the 2-dimensional editor. A editor window is made by creating new figures to the existing vector-graphics-window object in GRAPH. A sub-menu for the new editor is made with press buttons, again drawing the images with 2-dimensional editor. The menus for each object in the new editor are made using the menu-object of GRAPH. The necessary coding work for menus is to write fill-routines for the menus giving fields, the menu texts and default values. The actions of each menu choice are also written to a special file where actions of all menus are listed. The information from a vector-graphics-object will come to certain structures from which the data is read to the simulator or what ever is the purpose. In general, creating a new graphical editor with GRAPH is a matter of days.

REFERENCES:

[1] J.Jormakka-M.Negro-M.Butto:E-P308 Deliverable 2, EURESCOM (to appear).

[2] J.Jormakka: A model for SSP for dimensioning Intelligent Networks, IN'94 Heidelberg 1994.

[3] F.Langois-J.Regnier:Dynamic congestion control in circuit-switched telecommunications networks, teletraffic and datatraffic, ITC-13, Elsevier,1991.

8

Service prototyping in the OVOPS environment

Panu Puro[a], Jarkko Sonninen[b]
[a]Lappeenranta University of Technology, Data Communication Laboratory, P.O.Box 20, FIN-53851 Lappeenranta, Finland
Tel. +358 53 574 3613, Fax. +358 53 574 3650
[b]Systems Software Partners, Laserkatu 6, FIN-53850 Lappeenranta, Finland
Tel. +358 53 574 3612, Fax. +358 53 574 3650

Abstract

This paper will describe the OVOPS environment. The OVOPS environment is being developed by Telecom Finland and Lappeenranta University of Technology. It supports the design, implementation and debugging of distributed applications. Programs can be implemented operating system independently so that applications developed or embedded in the OVOPS environment can be ported to any system supporting OVOPS.

OVOPS can be used as a platform for implementing intelligent network services. The idea of using OVOPS in service creation is that it provides a way to design and implement services in an object-oriented and modular way. It encourages implementing software in modules (tasks) and provides facilities for message-based programs. Provided facilities are, for example, a communication channel between tasks with messages, scheduling of tasks, a base class for a task, other useful classes and a user interface for debugging.

Service creation is considered from a point of view of a programmer. OVOPS itself can be seen as a bottom-up approach for implementing message-based programs like intelligent network services. The bottom-up approach is good for fast prototyping and with it we can see the actual needs and problems of a service easily. A programmer may also affect the service implementation to make it more effective, which is usually impossible when high-level tools are used.

INTRODUCTION

Since the term "intelligent network" has been introduced we have been talking about fast service creation. It is true that with flexible architecture and network elements it is possible to offer all kinds of services without modifications to hardware. This is fundamental if we want to continue to develop intelligent networks.

Before any company producing intelligent network services can say to have a fast service creation cycle there is a lot work do to. Some work must be done because you cannot buy a ready system suitable for your needs. This can be said to be the situation at the moment but it is just matter of time when reasonable systems appear due to huge market in intelligent network business.

For the need of software tools the development of OVOPS was started in the summer of 1993. The goal was to implement an environment for event-based applications and especially for protocols. OVOPS consists of the OVOPS library and user interface. The user interface provides features to debug tasks in the OVOPS system, which is essential help in the development phase of software.

In the future we are planning to research the possibility to extend the OVOPS environment for intelligent network creating an environment for testing and developing intelligent network services.

SERVICE CREATION

Service creation can be seen as software engineering in a special environment, intelligent network. It has many problems to be solved before all benefits are exploited. Problems are raised by group work as well as other usual issues in software engineering.

An idea behind service creation is to produce services for intelligent network architecture. However, we are not just producing services but we want to produce them using as little effort as possible. This is an ideal goal that needs planning to be successful. Software tools are often introduced when we are talking about efficient software engineering although other possibilities exist as well.

A good way to minimize the time used in software development is to reuse code. This is one of the goals of OVOPS. The OVOPS library contains classes that can be used in all protocol implementations, which reduces the implementation time and number of programming errors. The needed code is in the OVOPS library that is ready and heavily used in other programs.

To produce a good piece of reusable software is quite difficult because it should be as general as possible or it should be easy to modify for different needs. Usually reusable software is not done because it takes extra effort and planning. This is effective but it could be more profitable in the long run to use this extra effort. Otherwise, the wheel is reinvented many times.

Software libraries are logical places to store general and useful software that should be available for every one in your organization. For example in an intelligent network service one library could be a group of functions for accessing a database. In this way you have a standardized way to do it and all of your services may use this same interface, which is a great help when you want to use another (different) database system for your services.

Unfortunately, using libraries is not so simple when managing large software projects. Some guidelines should be given for how to correct errors in the library and who will test the library. Addition to that every library should have a librarian who is responsible for maintaining the status of the library. The librarian is the person who answers all questions asked about the library by users.

Having useful libraries increases possibilities to prototype services because you have ready and useful code ready and available. Libraries have one good feature. Using libraries reduce the amount of code that is copied. This is good because if the copied code contains an error, it must be corrected only in one place increasing the quality and consistency of software and decreasing the amount of code needing maintenance.

All these issues mentioned above should be guided for programmers. Guideline helps to achieve all benefits of good software engineering.

After all, software tools are essential for productive software engineering. A tool should be flexible enough providing a possibility to employ it as you wish. It is too typical that the tool is not designed to work with other tools. Flexible tools, otherwise, are often complicated to use making it hard to get familiar with them.

OVOPS

Idea

OVOPS is designed to be a platform for other software tools. It is flexible enough with only few restrictions. One of the main benefits of OVOPS is that it provides software library that contains useful classes often needed in your software. Programming in the OVOPS environment modular architecture is encouraged improving compatibility and interoperability of your programs.

Shortly the idea behind OVOPS is to provide:

- Scheduling of tasks
- Message passing method between tasks
- An interface to devices (the operating system)
- User interface for debugging
- Timers, frame and other useful classes

OVOPS is meant to be suitable for different areas. This means that the tool must be flexible. OVOPS does not restrict the structure of the program. The system may consist of as many tasks and schedulers as wanted. The number of the channels between tasks can be specified by the user.

Overview

OVOPS is a practical approach for developing event-based applications. It supports the design, implementation and testing of event-based applications by providing classes for tasks, messages, schedulers, ports and devices. These classes are operating system independent and communicate by given message structures so that applications developed or embedded in the OVOPS environment can be ported to any system supporting OVOPS.

OVOPS is an environment where different applications, databases and protocols can be implemented and interfaced. There are no formal restrictions from the OVOPS side to the specifications and languages used in the interfaced applications. Foreign environments can be supported by specifying an embedding OVOPS task class for them.

In the OVOPS design performance was one of the most important aspects. So, the OVOPS library provides basic objects, or services. Of course, the services of the OVOPS library can be enriched by adding or writing specialized libraries.

OVOPS is suitable for being an environment in which tools and formal description techniques can be used. OVOPS can be said to be a practical approach to problems in large distributed applications and it should be used as a platform for software engineering.

An OVOPS system consists of tasks and channels. The channel connects two tasks to each other. Tasks communicate through a channel by sending messages. The message passing is asynchronous.

Tasks in OVOPS are independent from other tasks. The only thing a task knows about another task is an interface. The interface consists of a port and messages that can be send through the channel. A developed OVOPS system can be divided into several processes that can run in different computer systems.

The operation of tasks is event driven. All tasks have to perform their work gradually because the system is not pre-emptive. Only one task is executed at the same time. The task must release the control voluntarily. If not, all the other tasks inside the same process will be waiting the end of the execution of the running task.

The executions of tasks inside each process are scheduled by a user written scheduler, or a scheduler provided by the OVOPS library. Schedulers can be scheduled by other schedulers. So, it is possible to build a complex hierarchy of tasks. The top level scheduler is executed by the main loop of the process. There is no so called kernel in the OVOPS system but the functionality of OVOPS is based on contracts between classes.

Complex OVOPS systems with multiple schedulers have to be designed carefully and formal tools to prove and simulate complex behavior are needed if you want to be absolutely sure that you get benefits using a complex structure.

Usually, all processes need access to devices, a terminal for example. Physical devices are handled by driver tasks. The driver task is a task that uses a device class as an interface to the physical device and forwards data from or to the device by sending messages. The driver task is needed due to the nature of the system and the tasks are wanted to be kept independent from the operating system used.

CONCEPTS

VOPS

At first we could consider what is a VOPS (virtual operations system or virtual operating system). Why do we need the VOPS? The VOPS can be seen as a layer above the operating system. It helps a programmer to implement more modular programs, which are usually event-based applications, that communicate by using messages.

With the term VOPS we understand the platform that gives additional help and services for the programmer to implement event-based applications. OVOPS provides services for debugging and tracing system events, scheduling of tasks and transferring messages. These are same kind of services that the operating system usually provides but these are intended to be independent from the operating system improving portability between different operating systems. No assumptions are made if the operating system provides help for multi-tasking or not, which can be very true for embedded systems. The VOPS style architecture may sometimes be more efficient than the application that deploys multiple processes and interprocess communications for the same purpose because communication is lighter and there is no need for context switches, for instance.

OVOPS System

With an OVOPS system we mean the group of user written tasks and the system tasks that the OVOPS provides. It is the entire system that performs its work, for example, a protocol stack.

The term is later used to express the OVOPS library and the user interface, in some cases.

OVOPS Class Library

The purpose of the OVOPS class library is to provide standard services for applications implemented with OVOPS. The library is not intended to cover all the needs that a programmer has but it provides some fundamental services needed in event-based applications.

The library cannot support all needs in different software areas. So, all additional services are supported by specialized libraries or tools for a particular purpose. In this way the OVOPS library can be kept as standard and stable as possible.

User Interface

OVOPS can be divided to two parts, the OVOPS library and the user interface module. The user interface provides help for tracing and debugging of user defined and implemented tasks. The main purpose of the user interface is to provide help for tracing messages, i.e. events happened, in the system. This kind of help lacks from the debuggers that are also very useful in programming beside the user interface. That means that the user interface is in principle only used when you are developing an OVOPS system consisting of OVOPS tasks and channels interconnecting them. The user interface is an important part of the OVOPS system although it is used only in a development phase.

The user interface can be said to be some kind of debugger because it provides same services and features. However, the user interface needs help from the user because it does not use any debugging information on object files but it needs hooks to the user tasks. Hooks are C preprocessor macros, which are provided by the user interface and contain function calls that inform the user interface about variables in the user task. The hooks are for getting information and modifying variables in the tasks by using the user interface. Only hooked variables can be seen and set by the user interface.

There are no hooks in the code in the OVOPS library, which means that the classes of the OVOPS library are the exact code that will be in the developed product without any preprocessor's macros. In other words the OVOPS library can be linked with the development phase code and the product code. However, the hooks in the user written tasks should be able to be removed when the user interface is no longer wanted.

Any printing and setting routines are not wanted to implement to the classes of the OVOPS library, i.e. the code needed only in a development phase, for example print functions, is kept separately. This is the reason why hooks are needed. The actual printing and setting routines, which have access through hooks, are in the user interface, not in classes in the OVOPS library.

Tasks

Tasks are executable classes in the OVOPS. Because the system is not pre-emptive tasks have to perform all processing gradually. In other words a task has to release the control to the other tasks. Otherwise, the other tasks are and will be waiting execution forever, and only one task is running.

It is natural to model a task by using a finite state machine because tasks are asynchronous and event-driven. The task usually has certain messages that are input events and they produce transitions and output messages when processed. This model suits well for the nature of tasks in OVOPS. However, OVOPS itself does not provide any help for modeling the finite state machine. It is supposed to be provided by other tools.

Tasks do not have any predefined structure, except some variables needed by the system and message queues. The user's job is to build interfaces, any number of interfaces, to the task and to divide the application into modules, or tasks, and, of course, to write the functionality of the task to the run function. The run function is the function that is called by the scheduler when the task is decided to be executed.

In OVOPS there are two kinds of tasks, Otasks and EventTasks. The Otask is the base class of all executed tasks. The EventTask is a task that has a message queue and it is usually inherited by the user written tasks. EventTasks have predefined run function that takes a message from the message queue, if present, and calls an execute function with the message. The execute function should have a user-written body. Only special kinds of tasks do not need message queues, i.e. they are inherited from the Otask directly. These are, for example, system tasks, schedulers, I/O handler and timer task.

Message Passing

The message passing is one of the most important features in OVOPS. It is handled by port classes. The user makes instances, as many as needed, of the port and connects them to other ports. One port can be connected to only one port at the same time. In other words there are not predefined ports in any task.

The port has to know to which task it belongs because it has to be able to store the received message to the message queue of the task.

The OVOPS library contains two different ports. One is just for use of inside the OVOPS system in the operating system process. The other is supposed to be used for communication between two or more operating system processes. It uses UNIX or Internet sockets as a transport service. The user may use these two ports as base classes and build specialized (function) interface for messages sent through the port. This is one way to simplify and to hide the sending of messages improving the readability of the program.

All message passing in OVOPS is asynchronous. A task never blocks waiting on the actions of the recipient to which it sends a message. This is a part of the nature of the system, no task can block unless it is in its own process. Asynchronous message passing makes it possible to implement efficient tasks, but it makes it also much harder than doing it in an easy way. With a complex code that models a finite state machine you are able to handle many events concurrently, instead of handling events in particular order sequently without many concurrent contexts.

Ports are designed as light as possible making the over-head of message passing low, but of course there is always a little over-head when sending messages. You should be careful when giving parameters to messages trying not to copy parameters, which may easily low down the performance of your system. Large data structures as parameters should always be passed by using pointers and references.

In practice almost all the messages are inherited from the message class of the OVOPS library because it does not contain any members that can store user's data. The OVOPS message contains just port addresses, the ports that the message has gone through.

Distribution

By the term distribution we mean in the OVOPS world how to connect tasks in different operating system processes. As described before the message passing is provided by ports and by using socket ports data can be transferred between two processes. UNIX or Internet domain sockets can be used.

Unfortunately communication between two processes is not so easy. The programmer must provide her own encoding and decoding functions that take care of transforming the message to a stream of bits for all the messages sent through the socket ports. Of course, you can use existing coding tools. But anyway, there is some over-head transferring messages to another process, or computer, both in development phase and run-time. In development phase you should provide the coding functions somehow and in run-time the messages have to be coded and decoded before and after sending through the socket.

INTEROPERATION OF CLASSES

OVOPS does not have so called kernel because all the classes are taken in use by the user. The user has to even construct the main loop of an OVOPS system for correct operation by herself. This all means that the operation of the system is based on contracts between classes of the OVOPS library. The following text describes these contracts and the functionality of classes.

Of course, if you are writing your own classes by inhering from classes of the OVOPS library you can change functionality, but you have to be sure that your classes will work with existing classes. Normally, only relationship between a port and event task is interesting for the user that implements software using OVOPS. Other relationships are useful for OVOPS developers, persons who write their own schedulers, for example. However, sometimes detailed information helps to understand the system better.

Port

Between two ports are only few contracts that are only used when the port is the basic port (Port). Your own modified ports may operate differently, but the interface of ports remains the same.

When a message is sent by calling putMessage function the port checks if the port is connected, that is the port has a reference to another port to which it may want to send messages. If the port is connected the port calls the getMessage function of the port of the other end and causing it to receive the message. A message can be transferred only if the port is connected to another port. The getMessage stores or handles the message, or whatever. We will describe this later.

Because ports have references to other ports they have to communicate with each other when a port is disconnected. This confirms that a port does not have a reference to another port that does not exist.

Ports can be disconnected by calling disconnect member function, or ports are disconnected automatically when they are deleted. The disconnect function clears the reference. So, the port is no longer connected.

When a port is connected or disconnected a short handshaking is performed. When the connect is issued the port calls the connectRequest function of the port of the called end that calls the connect function of the task. After this the connect function of the task of the caller end is called then the connection is ready. If the connection was refused the disconnectRequest is issued to the port of the called end.

And when disconnecting the disconnect function of the task is called first. Then the disconnectRequest is issued to the port of the called end and the connection is down. The disconnectRequest function will call the disconnect function of the task.

EventTask and Port

A port has to belong to an event task, which is a task with a message queue, if the default port class is used. Otherwise, the port cannot do anything with the message that was received. The task must be known because the port calls the save function of the event task to store the message into a message queue to be handled later.

There are also two other functions in event tasks that ports call. When a task connects a port the connect function of the task is called in both ends. In both ends it can be used to monitor connections or it can be used for accepting or refusing connections.

Almost the same situation is when a port is disconnected. The disconnect function of the task is called, but it cannot deny the disconnection. The connection is always disconnected, anyway.

Scheduler and IoHandler

An I/O handler is a task but it has some special functions. That is why the scheduler has to know which task is the I/O handler. The special function that the scheduler has to know is the block function. It is called when the scheduler asks the I/O handler to block until something is happened in devices, data can be read or written or a time-out has occurred. The scheduler can call the block function only when there are no messages waiting in any message queue in any task, that is the load of all tasks is zero. Normally when there are messages to be handled in tasks the scheduler calls the run function of the I/O handler to process I/O requests of devices.

Nothing can be said about the calling frequency of the I/O handler. The current scheduler of the OVOPS library calls the I/O handler when it finishes executing tasks with certain priority. This makes the frequency quite undeterministic because the number of executed tasks varies. The I/O handler is not called before every task is executed because it is not usually needed, of course it depends on how much you communicate with the outer world. Undoubtedly the frequency affects to the performance of the system. If it is called too frequently it is extra overhead and if it is called too rarely the throughput of the system may suffer.

Scheduler and Otask

Every task has the knowledge of its scheduler if it is wanted to be executed. When the scheduler is decided to execute a task it calls the run function of the task. The run function is assumed to the actual task's job.

The task tells the scheduler when it should be executed by calling the request function of the scheduler. It usually happens when the task receives a message and the request function is called by the save function of the task. The scheduler is responsible to take care of executing the requested task as long as the load of that task reaches zero, that is the task has nothing more to do. This minimizes the need of calling the request function repeatedly. We do not assume that every task has a message queue and that is why the Otask class does not have it and we are talking about the load of a task.

When a task is created the constructor of the task calls the inform function of the scheduler to tell that a new task has been created and it should be scheduled.

Analogically, when a task is deleted the destructor of the task tells the scheduler by calling the forget function of the scheduler that the task does not need scheduling.

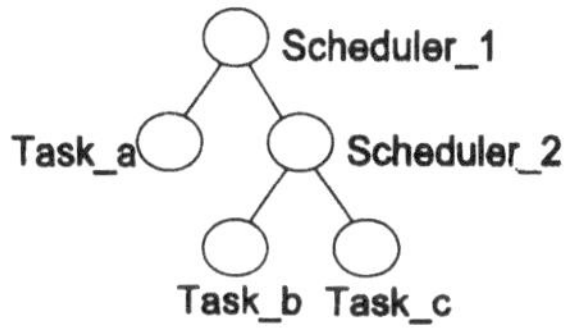

Figure 1. Multiple schedulers. One example of a very simple system structure.

The scheduler itself acts as a task, which makes it possible to a scheduler to have tasks and schedulers to be executed. It should be clear that the use of multiple schedulers should be researched carefully because scheduling takes a certain amount of overhead and the execution of the system becomes easily heavier than expected. But it can also be a great benefit if you are developing a very specialized scheduler for a particular purpose. As you may notice you do not need necessarily the top most scheduler but you can call the second level schedulers yourself reducing the number of schedulers in the system.

The scheduler performs its work gradually, executing only one task at the same run function call. And of course, the run function must not block because all the other tasks are waiting for execution.

IoHandler and Device

The relationship between an I/O handler and devices is much like the relationship between the scheduler and tasks.

A device has to know the I/O handler to which it belongs because it has to tell the I/O handler when it is created and deleted. This is done by calling the inform and forget functions of the I/O handler and the reason is that the I/O handler has to know all the devices in the system because it is monitoring them.

The device changes the status due to functions called by the user. The I/O handler monitors the status and checks if it is possible. For example, when the user requests data to be read the device changes the status for reading. Later the I/O handler checks if it is possible to read the file descriptor associated with the device. If it is possible the specific callback routine is called to handle the actual reading.

INTRODUCTION TO USER INTERFACE

OVOPS User Interface is a separate library. In the core OVOPS there are no support or dependencies for debugging. User Interface Library can be seen as an application or utility library that uses OVOPS.

The final version of the user program is usually meant to use only the core OVOPS library. In debugging phase the programmer could use the user interface library with certain C preprocessor macros. In that way it may be easier to disable the user interface afterwards.

Since there is no debugging information or functions in core OVOPS classes, user must explicitly give to the user interface a reference to her objects. With this technique, the user interface gets a picture of the user system, and can manage it.

In practice the user is mostly interested in running her application and getting information about its status. The user usually initiates system by sending messages through interfaces. Messages start traveling between tasks and system changes its state. The user interface library helps keeping track of what happens by displaying information about tasks and the message before and after its execution.

Commands

The OVOPS user interface is character based. User is prompted for commmand words that are interpreted by the user interface library.

```
Welcome to OVOPS (Object Virtual OPerations System)
Copyright (C) 1993 Telecom Finland

( log macro system Exit alias unalias shell ? )
```

In the topmost level of command hierarchy there are the following commands.

- Macro
- System
- Log
- Alias and Unalias

- Other

At the startup of system a special file ovops.rc is read, and commands in it are executed, just as user would have written them. With this file user can preset variables, aliases and so on to make debugging session more comfortable.

In most cases commands can be abbreviated simply by giving first characters of word. The command that first matches to given word is executed. No ambiguities are tested. OVOPS user interface will give you list of possibilities, if you don't give enough input, or it does not accept your data. You may quote longer texts containing special characters and blanks using single or double quotation marks around text.

Example Session

Here is a short piece of session using OVOPS interface. In the session, first the structure of system is displayed, then the variables of task "a" is looked at, and finally a message is send through upper interface of task "a."

```
net/project/ovops-1.0.1/ui/test:1>./amain

Welcome to OVOPS
Copyright (C) 1994 Telecom Finland

( log macro system Exit alias unalias shell ? )
sys str


Task 'ui' = {
    Port 'output' -> Unconnected
}
Task 'a' = {
    Port 'down' -> c:up1
    Port 'up' -> ui:output
}
Task 'b' = {
    Port 'down' -> c:up2
    Port 'up' -> ui:output
}

Task 'c' = {
    Port 'up1' -> a:down
    Port 'up2' -> b:down
}

( log macro system Exit alias unalias shell ? )
sys a print


Task 'a' = {
    Var 'a' = 0
    Frame 'b' [0]  =  {
        31 31 12
    }
    Timer 't' = 10.000000
    String 'name' = "Task A"
```

```
    Text 'hmm'
    MessageQueue: 0  SystemQueue: 0  Load: 0
    Trace: Message
}

( log macro system Exit alias unalias shell ? )
sys a up

( CrMsg CcMsg DtMsg DrMsg show set print trace struct ? . )

CrMsg CR 123

*** Message to a:up from Outer Space ***
Message 'msg_1' = {
    Enum 'typ' = CR
    Var 'qos' = 123 (1918981664)
}

Got Message: CR
*** Message to c:up1 from a:down ***

Message 'msg_1' = {
  Enum 'typ' = CR
  Var 'qos' = 123
}
```

SIMULATION SUPPORT IN OVOPS

Intelligent Network services are complex program entities and it is difficult to analyze them in other ways than simulation. Simulation provides also a possibility to test various configurations and values of parameters easily.

As OVOPS is a generic programming tool it is relatively easy to build simulation support above it. The OVOPS way to structure functionality as communicating tasks fits well for modeling simulated systems.

Use of simulation

Simulation of IN services can be performed in many levels of abstraction. User can construct an OVOPS system for modeling interactions between network elements such as SCP, SSP and SDP for analyzing performance issues of IN architecture. Such a system could for example give information about overall performance of telecommunication system. There could be certain delays in between entities and amount of traffic may vary.

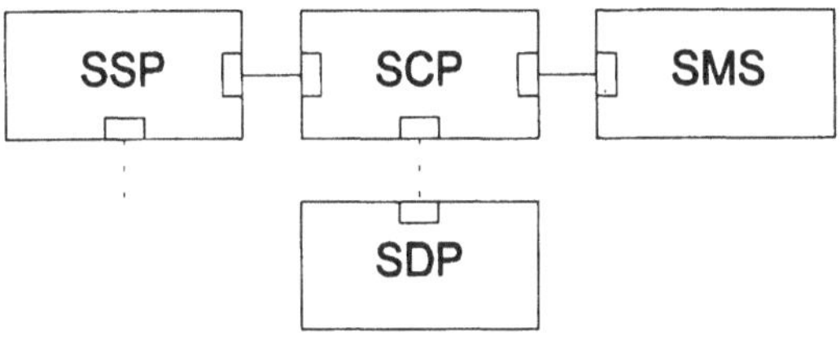

Figure 2. Example simulation system for investigating intelligent network architecture.

Simulation may help also the actual implementation of service. After the service and its environment are specified, a simulation environment for prototype can be used to test the first version. In this kind of environment simulated resources such as databases and SSPs can be used. It is faster to make prototypes without needing to use the real environment, for example telephone exchanges. This applies also for the first testing of service, though the final testing must always happen at the real system.

The interface between the service and execution environment can be kept similar in the simulation environment and real environment, so in principle there are no changes between the simulation version and final version of the service.

The OVOPS system is based on modules, so it is possible to gradually replace simulation modules with real service environment modules and to test at every phase.

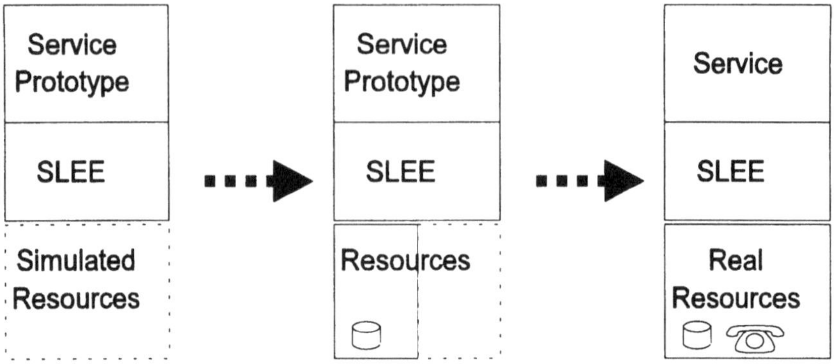

Figure 3. Modularity of OVOPS environment gives a possibility to replace gradually simulation modules with real modules.

OVOPS Simulation Environment

A support for virtual time and scheduling belongs to the OVOPS simulation library. Another important part is statistical functions.

For the IN services the code for simulated resources is also needed. SLEE has to be customized to be useful in simulation. Since the SSP part of service is provided by simulation it is easier to generate service traffic by traffic generator modules.

Logging events for further study and comparison are an obvious and essential feature and support for that is part of the core OVOPS user interface. Ability to control and display statistics is also needed.

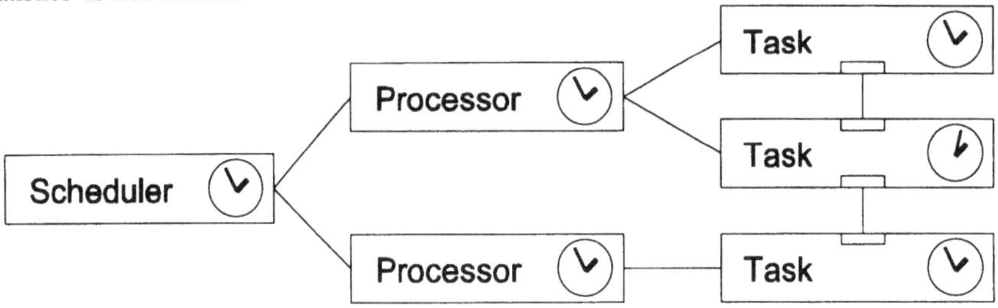

Figure 4. Example simulation system hierarchy

Basic concept in simulation environments is often an *event*, which is an input appearing at a certain moment. The event is mapped in quite a straightforward way to OVOPS Message class with a time stamp.

Usually the OVOPS system consists of tasks connected together with ports and schedulers running tasks. For virtual time, a new concept called Processor has to be introduced. New, modified Task and Scheduler are also required.

Processor models a real world process or a processor. It can be thought as one kind of scheduler that can execute one task a time. The main level scheduler controls processors to execute events in their time stamp order.

Virtual time has to be added to the main OVOPS classes. That is for keeping track which event has to be executed next and to update statistic variables.

The main functionality of the system is implemented in tasks. There are no major logical changes to the tasks in the simulation environment compared to the real environment, since execution and scheduling in a virtual time system matches to the real situation.

The system described in this chapter should fasten the development of service prototypes, aid and speed up testing in the way from prototypes to the final product.

BIBLIOGRAPHY

/1/ Margaret A. Ellis, Bjarne Stroustrup: The Annotated C++ Reference Manual, New Jersey, Addison-Wesley, 1990, 447 p., ISBN 0-201-51459-1.

/2/ CCITT Recommendation I.312/Q.1201 - Principles of Intelligent Network Architecture.

/3/ CCITT Recommendation X.701 - Systems Management Overview (ISO 10040).

/4/ CCITT Recommendation X.722 - Guidelines for the Definition of Managed Objects (ISO 10165-4)

/5/ Tapani Karttunen, Intelligent Network Service Creation Process, Workshop on Intelligent Networks - Proceedings, Lappeenranta, Lappeenranta University of Technology, 1993, 8 p.

/6/ Olli Martikainen, Valeri Naoumov, Konstantin Samouylov, Portable Intelligent Network Software Implementation, Network Information Processing Systems - Proceedings, Sofia, 1993, 6 p.

/7/ OTSO user's guide, Espoo, VTT/TEL, 1992, 132 p.

/8/ Object Virtual OPerations System Manual 1.0, Lappeenranta, Lappeenranta University of Technology, Telecom Finland, 1994, URL: http://shagrat.it.lut.fi/ovops/

9

The local management for a service control point

Tang Haitao[*], *Esa Kärkkäinen*[**]
Telecom Finland Ltd
Telecom Research Centre
P.O.BOX 145
FIN-00511 HELSINKI
FINLAND
Telephone: [*]*+358 2040 2513,* [**]*+358 2040 2990*

Abstract

Suitable local managers seem to be essential for increasing the performance, availability, and flexibility of the Service Control Point (SCP). In this paper, the main points of designing one local manager for an experimental SCP are introduced. The study shows the local manager is suitable for the purpose.

1 INTRODUCTION

A Service Control Point (SCP) is a key component in the Intelligent Network (IN). SCP works as a central server for customer services. Any problems in the SCP will greatly affect the performance of IN. Even if the hardware and software products used for IN would be more reliable in future, the environment of the SCP and other harmful factors will continue to threat its performance. Therefore, it is required to maintain the SCP with real time, full flexibility, and high efficiency to minimize the impact of failures on the system performance and to introduce quickly new services as well as change existing services. A local manager seems to be the solution to most of the requirements.

1.1 The SCP system and its local manager

The experimental SCP developed in Telecom Finland Ltd has the principal structure shown in figure 1 [1]. The SCP consists of two similar SCP-supporting systems, one active and the other passive at the time.

The SCP is implemented on an usual computer platform, which has an unix-like operating system. Each intelligent network service acts as an own service logic execution environment

(SLEE), which is a separate process in operating system. This makes it possible to have many different services in one machine. Services are also easier to manage because one service does not affect other services. Common commercial relational databases, like Sybase or Oracle, are used. Queries and updates from the services are centralized through the scheduler process. The scheduler is responsible for data integrity in the database. The SCP is connected to Service Switching Point (SSP).

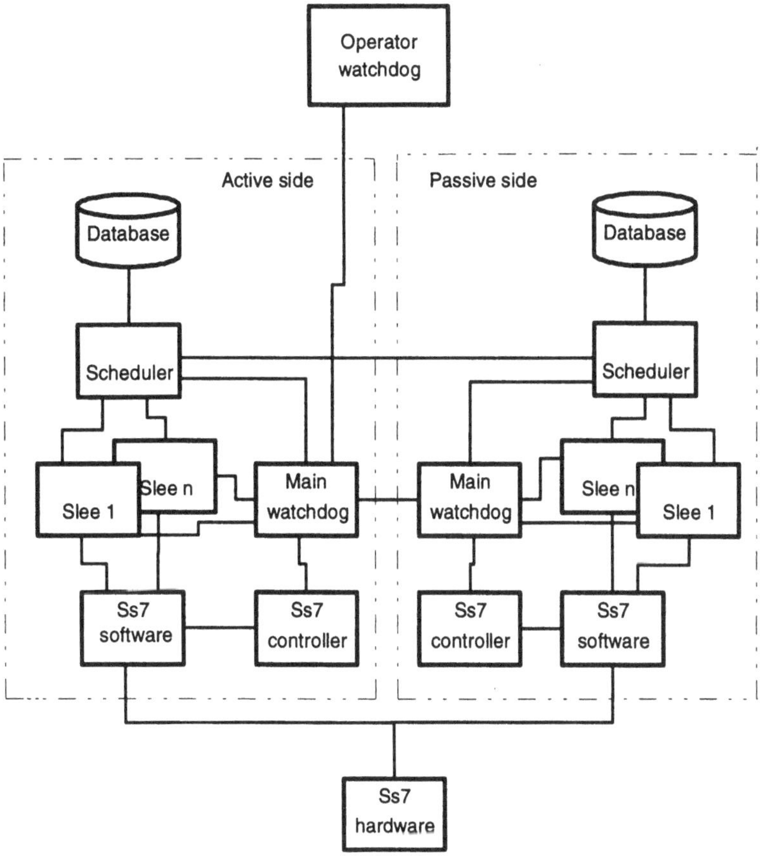

Figure 1.

The local manager inside the SCP is implemented with two main watchdogs and one operator watchdog. The local manager is used for maintaining the SCP and increasing its availability. It should be able to make automatically all parts of the SCP ready for the goal needed by the SCP.

The two main watchdogs connect each other and their information exchanging makes it possible for one of the SCP systems to be active and the other passive dynamically. The two main watchdogs decide together the roles of the two SCP systems.

The operator watchdog is mainly an interface between the SCP and SCP operator or other possible managers. The operator can control the SCP through this interface, even the local manager works independently and automatically. The reason for using this kind of operator watchdog is the need for the reliability, structure simpleness, and safety of the SCP.

From the system point of view, the local manager should also make it possible for the operator to introduce new services and change existing services quickly and without affecting other services in the SCP. The local manager is not a part of the real services in the SCP. From the efficiency point of view, it should not use system resources too much and be as simple and as reliable, as possible.

1.2 CVOPS

Local managers are built on CVOPS (C-based Virtual Operating System), which is a portable protocol development and run-time environment. CVOPS has been developed at Technical Research Center of Finland (VTT). The main reason for selecting CVOPS as a development tool for local managers is that CVOPS has been used successfully in other parts of the SCP.

The portability of CVOPS is based on a virtual operating system concept, which means that the operating system sees the whole CVOPS tool with all the protocol entities as a single process [2]. The protocol entities are implemented in different CVOPS tasks called vtasks. CVOPS provides FIFO (First in, First Out) scheduling and communication between the vtasks in a same process.

CVOPS gives basic framework to build an application and also several support services for developer. Logic actions are described as an extended finite state automaton (EFSA). CVOPS can work in many operating systems. If applications use only CVOPS services, they are portable to the operating systems too.

2 DESCRIPTION OF THE LOCAL MANAGEMENT

The operator watchdog and the main watchdog are the important parts of the local management. The operator watchdog acts as a manager for the SCP, since the operator watchdog sends certain requests to the main watchdog and receives event messages from the main watchdog. The operator watchdog can also act as an agent for another network manager. The main watchdog is an agent for the operator watchdog and a manager for all other parts in the SCP system. It can send certain requests to any part of the SCP system and receive events. Two main watchdogs are a limited agent and manager for each other at the same time. Their structure, working logic, and functions are essential for the performance and the extendibility of the SCP as a whole.

2.1 The protocol of the local manager

The protocol used in the local manager is a special one, which is not any of the standard protocols like Simple Network Management Protocol (SNMP). The reason for using the special protocol is that it is simple and efficient enough for the local management case. The simplified message structure of this protocol is shown in figure 2.

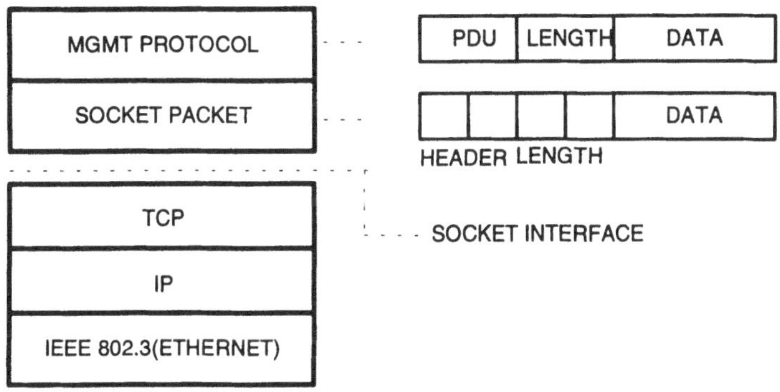

Figure 2.

2.2 The internal structure of the main watchdog

Figure 3 shows the internal structure of the main watchdog.

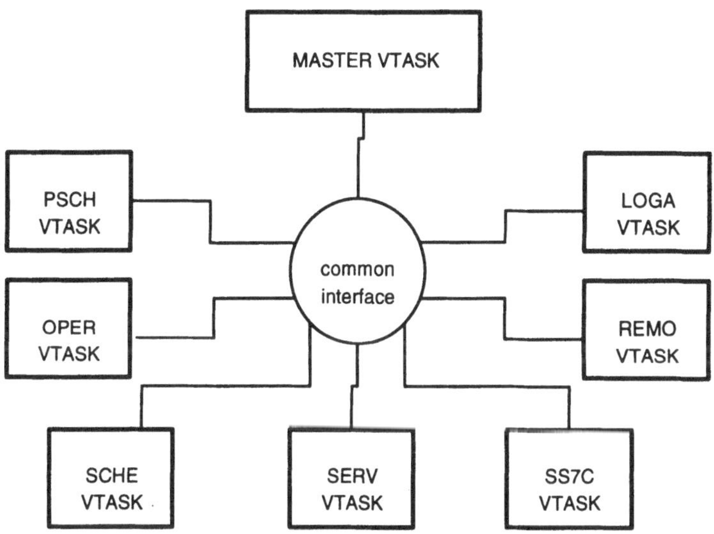

Figure 3.

The master vtask implements the control logic of the main watchdog. The slave vtasks can gather information and execute actions. All the vtasks have a common interface with each other and share common information. By this way, the structure of centralized controlling and distributed message processing is built up. One advantage of using this structure is that the overall logic is clear and the function groups are relatively independent. It is like using simple blocks to build a complex one. Another advantage is its flexibility, it means one can add new function group or modify the existing function groups without affecting the other parts too much.

Table 1. Main functions built in each vtask

Vtask	Managed part
master	global decisions
remo	Connection with the main watchdog at the other side
ss7c	Signalling System number 7 (ss7)
serv	Slees
sche	Scheduler and databases
oper	Connection with the operator watchdog
loga	Logging
psch	Checking and killing processes.

One very important thing for designing the main watchdog is the way of making certain decision from the asynchronous events gathered into the main watchdog [3]. To do this, one global data structure and certain algorithm for correlating the asynchronous events are used. The principle is shown in figure 4. It is better to use the following simple example to interpret it.

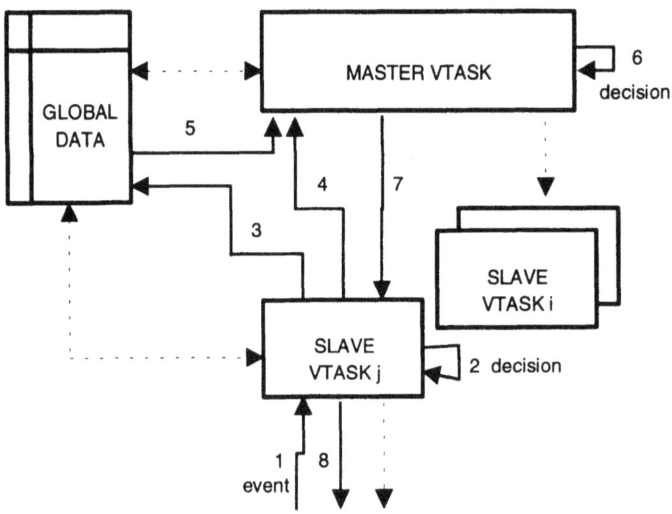

Figure 4. The numbering means the order of actions.

When some event comes, a slave vtask will make local decision and update the global data. Then, it will inform master vtask about this event. When master vtask receives this event, it will make a certain decision and give a certain action command to the slave vtask or other slave vtask(s). In turn, the slave vtask or other slave vtask(s) will execute the command.

2.3 The functions given by the main watchdog

Table 2. The offered services by main watchdog to the SCP

Service	Actions
starting	Start the system
maintaining	Keep the system alive
closing	Stop the system gracefully
switch over	Move services from one side to another side
interacting	Exchange information with the operator watchdog.

When the main watchdog has been in maintaining state, it keeps checking its own and the other side of the SCP system according to its negotiated role. If the main watchdog finds something wrong, it will try to fix the problem. Sometimes, it will go through switching over. There are two kinds of switch over cases, normal switch over and immediate switch over. If the main watchdog goes on normal switch over, it will first check if the other side is able to accept the possible active role. If the answer is YES, the main watchdog goes on closing and then, with the finishing of closing, this main watchdog informs the other side to take over service. If the main watchdog has to choose immediate switch over, it will inform the other side to take over service role or be active itself immediately.

In closing, main watchdog will go through the system shut down, which includes shut down of the slee(s), deactivating the ss7, and shut down of the scheduler. The main watchdog will also inform the other side to take over the service.

It is also important for the main watchdog to support the interactions between the SCP system and an operator or other managers. There is no any direct connection between the service stack of the SCP and the operator or other managers. The SCP may need certain information from the operator or other managers. For this reason, the main watchdog has to serve as a coordinator between them too.

2.4 The operator watchdog

As an interface between an operator and main watchdogs, the operator watchdog gives a way for them to control the main watchdog and other SCP parts by the following services:

Table 3. The offered services by main watchdog to the operator

Service	Actions
connecting	connect to the main watchdog
disconnecting	disconnect from the main watchdog
shut down	shut down one side or both sides
look	look the running situation of the SCP
change log	change the log of the scheduler, the slee,or the main watchdog

start	start any of the slees
stop	stop any of the slees
reset	let the slee to read its configuration

If one looks the local manager as a part of Telecommunication Management Network (TMN), the principal structure is shown in figure 5.

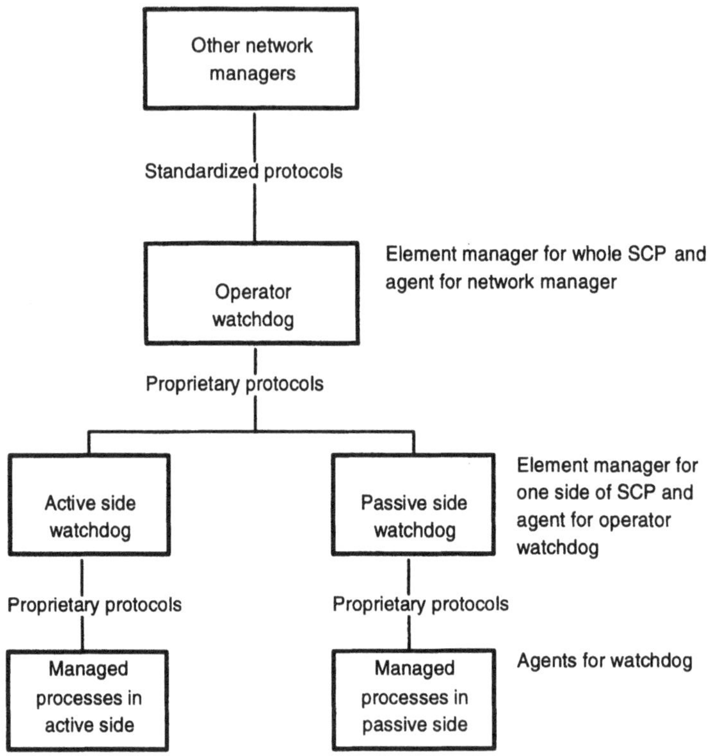

Figure 5.

A remote manager can connect to the operator watchdog. The remote manager may use some standard management protocol like CMIP (Common Management Information Protocol) or SNMP. The operator watchdog can act as a protocol converter, which translates messages between the remote manager and the SCP. The protocol itself is not enough since the remote manager and the managed process must have common understanding of the managed data.

In principle, TCP/IP (Transmission Control Protocol/Internet Protocol) -network, SS7 (Signalling System number 7) -network or X.25 network can be chosen for remote control. SS7 network is already used for the connection with SSP, but it is expensive and quite difficult to use. The application programming interface (API) for SS7 is not standardized. In contrast, the support for TCP/IP network is included in almost every unix operating system and the API is standardized as sockets. Furthermore, TCP/IP hardware is cheap. Currently, X.25 is not used in this SCP. For these reasons, TCP/IP-based management seems to be natural. Because SNMP is in wide use for TCP/IP networks, SNMP seams to be more natural choice than CMIP for this SCP management.

Protocol conversation is needed because it is not useful to use complex network management protocols inside the SCP. Simple and efficient enough protocols are suitable to be inside the SCP. The operator watchdog can be situated in a separate machine to remove protocol conversation load from the service machines.

2.5 The mib for the local manager

The MIB of the local manager is limited and special in its information scale. The principal structure is shown in figure 6.

```
SCP
|
|__ SYSTEM 1
|    |
|    |__ MAIN WATCHDOG
|    |
|    |__ DATABASE
|    |
|    |__ SCHEDULER
|    |
|    |__ SLEE 1
|    |
|    |...
|    |__ SLEE n
|    |
|    |__ SS7 SOFTWARE
|
|__ SYSTEM 2
     |
     |__ ( same as those under system 1 )
```

Figure 6.

Table 4. The attributes of the SCP object

Name	Type	Access	Status
scpStatus	ScpStatus	Read-only	Mandatory

scpStatus has values (not started, starting, single, double, going double, going single, not OK, going down, unknown)

Table 5. The attributes for the system object

Name	Type	Access	Status
systemStatus	SystemStatus	Read-only	Mandatory

systemStatus has values (not started, OK, not OK, active, passive)

Table 6. The attributes for the main watchdog object

Name	Type	Access	Status
wdStatus	WdStatus	Read-only	Mandatory
wdToScheduler	ConnectionStatus	Read-only	Mandatory
wdToSlee1	ConnectionStatus	Read-only	Mandatory
...			
wdToSleeN	ConnectionStatus	Read-only	Mandatory
wdToSs7	ConnectionStatus	Read-only	Mandatory
wdToOtherWd	ConnectionStatus	Read-only	Mandatory
Go down	BOOLEAN	Write-only	Mandatory
Ask report	BOOLEAN	Write-only	Mandatory
Logging level	Integer (1..5)	Write-only	Mandatory
Change log	BOOLEAN	Write-only	Mandatory
Read configuration	BOOLEAN	Write-only	Mandatory
Trouble report	ANY	Read-only	Optional
Report ready	ANY	Read-only	Optional

WdStatus has values (not started, OK, not OK), ConnectionStatus has values (not connected, connected)

Table 7. The attributes for the SLEE object

Name	Type	Access	Status
Slee identification	Integer (0..255)	Read-only	Mandatory
sleeStatus	SleeStatus	Read-only	Mandatory
sleeConnectivity	SleeConnectivity	Read-only	Mandatory
Connect to scheduler	BOOLEAN	Write-only	Mandatory
Connect to ss7	BOOLEAN	Write-only	Mandatory

Go down	BOOLEAN	Write-only	Mandatory
Ask report	BOOLEAN	Write-only	Mandatory
Logging level	Integer (1..5)	Write-only	Mandatory
Read configuration	BOOLEAN	Write-only	Mandatory
Trouble report	ANY	Read-only	Optional
Report ready	ANY	Read-only	Optional

SleeStatus has values (not started, OK, not OK, overload, down ready), SleeConnectivity has values (not connected, connected to scheduler, connected to ss7, connected to scheduler and ss7)

Table 8. The attributes for the scheduler object

Name	Type	Access	Status
schedulerStatus	SchedulerStatus	Read-only	Mandatory
scheConnectivity	ScheConnectivity	Read-only	Mandatory
Connect to local db	BOOLEAN	Write-only	Mandatory
Connect to remote db	BOOLEAN	Write-only	Mandatory
Go down	BOOLEAN	Write-only	Mandatory
Change log	BOOLEAN	Write-only	Mandatory

SchedulerStatus has values (not started, OK, not OK, down ready), ScheConnectivity has values (not connected, connected to local db,connected to remote db, connected to both dbs)

Table 9. The attributes for the ss7 object

Name	Type	Access	Status
ss7Status	Ss7Status	Read-only	Mandatory
Connect to ss7	BOOLEAN	Write-only	Mandatory
Disconnect to ss7	BOOLEAN	Write-only	Mandatory
Activate	BOOLEAN	Write-only	Mandatory
Deactivate	BOOLEAN	Write-only	Mandatory

Ss7Status has values (not started, active, passive, unknown, not OK).

3 CONCLUSION

The local manager is designed for the experimental SCP and has proved to be successful for the goal. With some modifications, it could used in the SCP product.

4 REFERENCES

[1] Kärkkäinen E. Telen SCP. Seminar work, Lappeenranta University of Technology. 1994.

[2] CVOPS User's Guide. Version 1.1, Technical Report, Technical Research Centre of Finland, 1991.

[3] Jakobson G. and Weissman M. Alarm Correlation. IEEE NETWORK MAGAZINE. November 1993. pp 52-59.

10

An advanced management architecture for IN

Dominique Gaïti
Columbia University
Center for Telecommunications Research,
Room 801, Schapiro Research Building,
New York, NY 10027-6699, USA
Tel: (212) 854 8928, Fax: (212) 316 9068,
Email: gaiti@ctr.columbia.edu

Abstract

To provide various advanced telecommunications services and to enhance their operations, it is necessary to introduce a flexible and adaptable architecture on the Intelligent Network (IN). IN may be seen as one of the most innovative concepts of that last decade, if this flexibility and adaptability may be achieved. One way to reach such a goal is to introduce some intelligence. In that way, we propose a new management architecture for IN introducing concepts from the field of Distributed Artificial Intelligence.

First, this paper describes a proposed generic architecture to support the IN management aspects. Then, to provide an intelligent architecture, we investigate the field of Distributed Artificial Intelligence, and a model adapted to these new concepts is proposed. Finally, we propose the integration of the generic IN management architecture and the intelligent model to provide an intelligent architecture for IN management.

INTRODUCTION

The objective of Intelligent Networks (IN) is to allow the introduction of new capabilities in telecommunication networks and to facilitate and accelerate, in a cost-effective manner, service implementation and provisioning in a multivendor environment. IN is a generic service-oriented architecture intended to be used for all kinds of services. The management aspects of IN are mainly focused on "service management", that is, management related to services on an IN architecture.

A conceptual model for IN was developed under the auspices of ITU-T (ex CCITT) to aid in the specification of IN. Functional and physical entities have been defined in this model. In particular, some functions are devoted to network management of IN services. A network management architecture is necessary to realize the functions described in these management entities. In this paper we provide an advanced management architecture well adapted to these IN concepts. This architecture uses some ideas developed in the Distributed Artificial Intelligence field.

The paper is in four sections. The first section describes IN concepts and the work focusing on IN management aspects. Section 2 proposes a generic architecture to support these Intelligent Network management aspects. To provide an intelligent management architecture, we investigate the field of Distributed Artificial Intelligence described in section 3. In section 4 the integration of sections 1 through 3 is realized to provide an intelligent architecture for IN management.

1. INTELLIGENT NETWORK MANAGEMENT ASPECTS

The IN Conceptual Model (INCM) was defined by ITU-T as a methodology for specifying the IN architecture. It is not an architecture in itself, but a framework for the design and description of the IN [1] [2]. The INCM consists of four planes (Figure 1), where each plane represents a different abstract view of the capabilities provided by an IN-structured network. The planes represent the service view, the global functional view, the distributed functional view, and the physical view of IN.

The service plane gives an abstract end-user view of the services provided by an IN-structured network. A service can be decomposed into one or more service features. A service feature is built from several Service Independent Building blocks.

The global functional plane defines the Service Independent Building blocks (SIBs) with the following principle: the network is viewed as a single virtual machine. It turns out that the distribution is transparent at that level.

The distributed functional plane models a distributed view of an Intelligent Network. This view consists of Functional Entities (FEs). A FE is a unique group of functions in a single location and a subset of the total set of functions required to provide the service feature supported by the SIB. A group of centralized or distributed FEs is mapped onto a SIB. Then a SIB is realized by the interaction between corresponding distributed FEs. The advantage of this representation is to have reusable FEs, i.e., a FE can be used by several SIBs.

Three types of management-related functional entities are provided: Service Management Function (SMF), Service Management Access Function (SMAF), and Service Creation Environment Function (SCEF).

- A Service Management Functions (SMF) supports deployment and provision of IN provided services and/or ongoing operations. Particularly, for a given service, it supports co-ordination of different Service Control Function (SCF) and Service Data Function (SDF) instances, e.g.:
    - billing and statistical information are received from the SCFs, and made available to authorized service managers through the SMAF;
    - modifications in service data are distributed in SDFs that keep track of the reference service data values.

The SMF manages and updates service related information.

- A Service Management Access Function (SMAF) provides an interface between the service managers and the Service Management Function. This service includes checking authority to access function.
- A Service Creation Environment Function (SCEF) allows services provided in Intelligent Network to be defined, developed, tested and made supportable by Service Management Functions. Output of this function includes service data definitions and service trigger definitions.

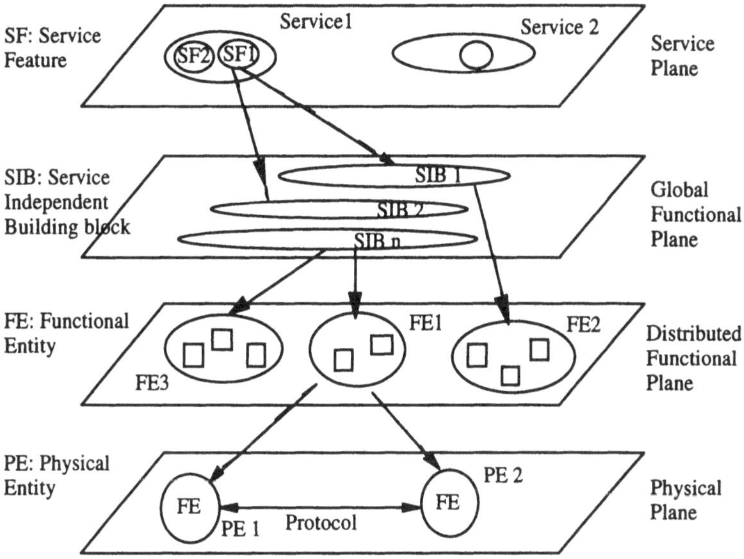

Figure 1. The Intelligent Network Conceptual Model.

The physical plane identifies the different Physical Entities (PEs). A Physical Entity consists of one or more Functional Entities. One or more FEs may be mapped onto the same PE, but a single FE cannot be split between two PEs [3]. This plane takes into account physical considerations, optimization aspects, and protocols.

In this paper, we are mainly interested in Service Management Functions (SMFs). They handle five activities (service deployment, service provisioning, service control, billing, and service monitoring). The SMFs are located in the Service Management Points (SMPs). Now the problem is how to design these SMFs to co-operate in an intelligent way? We have to define what resources are managed by what SMPs. How are the SMAFs realized given, that they are located in an intelligent terminal or a computer? What is the structure of an SMP?

In the next section we propose an architecture able to support IN management and to answer the different problems arising in this area.

2. MANAGEMENT ARCHITECTURE

We are going to define an architecture for integrating the management of all the resources of a distributed system. The architecture we have adopted has been defined in a previous paper [4] and is derived from proposals developed in the European Esprit or Race projects [5] [6] [7] [8] [9]. We defined a generic integrated management architecture through an object-oriented approach able to model all entities of a distributed management systems. This allows us to represent and to manipulate in a uniform way the information coming both from the attributes of the managed entities, and from management entities. An object machine permits basic services like creation and removing of types of objects and instances. This object machine should also support location-independent interactions between objects (invocation) and protection against wrong object use.

Management is modelled in the following terms:

- The resources whose activity produces a specific function for the treatment of the information;
- A manager that provides the three following functions:
    - A monitoring function to supervise resources and activities to insure that they are conform to a management policy , and to identify the possible conflicts;
    - A control function to select a control operation following an algorithm determined by the policy management;
    - A default function to describe what to do when the management of resources and activities cannot be fulfilled following the defined policy.

Let us note that a manager may be responsible for several resources, each of them subject to control and supervision.

The management is hierarchical to facilitate flexible implementation and future extensions. A manager may itself be managed by a higher level manager. It turns out that a manager can work with a certain autonomy depending on its own policy. At the same time, a manager may answer when it receives a control command. The basic model can be extended recursively to different levels of refinement and specialization.

The management is also multiple. It is necessary to share the resources that are managed simultaneously by several managers. It is also necessary to establish a negotiation between the managers to avoid conflicts due to concurrent control of the same resource.

Our management architecture defines the basic principles of organization of the IN world and the structures that describe it.

The domain is a first principle for the organization. A domain (from a management point of view) consists of resources to be managed in a distributed system and a specific management policy (or a component of this policy).

The second organizing principle concerns the Management Process (MP), which includes the hardware, the software and, the people executing the process. For example, we can group the objects to be managed according to the level where they are or the relationship between the different layers. The Management Process contains two components:

- The MP view, which is the abstract representation of the resources and the activities under control. It is possible to have a set of views that corresponds to different subsets of resources.
- The MP core that executes the management process. It manages and activates the resources through the MP view.

This distinction between the MP view and the MP core allows the separation of the management policy (namely, the goal of the management system) from the resources and the managed activities (namely, the managed objects). These principles are shown in Figure 2.

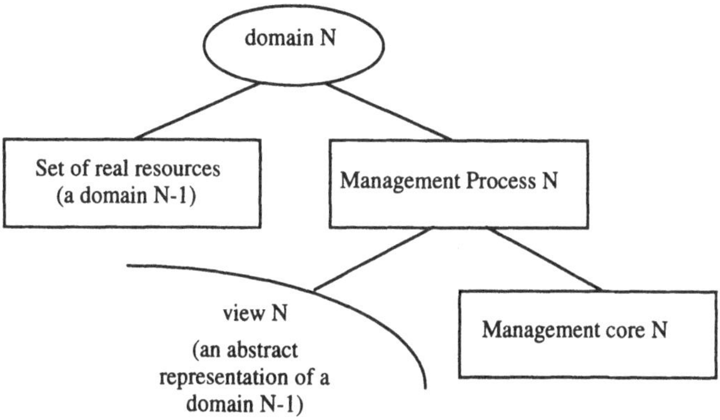

Figure 2. The principles of structure.

The concept of domain allows the description of a large number of management cases, from simple managed resources to very complex structures. Indeed, the domains may be structured to reflect a delegate administration in relation with another domain. This delegation may be applied in a recursive way. It means that a domain N may delegate to another domain N-1 the management of some of its real resources. The abstract representation of N-1 constitutes the part of the MP view of N. It turns out that N-1 represents a subset of N's resources. This decomposition may facilitate the management application.

This domain and MP-based architecture allows the domains to be completely autonomous since they are able to manage their own resources. This permits the description of a managed resource to be provided very easily. Moreover, the delegation of some management tasks by one domain to another allows the creation of complex schemes. Such an approach allows the modelling of layered structures with their dependencies.

The use of Knowledge Based Systems (KBS) lets one capture network management knowledge. One of the main features of a network management system is the large amount of information it handles. Looking at this knowledge as one big area would make it impossible to find a single representation formalism covering everything. But we can adopt a top-down approach and try to find some of the main objects in this large area.

We have found three objects on a very high level that we think cover a large part of our management system and can be fairly representative. These objects are the network, the services, and the users of the management system. For each object, we have proposed knowledge representation formalisms capturing the knowledge properly [10].

3. TOWARD AN INTELLIGENT MANGEMENT ARCHITECTURE

3.1. Distributed Artificial Intelligence Concepts

Real time and distribution problems are two fundamental aspects of network management. In fact, time constraints are crucial for telecommunication system features such as response time (from user point of view) and performance throughput (from the network-manager point of view). Network applications require very short response times when performing for example, routing table up-date or repairs in case of various network faults (e.g., network overload, hardware failure).

The aim is to keep the user unaware of problems occurring in the network. Consequently, the network manager should be able to detect immediately any network malfunction (fault detection), distinguish its real causes from the numerous side effects (fault isolation), and identify the problem (diagnosis). Then, he should ask for the required knowledge for building a solution to restore the network to an unfaulty behavior (effective repair). So, the required expert system has a difficult task to encompass all aspects of the above operations.

Moreover, a network is characterized by its heterogeneity, its layered structure, and the distribution of its resources. Thus, we have to formalize such a network by a system in which knowledge is distributed [11].

Distributed Artificial Intelligence (DAI) allows several processes, called agents, to solve a single problem. Each process uses local power and communicates with remote hosts. In the DAI field, the research aim is to propose a scheme of organization allowing the distribution of the resolution of a shared problem. This organization has to describe the mechanisms that can be used by agents to co-operate.

With DAI, problem solving involves the co-operation of decentralized processes. Agents may be small real processes or they can be complex real ones such as applications based upon large knowledge bases. The problem solving process is distributed, meaning that each agent has to share a common information set allowing the entire group to reach a solution. Decentralized processing means that data and control are often logically and physically distributed [12].

3.2. Distributed Artificial Intelligence Architecture

The implementation of a DAI architecture may follow different techniques [13]. In our project, the DAI architecture follows the principle of the blackboard architecture [14]. It incorporates three components (Figure 3) [15]:

- A structured knowledge base called "blackboard" that contains the current state of the solution for a given problem. This shared data area is analogous to the working memory accessed by many production rules in a rule-based system. It is divided into separate areas of semantic abstraction called levels.

- A collection of independent agents (or knowledge sources) which may read and write one or several levels. They can be seen as a collection of independent processes able to co-operate to solve a problem. Each agent is specialized in one particular sub-domain.
- A control system that insures the supervision of the actions of the different knowledge sources. The control system is also a collection of integrated knowledge sources.

This organization is not sufficient to realize a real multi-agent system. We need to develop a real architecture for the agent itself; it cannot be just a collection of rules, an entire procedure, or a whole rule-based system.

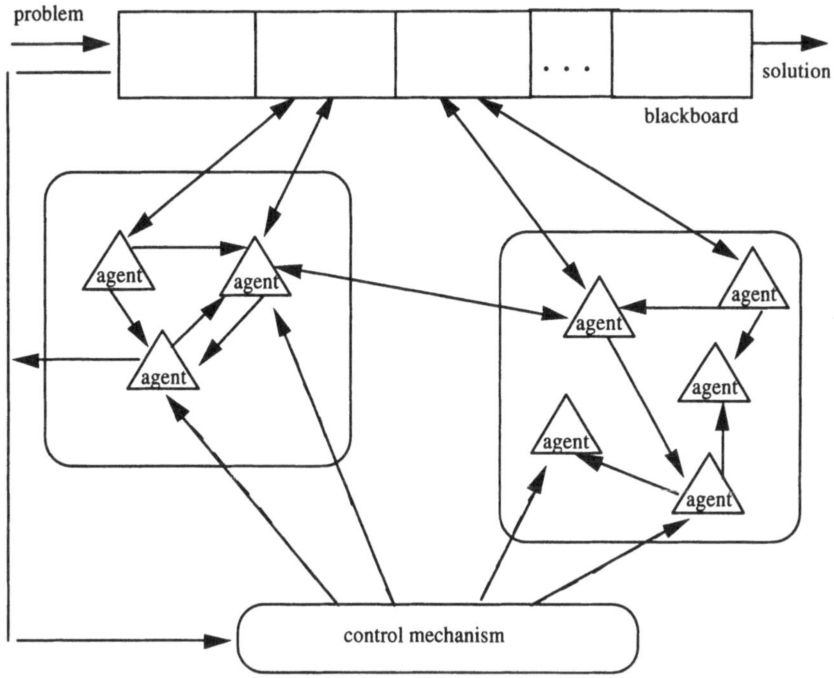

Figure 3. The blackboard architecture.

The agent itself is organized following the principles of the blackboard architecture. It has three components (Figure 4):

- a knowledge module: this relates to all the usual data and the knowledge of the domain;
- a control module: this defines the problem-solving strategy and the control mechanism to determine the next action to perform. It corresponds to the inference engine in the rule-based system;
- a communication module: this gives a model of the other agents in the environment. It also provides an interpreter of messages and a logic for communications with the other agents.

Each agent may be seen as an independent software processor, with its own resources. The agents communicate by sharing the information if we look at a high level, i.e., at the entire multi-agent system level. They communicate through the blackboard that contains, at the beginning, the facts of a given problem. The agents read the information in the blackboard and write in it to reach a solution step by step.

If we look at a lower level, the communication module in each agent is in charge of receiving and sending messages to the other agents to co-operate in building a solution if they do not have enough knowledge to solve the problem themselves.

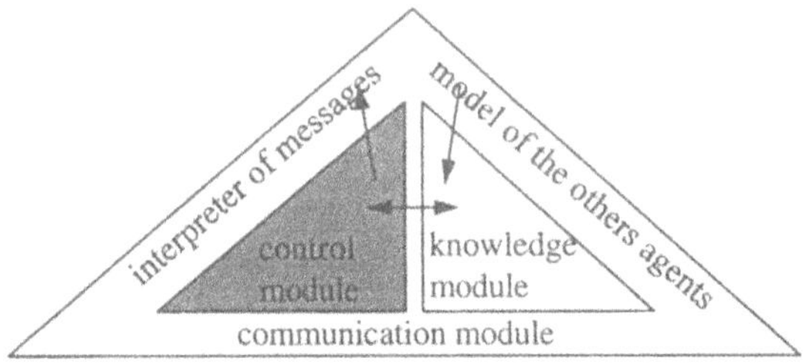

Figure 4. An agent.

3.3. An Intelligent Integrated Architecture

In § 1 and § 2, we defined a generic integrated management architecture. Now we want to apply the DAI model on this architecture to get an Intelligent Integrated Architecture.

First, we have to define the different domains. The domains are composed of a set of managed resources but also encapsulate the components performing management. Therefore, management components and resources may be grouped to build domains according to different criteria associated with the contained management resources (type, location, functionality, ownership, or objectives (management function, applied policy, authority)). For example, if we are interested in the management of N interconnected sub-networks, a domain can be a sub-network. But, a domain may be constituted with the machines of a manufacturer X or may be constituted with the software and hardware of the layer n of the OSI architecture (on all the network or on a sub-network or only on a part of the network).

The domain is managed by one or several agents that can be situated in any node of the network belonging to the domain. These agents work in a co-ordinated manner on a common data structure: the blackboard. This is situated in the MP core where all the management functions are provided. The agents may work in group on a given problem (in the blackboard) or may work alone or with some other agents of their environment to solve more trivial problems when they have the appropriate knowledge. The structure of the agents and the way they are organized are described in §3.2.

4. AN ADVANCED MANAGEMENT ARCHITECTURE FOR IN

In the IN, several scenarios for the physical architecture can be deployed. We propose to adapt our intelligent management architecture on the physical plane, namely the last plane of Figure

1. This plane is decomposed into domains. They are defined following different criteria, depending on the structure of the network, the needs of the operator, the user constraints, and the diversity of the infrastructure. An example of a domain is illustrated in Figure 5.

The domain is composed of one Service Management Point (SMP) that performs service management control, service provision control, service deployment control, etc. Examples of functions that the SMP can perform are database administration, network surveillance and testing, network traffic management, network data collection. Functionally, the SMP contains the Service Management Function (SMF) and, optionally, the Service Management Access Function (SMAF) and the Service Creation Environment Function (SCEF). The SMP can access all other PEs.

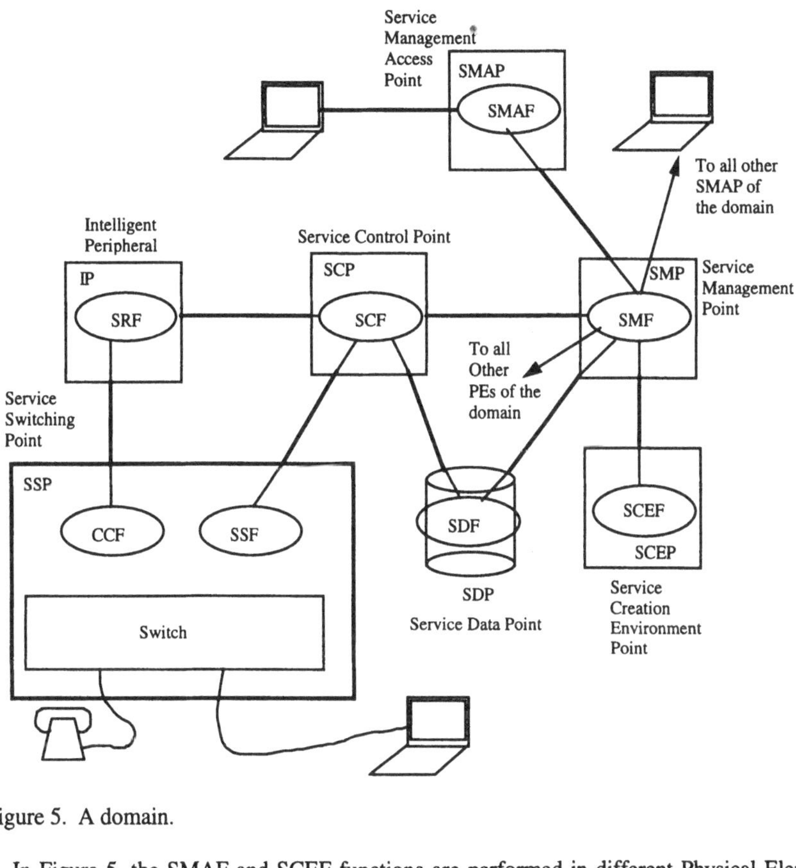

Figure 5. A domain.

In Figure 5, the SMAF and SCEF functions are performed in different Physical Elements. The SMAF is included in a Service Management Access Point (SMAP). This SMAP may be used as a service independent front end to the SMP or to all SMPs of the system. This simplifies the user's procedures (customer or network operator) as there would be only one entry point for the management of all IN-structured services. In addition, all security-related data for user identification, for example, may be specified for each SMAF. These functions need not be grouped together as it happens when the SMAF is located in the SMP. The SCEF is included in the Service Creation Environment Point (SCEP) that is used to define, develop and test an IN service, and to input it into the SMP.

Other Physical Entities can be Service Switching Point (SSP), Service Control Point (SCP), Service Data Point (SDP), Intelligent Peripheral (IP), Service Switching and Control Point (SSCP), Service Node (SN) and so on.

For each domain a Service Management Point (SMP) is defined. The MP core is situated in this SMP and is in charge of the Service Management Functions. At each IN Functional Entity there is an agent that can communicate with other agents and with the blackboard situated in the MP core. The management relationship between the SMF and any FEs is performed through the communication between the agents and the blackboard. The information flows between the SMF and the other functional entities are defined by the agent capabilities. Moreover, a pool of agents is provided in the Service Creation Environment Point (SCEP) supporting the Service Creation Environment Function (SCEF). For example, the SCEP may contain an agent for service specification, an agent for service analysis, an agent for service generation, and an agent for service testing.

The architecture we described in Figure 5 describes one domain. Now, a general topology of our global architecture is illustrated in Figure 6.

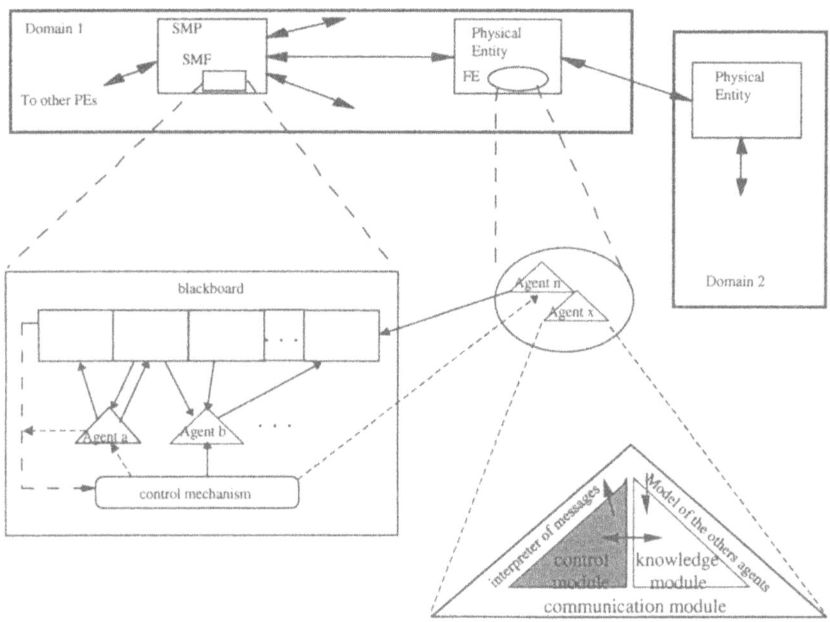

Figure 6. An intelligent management architecture for IN.

In this figure two domains are shown. Each domain contains one Service Management Point with one blackboard and a set of agents attached to this blackboard.

The blackboards and the agents of the different domains may co-operate to insure the global management of the system. The main problem is to organize the global system to define a general service management process able to take charge of a global service. This can be performed through a hierarchical structure of distributed systems.

We may define a main agent, situated in a central control node. This main agent has the same structure we described before with three components but not the same functions:

- A knowledge module containing the awareness of all the other agents in the global system. It means that the main agent knows exactly which agent is able to solve a particular task. It also has knowledge concerning the way to subdivide a global problem into sub-problems and the way to put together partial results to obtain a global solution for a global problem;
- A control module: it fulfills the same functions as in ordinary agents;
- A communication module: it gives a model of the different blackboards in the environment. It also provides a way to read and write in these blackboards.

When a general problem appears, the main agent subdivides it into sub-problems and looks for the agents able to solve a particular sub-problem. When it finds them, it writes sub-problem data into their associated blackboards. It can write in different blackboards.

The main agent reads from these blackboards after the agents finish their work and puts together all the partial results it found to achieve a global solution.

5. CONCLUSIONS

This paper proposes a framework for defining an intelligent and integrated model to be applied in the service management area of the Intelligent Network. The main idea was to reduce the complexity of the management process. We dealt with complexity by introducing agents and blackboards as intelligent concepts. In this paper, we worked on a low time scale corresponding to service management tasks. In a future issue, we would like to enlarge the applicability of the architecture by introducing new concepts in the agents to take care of real time considerations.

REFERENCES

[1] ITU-T I.312/Q.1201 Principles of Intelligent Network Architecture, October 1992, Geneva.

[2] José M. Duran and John Visser, International Standards for Intelligent Networks, IEEE Communications Magazine, February 1992

[3] W. Ambrosch, A. Malher, B. Sasscer, The Intelligent Network, Springer-Verlag, 89.

[4] D. Gaïti, I^2NMA: an Intelligent Integrated Network Management Architecture, International Journal on Network Management, vol 4, 3, 1994.

[5] The final technical Report of RACE project ADVANCE (R1009), 1992.

[6] The final technical Report of RACE project NEMESYS (R1005), 1992.

[7] The final technical Report of RACE project AIM, (R1005), 1992.

[8] ROSA Architecture - Final release, RACE project 1093, April 1992.

[9] M. Möller et al., DOMAINS Basic Concepts for Distributed Systems Management, Proceedings of the fifth RACE TMN Conference, 1991.

[10] I. Rahali and D. Gaïti, A Multi-agent System for Network Management, In Integrated Network Management II, pp. 469-479, Ed. I. Krishnan & W. Zimmer, Elsevier Science Publishers B.V. (North Holland), 1991.

[11] D. Gaïti and I. Rahali, Applying Artificial Intelligence Techniques to the Management of Heterogeneous Networks, In Network Management and Control, pp. 189-200, Ed. A. Kershenbaum, M. Malek, M. Wall, Plenum Publishing corporation, 1990.

[12] L. Gasser, Distributed Artificial Intelligence, AI Expert, July 89.

[13] E. Werner, The Design of Multi-Agent Systems, Decentralized A.I.-3, Elsevier Science Publishers B.V., 1992.

[14] R. Engelmore et al., Blackboard Systems, Reading, Mass.: Addison-Wesley, 1988.

[15] D. Gaïti, Distributed Problem Solving in Network Management, Proc. of International Conference IFIP/IEEE on Distributed System: Operation and Management, Long Branch N.J., USA, October 1993.

11

Secure IN internetworking

Alexander Herrigel and Xuejia Lai
R^3 Security Engineering AG[1]
Zurichstrasse 151, CH-8607 Aathal, Switzerland.
Fax: +41 1 932 66 60
Email: herrigel@r3.ch

Abstract

This paper presents a new approach for secure IN internetworking. Based on a threat analysis, an adequate cryptographic protocol is proposed to address the derived security concerns. The cryptographic protocol presented is based on the recently published standardization framework ISO/IEC CD11770-3.

1. INTRODUCTION

Confronted with the growing complexity of the public networks and the need for rapid development and deployment of enhanced services the telecommunications community developed a new concept, called the Intelligent Network (IN) architecture. This architecture has a modular structure, separates the call processing logic from the service intelligence, and is based on a centralized service control [ITU-T]. With respect to the network operators requirements for vendor and network independent solutions, the telecommunications community has developed for the IN architecture two standards, called the AIN and the ITU-T Q.1200 recommendations. The ITU-T standard is actually evolving to support service transparency in a multi-vendor environment (Figure 1).

[1]This work has been funded by the Swiss National Science Foundation under the SPP program.

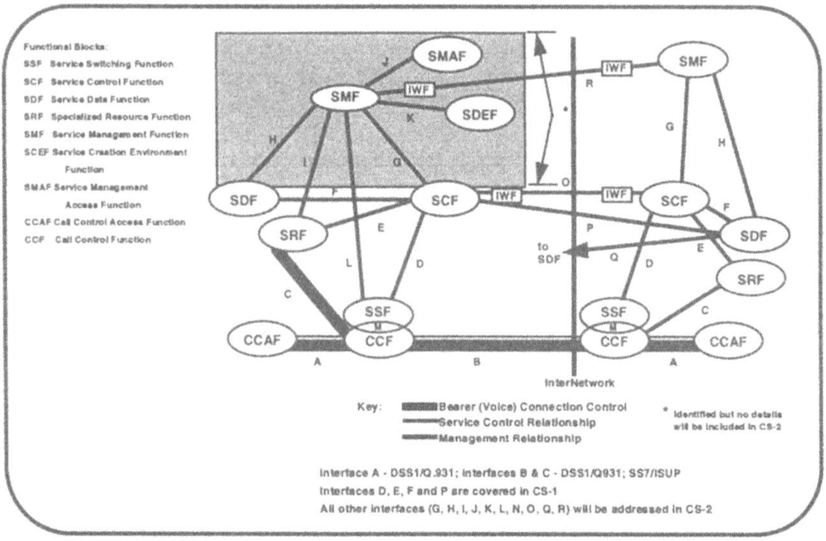

Figure 1: The modified distributed functional plane for CS 2 [Nagaraj].

The IEEE communications society and the ITU standardization body (SG11) are currently investigating an enhanced Capability Set, called CS 2, for the IN architecture. Due to the market demand and deregulation, IN Internetworking aspects between different Network Operators are addressed. Internetworking will provide third party access to an IN and enable the exchange of control and management information between different Network Operators running different Intelligent Networks. From a business perspective, IN internetworking is important in Europe for the Network Operators, since it provides the basis for offering European-wide IN services to the service users. Because the management and control data influence the service processing and billing, the messages transferred to exchange these data are very sensitive and security is a very critical aspect in this domain. In addition, some national regulatory bodies require to protect the privacy of service subscribers by additional means.

The following interfaces are actually investigated [Chatras, Nagaraj, ETSI]:

1. SCF - SDF,
2. SCF - SCF,
3. SDF - SDF and
4. SMF - SMF.

The importance of security issues for IN Internetworking has also been recently noticed by the EURESCOM project group P230 [Hulzebos]. Our paper addresses the Internetworking security aspects for the following interfaces:

1. SCF - SCF, and
2. SMF - SMF.

The paper is organized in 5 sections. The threat analysis for the Internetworking is described in section 2 to identify the security concerns. In section 3, notations and definitions are introduced to provide the adequate basis for a precise formal description of the cryptographic protocols. The derived protocols are presented in section 4. Finally, results and conclusions are summarized and discussed.

2. THREAT ANALYSIS

2.1. Involved parties

From our perspective, the following entities have to be considered for the IN scenario:

IN-Service: An IN-service is considered as a stand-alone commercial telecommunication service that is executed in an IN environment. This environment can be partitioned into several domains.

Network Operator (NO): A network operator is a commercial organization which runs and maintains an IN infrastructure. During the execution of an IN service, such an infrastructure can be involved for the origination, transfer or terminating call processing.

Service Provider (SP): A service provider is a organization that offers an IN service for subscription. The SP has the legal responsibility for the correct operation of the offered services. The NO itself may also act as a service provider.

Service Subscriber (SS): A service subscriber is a person or an organization that has a contract with a service provider for the supply of a service.

Line Subscriber (LS): A line subscriber is a person or an organization that has rented or bought an IN access device.

Service User (SU): A service user is the end user of a typically IN-service. In a business environment, it can be an employee of the service subscriber.

Regulatory Body (RB): The regulatory body imposes the legal rules for the different parties in a commercial IN environment to ensure fair competition.

2.2. Characterization of the IN architecture

The IN architecture is based on three different layers, namely the service creation and management layer, the service control layer, and the switching and network resource layer. These layers have different resource entities which are shortly described for an adequate threat analysis.

The Service Management System (SMS) represents the service creation and management layer. It supports the specification, the introduction and maintenance of all IN services. It holds the master copy of all network databases and maintains the SCP's tables and service logic. In addition, it maintains the service and control information in the IPs. A Service Creation Environment (SCE) is a an integral part of the SMS. Typically, new IN services are introduced via a file transfer to the SCPs. The SMS updates and retrieves management data from the Service Control Point (SCP), the Adjuncts, and the Intelligent Peripheral (IP).

The network service control intelligence is positioned at centralized nodes (SCP). The SCPs execute the service logic for call control. They support on-line transaction processing with high transaction rates and volumes. In addition, these nodes hold on-line local databases with respect to the IN services they support.

The SSPs provide the access to the service users. They trigger the calls which have been generated by an IN access device and send the imposed service requests to the SCP. The IN service is executed by the SSP on the basis of instruction guidance from the associated SCP within the framework of a Basic Call State Model (BCSM). The SSP is connected to other local or trunk exchanges on the basis of the bearer connection control and a non-IN call control protocol. In addition, the SSP is connected to the associated SCP on the basis of the SS7 protocol.

The Adjunct offers the same functionality as an SCP. It is, however, connected to a local exchange with high speed connections and supports specific logic programs of a service provider.

The IPs supports enhanced voice/data services like announcements, digit collection, voice messaging, speech response. An IP is shared between different SSPs. It is typically accessed by an ISDN interface.

There are two modes of interactions in this environment:

- Transaction messages for providing updates of management, control, and service information.
- File transfers, typically via the X.25 network, for the installation and modification of the service logic and management data.

IN Internetworking is defined as the process of executing a requested IN service across at least two different autonomous domains. Every domain represents a separate IN network with the above described involved parties and IN resource entities. The IN service can only be executed successfully, if the two different domains cooperate by exchanging management, control, and service data. Since the different domains are typically run by different Network Operators in competition, there is only a very limited trust relationship between these domains. Each domain enforces its own mechanisms to provide integrity , availability, and service user privacy. Depending on the number of different domains involved in a IN service processing, the following different types of domains are identified: 1. The origination domain, 2. the interconnection domain, and 3. the destination domain. From our perspective, the Interworking Function (IWF) of the enhanced distributed functional plane provides the necessary addition means to support secure IN Internetworking during the call processing or exchange of management information.

2.3. Derived threats and security services

The following threats have been identified:

Threat 1: Modification (replication, insertion, deletion) of service, control, and management data during the transmission.

Threat 2: Replay of old messages by unauthorized parties.

Threat 3: Impersonation of network entities (SMS, SCP) to penetrate the NOs databases and service (denial of service).

Threat 4: Illegitimate use of system resources (SMS, SCP) from authorized entities.

Threat 5: Repudiation of management information flow. If management data modification is requested, it should not be possible to deny later having requested the modification.

Threat 6: Disclosure of information from the transmitted data to gain competitive advantages.

Only threat 4 is covered by the access control policy of the operating systems (Unix, etc.) which are typically applied in a specific domain.

The following security services must be supported for a secure IN Internetworking. The reader is referred to [ISO] for a detailed definition of these security services.

1. Message Content Authentication
2. Message Origin Authentication
3. Non-Repudiation of Origin
4. Confidentiality

The co-operation of the different domains is only possible if specific legal contracts have been negotiated and agreed between the Network Operators. These contracts define a low level trust relationship with respect to the messages, that have to be exchanged.

3. NOTATIONS AND DEFINITIONS

The following definitions are introduced to distinguish the different cryptographic approaches which are applied in the protocols.

Cryptographic algorithm: A function f: (A x B) -> C from the Cartesian product set (A x B) to a set C is called a cryptographic algorithm if

1. the computational complexity CC to find for a given c = f(a, b) in C and a in A without the knowledge of b in B such that f(a, b) = c is infeasible,
2. it is easy to compute a for a given c from C and b from B.

The set A is identified as the plain text space and the set B is identified as the key space. The set C is called the ciphered message space.

Asymmetric cryptosystem: An asymmetric cryptoystem is based on a cryptographic algorithm that uses two related transformations, a public transformation (public key) and its inverse, a private transformation (private key). The two transformations have the property that, given the public transformations, it is computationally infeasable to derive the private transformation.

Symmetric cryptosystem: An symmetric cryptosystem is based on a cryptographic algorithm that uses two transformations t and l. They have the following properties:

1. Given one transformation it is possible to derive the other.
2. Given an a in A, a b in B, and a c in C the following relation holds: l(t(a, b), b) = l(c, b) = a.
3. It is computationally infeasable to find for a given c in C an a in A without the knowledge of the b in B such that t(a, b) = c.
4. It is computationally infeasable to find for a given a in A a c in C without the knowledge of the b in B such that l(c, b) = a.

Collision resistant hash function: A function f: A -> B from a set A to a set B is called a collision resistant hash function, if it fulfills the following conditions:

1. The argument of f can be of arbitrary length and the image of f has a fixed length of n bits.

2. It is computationally infeasable to find for a given b in B and a in A such that t(a) = b.
3. It is computationally infeasable to find a and b in A, a ≠ b, such that f(a) = f(b).

Message authentication code (MAC): A function f: (A x B) -> C from the Cartesian product set (A x B) to a set C is called a message authentication code, if the following conditions are fulfilled:

1. The argument of f can be of arbitrary length and the image of f has a fixed length of n bits.
2. For a given c = f(a, b) in C, it is computationally infeasable to find an a in A without the knowledge of the given b in B such that f(a, b) = c.

Token: A token is a message which is sent from one entity to another entity during the execution of a cryptographic protocol.

Ticket: A ticket is a message that is partitioned by a token and the associated certificate.

Certificate: A certificate C is a specific cryptographic transformation M of a token T. M is defined as the composition of a collision resistant hash function f_{crh} and a private transformation t_{pr} of an asymmetric cryptosystem ($C = M(T) = t_{pr}(f_{crh}(T))$).

The following notation is introduced to specify precisely the proposed cryptographic protocols. The protocols are described by relations which identify the objectives (goals), the different starting assumptions, and the communication or computation phases.

$R_{operation}$	:=	{ < x, y > | The process x executes the operation y.}
$R_{condoperation}$	:=	{ < x, y, (z) > | The process x executes operation y, if the condition z is fulfilled.}
$R_{condmessage}$	:=	{ < x, y, (m_1, m_2, ..m_n), (z) > | The process x sends the message m = (m_1, m_2, m_3, ..., m_n) to process y, if the condition z is fulfilled. The considered communication channel is untrusted.}
$R_{assumption}$	:=	{ < x, y > | The process x executes the operation y for initialization purposes.}
R_{goal}	:=	{ < x , y> | x is the name of the cryptographic protocol, y is the goal of the cryptographic protocol}
$R_{constraint}$	:=	{ < x , z> | Imposed constraint z for the cryptographic protocol x.}

4. THE APPROACH

4.1. Motivation

The proposed cryptographic protocol includes an asymmetric one-pass key transport scheme which supports a mutual authentication with key confirmation (ISO/IEC CD 11770-3). The scheme establishes in first phase a shared secret key between two entities. A ticket of the management data is generated in the second phase to support message content authentication, message origin authentication, and non-repudiation of origin. The ticket is then ciphered to support confidentiality.

Suppose the SMS A wants to send some management data MD to the SMS B over an interconnection domain. Then the protocol is described by the following scheme:

Phase 1:

1. A generates the key agreement token $KATo_A$:= < SN_t, ID_A, SK_{CIP}, $SK_{CIP}[CD_C]$ >, with SN_t as the sequence number[2] at time t, ID_A as the ID of the SMS A, SK_{CIP} as the generated secret key for the ciphering, and CD_C as the control data for the key confirmation. Then SN_t is increased by one and stored. If a specific threshold value has been reached, the sequence number is initialized.
2. A retrieves the public key V_B from the SMS B and ciphers the key agreement token with this key ($V_B[KATo_A]$).
3. A generates the token ATo_A : = < $V_B[KATo_A]$, ID_B >.
4. A hashes the token ATo_A with the collision resistant hash function f_{crh} and signs the resulting output with the private key P_A from the RSA key pair (P_A, V_A). The ticket Ti_A := < $P_A[f_{crh}(ATo_A)]$, ATo_A > is then generated and sent to B.
5. B receives the ticket Ti_A, retrieves the public key V_A from A and verifies the ticket (valid signature). If the signature is valid, B retrieves the key agreement token $KATo_A$ by deciphering the token $V_B[KATo_A]$ with its private key P_B from the RSA key pair (P_B, V_B).
6. B sends the key confirmation token < B, CD_C > to A, if the sequence number SN_t is not minor as the last stored one. SN_t is then increased by one and stored.

Phase 2:

7. A receives the key confirmation token and verifies the data. If the confirmation is valid, A sends the following token to B: < A, SK_{CIP}[< < MD, SN_t >, $P_A[f_{crh}$(< MD, SN_t >) >] >. The sequence number is then increased by 1 and stored. If a specific threshold value has been reached, the sequence number is initialized. B receives the token < A, SK_{CIP}[< < MD, SN_t >, $P_A[f_{crh}$(< MD, SN_t >) >] >. B retrieves the symmetric secret key SK_{CIP} and the public key V_A from A. The management data MD is generated by deciphering the token, i.e. $SK_{CIP}[SK_{CIP}$[< < MD, SN_t >, $P_A[f_{crh}$(< MD, SN_t >) >] >.].
8. The management data is accepted, if the sequence number SN_t is not minor as the last stored one and the signature $P_A[f_{crh}$(<MD, SN_t >)] has been verified. SN_t is then increased by one and stored. If a specific threshold value has been reached, the sequence number is initialized.

4.2. The protocol

4.2.1. Imposed constraints

goal(IN Internetworking, The SMS A wants to send securely some management data MD to the SMS B over an interconnection domain).

constraint(IN Internetworking, the secret key SK_{CIP} has at least a length of 64 bits);

constraint(IN Internetworking, HASH(M) is a collision resistant hash function computation with the message M as the input);

---

[2] A random number based challenge can also be used in this phase.

4.2.2. Initialization

assumption(A, generate asymmetric RSA key pair (P_A, V_A));
assumption(A, distribute authentically V_A to B);
assumption(B, generate asymmetric RSA key pair (P_B, V_B));
assumption(B, distribute authentically V_B to A);

4.2.3. Protocol execution

4.2.3.1. Key transport with implicit authentication

operation(A, generate SK_{CIP});
operation(A, retrieve SN_t);
operation(A, compute $SN_t := SN_t + 1$ and store the result);
operation(A, retrieve ID_A);
operation(A, generate control data CD);
operation(A, compute SK_{CIP} [CD]);
operation(A, generate key agreement token $KATo := < SN_t, ID_A, SK_{CIP}, SK_{CIP}[CD_c] >$);
operation(A, retrieve public key V_B from B);
operation(A, compute V_B[KATo]);
operation(A, generate token $ATo := (V_B[KATo]), ID_B$);
operation(A, generate certificate $Sig := P_A[HASH(ATo)]$);
operation(A, generate ticket $Ti_A := (Sig, Ato)$);
operation(A, compute $SN_t := SN_t+1$ and store the result);
condoperation(A, initialize SN_t , (Sn_t > range));
message(A, B, Ti_A);

4.2.3.2. Ticket verification

operation(B, retrieve public key V_A from A);
condoperation(B, stop execution, ($V_A[Sig] < > HASH(ATo)$));

4.2.3.3. Key confirmation

operation(B, retrieve private key P_B);
operation(B, generate KATo by computing $P_B[V_B[KATo]]$);
operation(B, extract SN_t);
operation(B, extract SK_{CIP});
operation(B, extract ID_A);
condoperation(B, stop execution, (SN_t is minor as the last stored one);
operation(B, compute $SN_t := SN_t + 1$ and store the result);
condoperation(B, stop execution, (identified party does not match with ID_A));
condoperation(B, stop execution, ($SK_{CIP}[SK_{CIP}[CD]] <> CD$);
operation(B, store SK_{CIP});
message(B, A, (B,CD));

4.2.3.4. Sending management data

condoperation(A, stop execution, (CD is not valid));
operation(A, retrieve or generate management data MD);
operation(A, retrieve SN_t);
operation(A, compute $SN_t := SN_t + 1$ and store the result);
operation(A, generate $T := < MD, SN_t >$);
operation(A, generate $HT := HASH(T)$);

operation(A, retrieve private key P_A),
operation(A, generate Sig := P_A[HT]);
operation(A, generate MT := < T, Sig >);
operation(A, retrieve SK_{CIP});
operation(A, generate CMT := SK_{CIP}[MT]);
message(A, B, CMT);

4.2.3.4. Receiving management data

operation(B, retrieve SK_{CIP});
operation(B, retrieve public key V_A);
operation(B, generate MT by SK_{CIP}[CMT]);
operation(B, extract SN_t from T);
condoperation(B, stop execution, (SN_t is minor as the last stored one);
operation(B, compute $SN_t := SN_t + 1$ and store the result);
condoperation(B, initialize SN_t , (Sn_t > range));
condoperation(B, stop execution, (V_A[Sig] < > HASH(HT));
operation(B, extract MD from T);

6. CONCLUSIONS AND FUTURE WORK

We have presented in this paper a new approach for secure IN Internetworking. The approach has the following advantages:

1. The key agreement is based on an asymmetric technique. The imposed trust relationship between the involved parties for this scheme is adequate for the scenario. In addition, such a scheme provides a low key management complexity.
2. The scheme is efficient, since it needs only one pass for the key agreement.
3. The scheme supports key confirmation, i.e. the assurance that the SMS B is in the possession of the correct key.
4. The scheme provides explicit authentication from SMS A to SMS B through the signature and implicit authentication of B to A, since SMS B is the only entity that can compute the session key.
5. Since the session key is generated on a demand basis, there is no need to store any additional keys for the security services.
6. The described protocol uses a sequence number to prevent replay attacks. This sequence number can be generated by a counter. There is no need that both entities have to share a synchronized clock.
7. The same scheme can be applied for the following additional interfaces:
8. SMAF - SMF, and 2. SMF - SCF.

We are actually investigating the Nyberg-Rueppel key agreement protocol [Nyberg]. In contrast to other asymmetric key agreement schemes, this scheme supports also message recovery.

REFERENCES

[ITU-T] "General Recommendations On Telephone Switching And Signalling Intelligent Network, Introduction To Intelligent Network Capability Set 1", ITU-T Recommendation Q.1211, March 1993.

[Chatras] Bruno Chartras and Francois Gallant, "Protocols For Remote Data Management In Intelligent Networks CS1", IEEE Intelligent Network '94 Workshop, Heidelberg, Germany, May 24-26, 1994.

[Hulzebos] Raymond Hulzebos and Steve Reeder, "Pan European IN Reference Architecture ", IEEE Intelligent Network '94 Workshop, Heidelberg, Germany, May 24-26, 1994.

[Nagaraj] Kesavamurthy Nagaraj, "CS-1 Refinements and CS-2 Scope", IEEE Intelligent Network '94 Workshop, Heidelberg, Germany, May 24-26, 1994.

[Nyberg] K. Nyberg and R. A. Rueppel, "Message Recovery for Signature Schemes based on the Discrete Logarithm Problem", Proceedings of Eurocrypt'94, Springer-Verlag, 1994.

[ETSI] Network Aspects (NA), Security Requirements for Global IN Systems, DTR/NA-61201, Version: 1.0.5, date: 03.05.94, ETSI.

[ISO] ISO International Standard 7498-2: Open Systems Interconnection Reference Model - Part 2: Security Architecture, 1988.

12

Charging generalised: a generic role model for charging and billing services in telecommunications

Donald Ranasinghe
BT Laboratories, Martlesham Heath, Ipswich IP5 7RE, United Kingdom
Tel +44 473 642849, fax +44 473 637400
E-mail: ranasinghe_d_w@bt-web.bt.co.uk

Jørgen Nørgaard,
Tele Danmark Research, Lyngsø Allé 2, DK-2970 Hørsholm, Danmark.
Tel: +45 4576 6444, fax: +45 4576 6336
E-mail: jnp@tdr.dk

INTRODUCTION

In current IN charging has a mixed status which obscures the essence of the concept. It is considered 1) as a separate charging SIB and 2) as a service in its own right, such as (credit) Card Calling Service. On examining the service creation in an open broadband environment, charging had to be reconsidered because the card calling service is not a service on its own in the EURESCOM (European Institute for Research and Strategic Studies in Telecommunications, Heidelberg, Germany) project P103 context, but is seen merely as a way of modifying how charging is done for any kind of service. Thus initiating for example, a multimedia conference in a broad band network and have it charged to a calling card [6].

The motivation for this presentation and creation of a generalised charging model will be discussed in relation to EURESCOM Project P103. In this presentation, P103 service creation and composition guidelines will be used to describe a charging service constituent. These guidelines are based on object-oriented concepts from OOram [7, 8], and a service constituent is a reusable unit within service creation.

The work reported in this paper is ongoing in EURESCOM project P103 - Evolution of the Intelligent Network, therefore not yet finalised, but represents a snapshot of the more mature parts of the project.

The paper is organised as follows, first a short description of the charging problem and a brief introduction to the notations used in the paper. Then the charging service constituents are described in detail and finally a brief discussion of the results.

CHARGING SERVICES IN GENERAL

The scope of P103 is to define a service framework to support creation of telecommunications services using reusable components and to demonstrate how this service framework can be realised using evolving technology. Background to the project and earlier results can be found elsewhere [9].

The charging service constituent developed in P103 relies on [1], [2], [3], [4] and [5]. The aim of this paper is not to discuss new charging principles, instead it provides a service constituent with parameters and mechanisms for charging which can be applied to any service available in a broadband or in a narrowband network, such as telephony. The starting point for generalised charging in telecommunications services was trying to model the service Card Calling. However, Card Calling service on its own did not make any sense in an environment aiming at reuse. This is because within P103, any service subjected to the core of Card Calling service, should have the usage charge directed to a particular account which may be different from the default one. By generalising charging to be a service constituent that is used by all services, and then leaving it to the charging constituent to actually handle details of charging, it is possible to model services that may be subject to a Card Calling facility.

Charging referred to in this paper is applicable to network operators, service providers, service subscribers and service users for the usage of services.

During the development of the service constituent, no attempt was made to distinguish between charging related to (a) parameters related to subscription (access component) i.e. usage independent parameters and (b) parameters related to the connection (utilisation component) i.e. usage dependent parameters.

EURESCOM PROJECT P103 NOTATION

In P103 guidelines for the analysis and design of services and service constituents have been defined. The guidelines are based on the object oriented OOram role modelling [10], extended in several ways to meet the project needs. Notations selected are ITU-T Message Sequence Charts (MSC) [12] and ITU-T SDL-92 [11]. The result of applying the guidelines is a set of object types for a service application. The object types can be based on existing role models (*service constituents*) or on new role models.

In P103 project the term *service constituent* is used to denote the reusable components which are used to build the service specifications. A service constituent is, in order to be reusable, independent of any particular service but may be applicable to a number of services because it solves some recurring problems in service specifications.

The guidelines consists of three main steps as follows.

Step 1:

Study the service application and separate it into loosely coupled (independent) *Areas of Concern*. An area of concern can be viewed as an elementary mechanism with one particular aim. Describe each area of concern in relation to the service application as a whole.

Step 2:

Specify a *role model* for each area of concern by the use of MSC and SDL. The role model should be a generalised solution to the specific problem to increase the possibility of reuse later. If a similar area of concern already has been specified in terms of a role model this specification can be *reused* as it is, or as a basis for a new role model.

Step 3:

Map all role models that are necessary in the service application onto a set of object types so that each object type plays one or more roles. The interface and behaviour of each object type is derived from the interfaces and behaviours of each of its roles.

Area of concern

As with other analysis and design methods, structuring of a service description starts early in the analysis phase and is then used throughout the analysis and design work. The main purpose of this is to break the service down into smaller and easily manageable sub problems. The notion for sub-problem in OOram is *area of concern*. The way the service is divided into sub-problems is perhaps the novelty of OOram. Instead of looking at individual parts or objects of a service, like it would be done with other object oriented methods, the analyst structures according to (part-) *functionalities* of the service.

An area of concern does not have to be located in one object, instead it may spread over many objects. In fact, objects are implementation related and therefore they are not used in the analysis and design phases of the work.

As a consequence, structuring that originates from a logical understanding of the service and is unlikely anything that may resemble implementation. Thus shifting the attention from what is basically (one) implementation technique to functionality bound structuring.

Stimulus and Response

When the areas of concern has been determined, the next step is to identify *stimulus* and *responses* for each area. Initially an area of concern is characterised by a (prose) description of the area (intentions, purposes, etc.) and a set of *stimuli/responses*. A *stimulus* is an event that is not accounted for in the area of concern. However, if it happens, then a certain behaviour will follow, as a *response*. The response can be anything from a single message being sent to a whole sequence of behaviour.

An example of stimulus for modelling a switch is shown in Figure 1, off-hook, dialling and on-hook and appropriate actions for each one. This gives a way of stating a stimulus the switch model accepts, but not how it is eventually provided. The responses are not shown here.

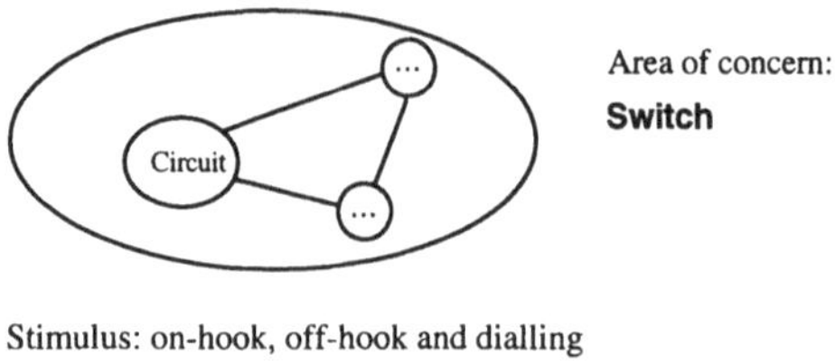

Figure 1. Some stimuli for a simple switch.

During the development phase, other role models providing the cause of the stimuli, may be composed for the switch model, thus giving a more comprehensive role model.

One important aspect when getting an input to the role models in this very open manner is that it stresses the externalness of the stimuli which causes an action. In a sense there is essentially no difference between coming from a human user or from another part of the service. It emphasises an autonomous area within in the area of concern with respect to other areas of concern that may exist.

Scenarios

To get a better understanding of the service being analysed 'use case' scenarios need to be developed, expressing typical communication sequences between entities. The purpose is to get a better understanding of what is going on in the area of concern and what roles are needed. As in many other activities this is an iterative process that may be visited several times.

ITU-TS Message Sequence Charts is a formal notation used for writing down the scenarios. Because MSCs are closely related to SDL-92 this allows easier comparison of MSCs and SDL-92 descriptions.

Role Models

In general, for each area of concern a role model is built. A role model is a collection of roles (each role having a group of functionalities) and their collaborating structure. A role model containing such a collection of collaborating roles fulfils the requirements in the area of concern.

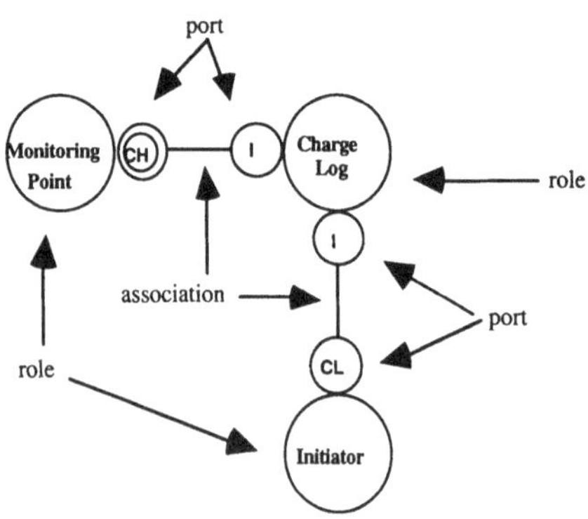

Figure 2. An example role model.

Roles are *associated* with each other if they need to communicate. In that case a number of *contracts* (set of signatures) are specified and positioned in the *ports* of the roles. In figure 2, the Initiator has a port CL which holds the contracts it has with the role Charge Log. The contract contains the signatures of the operations it expects the Charge Log role to provide, similarly for the Charge Log with its port I.

A *role* acts as an efficient tool during requirements capture, where the requirements on different parts of the service (-model) are being investigated. It allows "looking" at the rest of the role model from a specific roles's (functionality's) perspective. The role can then state requirement of *other* roles needed, as *collaborating* roles. This in reality is a different approach in trying to find requirements by implementing entity's point of view where *functionality required* is expressed, not the functionality that appears to be necessary. If the roles does not require functionality, it is not written down, even if it appears necessary to the implementor.

During development of the role model, by identifying the roles, the collaborating structure is also determined. It captures the information such as which other role needs to know about a particular role .

The functionality requirements are expressed in *contracts* between roles and is written down in SDL-92 syntax as signatures for *signals, remote procedures* and *datatypes*. SDL-92 is used to avoid the introduction of an intermediate notation (and a mapping problem) before starting to specify the role. Behaviour is specified as a partial process type describing the handling of stimuli and communication, using all available constructs of SDL-92. Such a process type may be specialised.

The goal is to provide service constituents that are not changed later, but only specialised. To achieve this, each transition that may cause a stimuli, has a virtual procedure declared (in the process type) which will be called (in the transition). Thus necessary openings in the behaviour are provided, and none of the already specified behaviour can be accidentally overridden by a specialisation.

Composition: Synthesis

Having fully designed and specified role models for subproblems of the service, the areas of concern need to be composed. The role models are the result of breaking the service into smaller and more manageable pieces, perhaps recursively if needed. Because the splitting of a service is done differently from most other analysis and design methods, composition will also have to be different. Considering that an area of concern is more than an object, it is obvious that working with usual interfaces alone is probably not enough to complete the job.

Composing means identifying the relations between involved role models and making sure that the *causes of stimuli* are put together with the parts requiring the stimuli. Because causes of stimulus can come from different places this needs careful consideration. The process of doing composition is called *synthesis*. Bringing two role models together in synthesis means identifying the roles in the role models where the caused stimuli have to be introduced.

A simple access role model as shown in Figure 3, will cause the stimuli off-hook, on-hook and dialling.

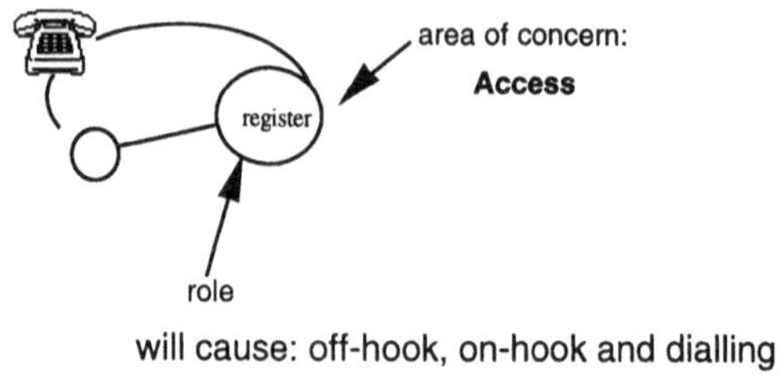

Figure 3 Synthesis.

In P103 the contracts and behaviour is specified in SDL-92. For roles to be synthesised the starting point is the partial process graphs that describe the behaviours of the roles. Because SDL-92 is the formal language being used, communication is either sending messages or making remote procedure calls. This determines the possibilities for interaction. Stimuli for a role model is treated as receiving messages or accepting remote procedure calls under the assumption that any particular stimulus is only accepted in one role.

After having identified the roles that cause stimuli, what remains is to specialise the behaviour to actually generate the stimuli. Each potential cause has its own virtual procedure declaration attached. This declaration may be specialised if the stimuli of some kind have to be generated.

To show how this works in SDL, assume that we have a (partly specified) PROCESS TYPE for the Telephone and we want to specialise that to be the cause of stimuli in the switch.

```
PROCESS TYPE Telephone
VIRTUAL PROCEDURE OffHook;
    RETURN;
   .
   . (* OffHook is called somewhere in here*)
   .
ENDPROCESS TYPE

PROCESS TYPE SwitchTelephone
INHERITS Telephone
REDEFINED PROCEDURE OffHook;
OUTPUT OffHook(...) TO Switch;
RETURN;
ENDPROCESS TYPE
```

The stimulus is sent to the *Switch* and the off-hook interaction is resolved by specialising the existing complete specification. There are some technical details left out here, but they are not important for the example illustrated.

CHARGING COMPONENTS

This presentation discusses a generic charging role model consisting of five roles as shown in Figure 4. Each role in this model will then be decomposed as shown in Figure 6, into a set of relatively low level service constituents related to charging and billing done with a card account. The aim of the decomposed role model is to provide necessary and sufficient details to provide requirements to simulate the charging and billing service constituents (with SDL) and check the (behavioural) correctness of the specification. With appropriate test cases, this would provide an opportunity to prove the behavioural aspect of the model.

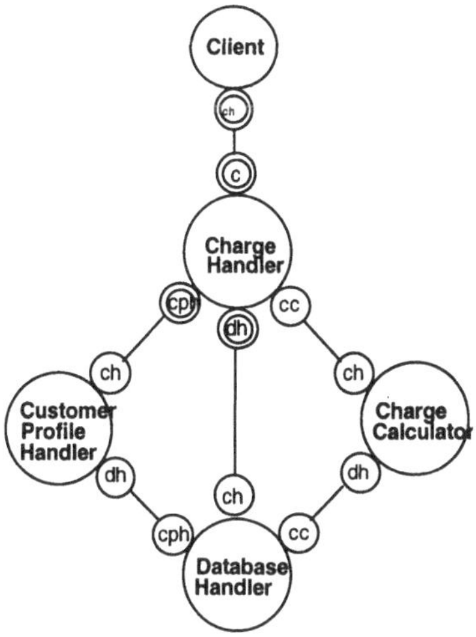

Figure 4. Charging Role model

c - Client
ch - Charge Handler
cc - Charge Calculator
dh - Database Handler
cph - Customer Profile Handler

Table 1. shown below is a typical stimulus-response for the charging area of concern

Stimulus	**Response**
Transfer the cost of using a service to an account	Check that such an operation is permitted Ask for the service usage parameters Store the cost of service Establish contact with relevant entities Get parameters for charging purposes Measure any charging parameters that affect the quality of service Give feedback
Calculate the cost of using a service and give feedback	Get parameters for calculation Calculate cost Give feedback

Calculate the potential cost of using a service (hot billing), given a set of parameters	Get parameters associated with the requested service Calculate cost Give cost value
Store the cost of using a service	Store the cost of the service in an appropriate format.
Give information on service charging activity	Generate the information requested and give it in the correct format.

Roles and Collaborating structure for Charging Aspects

Based on the stimulus-response described above, a role model for charging was produced as shown in Figure 4.

Client {Role}

The Client role represents the stimulus for the role model. The Client role can be replaced with any service, thus making the charging role model adaptable to charge any service, in general.

Charge Handler {Role}

The Charge Handler contains a set of actions and procedures which are associated with a service, in order to determine all the parameters required for calculating service charge. Following functions are its main activities.

- Method(s) for measuring charging information.
- To measure parameters related to charging that affect the quality of service.
- Dispatching the information related to charging activity to those entities that request it.

Charge Calculator {Role}

The Charge Calculator contains method(s) or procedures for calculating service usage.

Database Handler {Role}

Three key activities of the database handler are as follows:

- a storage function, required for the storage of charging information
- to receive charging information.
- to hold information on tariffs applied to charging.

The database handler contains the subscriber account number, CCITT - E.164 address, and the physical locations associated with these addresses. This information is stored and mapped into an address inside the database handler during the whole life of the subscription, from registration to the removal.

The database handler contains charging information related to network usage, service usage and access.

Customer Profile Handler {Role}

This role contains the information of the customer (Subscriber instance) in relation to:

- Administrative address,
- Physical location,
- Privileges, e.g. service user privileges,
- Services Allowed,
- Customer type,
- Transmission related data,
- Access charging amount,
- PIN code,
- Service access codes and
- authorisation codes.

Figure 5 shows one possible scenario for the charging role model.

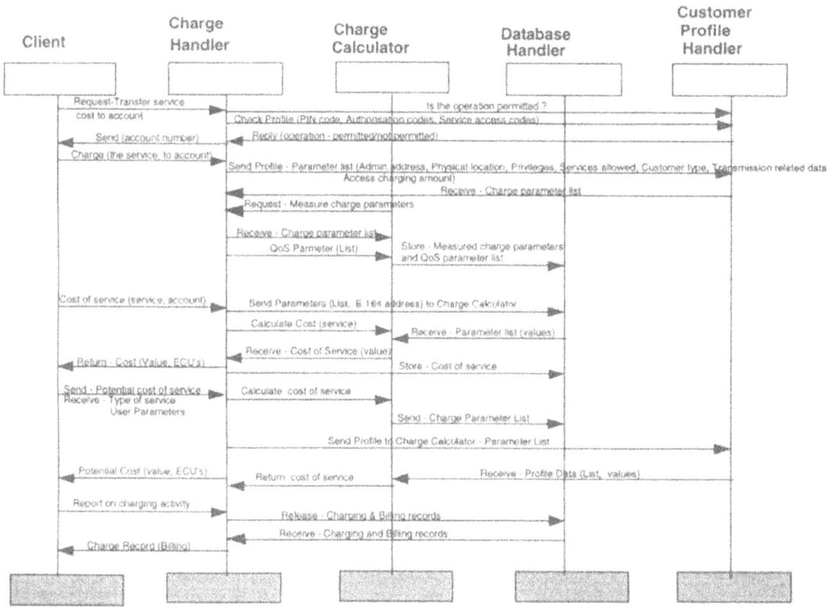

Figure 5. Charging MSC - Message Sequence Chart

Abbreviations used in Table 2 are shown below:
CH - Charge Handler,
CC - Charge Calculator,
DH - Database Handler and
CPH - Customer Profile Handler.

Table 2. Shown below gives a description of Messages and the Action sequence shown in Figure 5 (Charging MSC)

Message from Client (a User or another Service)	**Action by CH, CC, DH & CPH in response to message from Client**
Request - transfer service cost into account	CH to CPH - Is the operation permitted ?
	CH to CPH - Check Profile (PIN code, Authorisation codes, Service Access Code)
	CH to Client - Send account number
	CPH to CH - operation-permitted/not-permitted authorisation - verify account has funds (black/white list)
Charge - the service (Service Provider) - to account (Network Operator/ Service Provider/ Service User)	CH to CPH Send Profile - Charge parameter list (Administration address, Physical location, Privileges - service user privileges, Services allowed, Customer type, Transmission related data, Access charging amount)
	CPH to CH - Receive charge parameter list
	CH to CC - Calculate charge
	CC to CH - Request "measure charge parameters"
	CH to CC - Receive "charge parameter list"
	CH to CC - Receive "QoS parameter list"
	CC to DH - Store "measured charge parameters, QoS parameter list and calculated cost of service"
	CH to Client - Allow access to list of parameters related to service cost and tariffs associated with service
	CH to CC - Receive measured charge parameters
	CH to CC - Calculate cost of service
	CH to CC - Receive QoS - list of parameters
	CC to DH - Store measured parameters & calculated cost
Cost of service - to pay a service in real-time / debit to an account	CH to DH - Send charge parameters (list), E.164 address, and tariffs to CC
	DH to CC - Receive charge parameter list and tariffs
	CH to CC - Calculate the cost of service
	CH to DH - Store calculated cost of service at E.164 address
	CC to CH - Return cost of service to Client - Billing - value (ECU's)
Potential cost of service usage (for paying a service) - Type of service / user parameters to establish cost (real-time usage of service)	CH to CC - Calculate the cost of service usage
	CH to CC - Receive Tariffs, Access charge and User parameters
	CPH to CC - Receive profile data (list, values)

6. Provide report on charging activity - Billing with user parameters	Request from CH to DH - Release records of charging activities to Client
	CH to Client - Receive charging records

Further Subdivision of the Charging Constituent

The overall charging role model shown in Figure 4, is decomposed into a number of smaller constituents, which focus on particular aspects of functionality.

Figure 6, shows role models for the aspects of Usage Charging, Account Handling, Account Selection and Payment Selection.

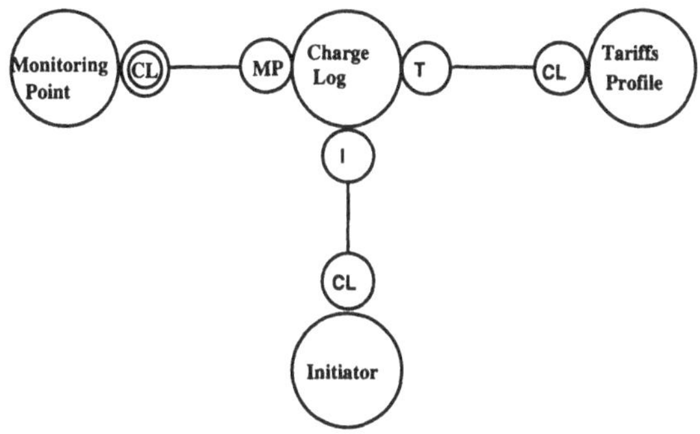

Figure 6(a). Usage charging role model

Figure 6(a) shows the role model concerned with charging a user of a service on behalf of the owner of the service. In general, a user can be a person or another service.

Following are the stimuli which occur in this role model.

1. The initiator decides to start charging the user. The initiator sends a start_charging message to the charge log. The charge log starts to register all charge related events. If the charge log is unable to execute this task, then a failure message is sent to the initiator.
2. The initiator stops charging the user, and sends a stop_charging message to the charge log. This terminates the action of the charge log i.e. suspend logging of all charge related events.
3. A monitoring point detects a change in the charge state of a resource and reports the change to the charge log. This causes an update in the charge register.

Figure 6(b). Transaction handling role model

Figure 6(b) shows the role model concerned with the registration of a payment transaction on the account of a service user (e.g. a credit card account). The transaction must contain an identification of a receiver of the payment (normally the owner of the service) and the amount to be paid.

Following are the stimuli which occur in this role.

1. The charged service decides to start the transfer of charge to the account handler. This follows an initialise message from the charged service to the account handler. The account handler receives a credit_limit from the account and returns a ready message to the charged service.
2. The charged service sends charge state information to the account handler. The account handler compares accumulated charge with the credit_limit. If the credit_limit is exceeded, account handler withdraws the credit limit from the account and closes the account. This follows the message credit_limit_exceeded from account handler to the charged service.
3. The account handler detects (based on the current charge rate) that the credit_limit is likely to be exceeded and sends the credit_limit_exceeded message to the charged service.
4. Termination of the charged service - The charged service sends a stop message to the account handler, withdrawing the accumulated charge from the account followed by the closure of the account.

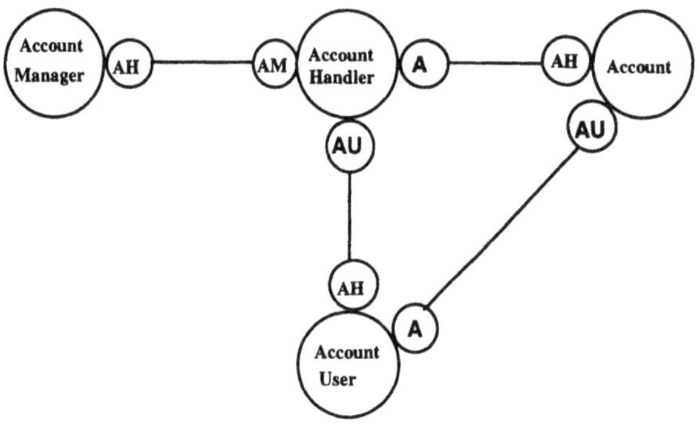

Figure 6(c). Account selection role model

Figure 6(c) shows the account selection role model. This role model is concerned with the selection and initialisation of an account for an account user. An account manager is responsible for selecting the account to be used. An account manager identifies the account to be used. The selected account is then initialised. If the account cannot be initialised a failure message is sent to the account manager and the account user.

Following stimulus occurs in this role model.

1. The account user needs an account reference. The account user sends a get_account message to the account initialiser. The result of the account initialisation is returned to the account user together with a reference to the account.

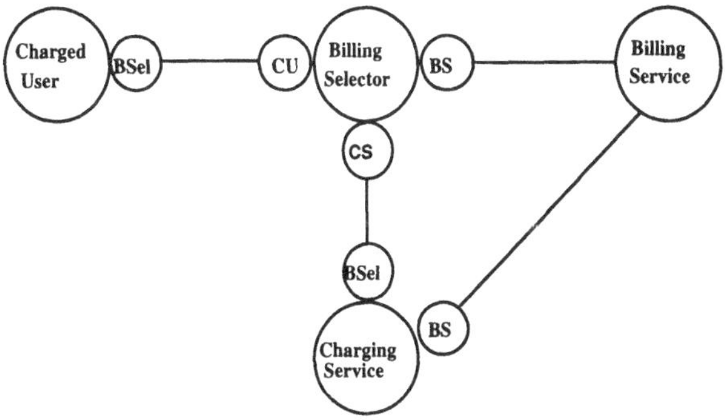

Figure 6(d). Billing selection role model

Figure 6(d) shows the billing selection role model. This role model is concerned with the selection of a billing service for a charged user. The user who is charged for the usage of a service is given the option to select the type of billing service to be used e.g. coin box billing, credit card account, standard method by the owner of the charging service, prepaid telecom cards ect. It is assumed that different ways of paying for a service may require different software implementations. Therefore, it would be necessary to invoke the kind of billing service chosen by the user.

Following stimulus occurs in this role model.

1. The charged service requires a reference to a billing service. The charged service sends an initialise message to the billing selector. The billing selector will return the result of the billing selection to the charged service together with the reference to the selected payment service.

The service constituents described in this paper would be incomplete without the appropriate role models for user authentication and the service network charging. These role models are discussed in detail within the EURESCOM P103, PIR (Project Internal Report) 1.5, Issue 2. However, the synthesised role model for charging and billing shown in Figure 4, and the message sequence chart shown in Figure 5 has accounted for these two role models during its synthesis.

DISCUSSION AND FUTURE WORK PLAN

The discussion in this paper is centred around a proposal to generalise the concept of charging as a service constituent that may be used from any service in a network. By having a charging service constituent, the service specifications does not have to include aspects that have to do with charging. Thus, allowing any service to be charged to a calling card, as opposed to the card calling services currently in use. This is recognised as an important property that can be applied in networks, which may contain a great variety of services and many different charging policies. Future work include consolidating and detailing the descriptions of the constituents and to specify their behaviour.

ACKNOWLEDGEMENTS

The work reported in this paper is based on intermediate results from EURESCOM collaborative project P103, Task 1 where the following have contributed to the work Bengt G Jensen and Knut Lillegraven from Norwegian Telecom Research (Kjeller, Norway), Parminder Mudhar from BT Laboratory (Ipswich, UK), Lars Pedersen from Tele Danmark Research (Horsholm, Danmark) and Roberto Minetti from CSELT (Turin, Italy).

REFERENCES

[1] ETSI/NA5 technical report "Parameters and mechanisms relevant for charging in B-ISDN" DTR/NA-52204.

[2] EURESCOM P105 "European ATM network studies" (Deliverable No.3, Volume 2 of 4: Network Management - Parameters for Charging and Accounting).

[3] CCITT Recommendation D-Series (03/93), D.000, D.98, D.178, D.120

[4] The EURESCOM P103, Project Internal Report 1.5 Service Constituents, Issue 1, (Entity-Relationship Diagrams of the Card Calling Service).

[5] The 1988 CommEd Proceedings on Telecommunications Billing and Charging (London), CommEd Ltd, 137, Dulwich Road, London SE2 0NG, UK.

[6] EURESCOM P103, Project Internal Report, 1. 4, (July 1994).

[7] Presentations of the method at IEEE IN '94 Workshop Records Volumes 1 & 2. (May 24 - 26,1994, Ramada Renaissance Hotel, Heidelberg, Germany).

[8] EURESCOM P103, Project Internal Report, 1.1 - Scene Setting.

[9] "Object oriented IN service provision" R. Nilsen, J. Simons and P.Dellafera, The fourth Telecommunications information Networking Architecture Workshop, 1993, pp I-233 - I-245.

[10] Reenskaug, T et al. "OORASS", Seamless support for the creation and maintenance of object oriented systems" In: Journal of Object-Oriented Programming (Oct. 1992).

[11] Recommendation Z.100 - SDL, ITU-T, 1992, Revised Recommendation.

[12] Recommendation Z.120 “Message Sequence Charts, MSC,” ITU-T, 1992.

13

Database access in intelligent networks

Kimmo E. E. Raatikainen
University of Helsinki, Department of Computer Science
P. O. Box 26 (Teollisuuskatu 23),
FIN-00014 University of Helsinki, Finland

Abstract

Future telecommunication services will extensively exploitdatabase technology. Information needed in operations and management of telecommunication services will be organized as a logical entity. The world-wide nature of telecommunication dictates that the logical entity can only be realized through interoperability of autonomous databases that may have different internal characteristics. We review the requirements for database access in the IN architecture. Our premises are derived from the IN long-term architecture framework as specified in ITU-T Recommendation Q.1201.

1. INTRODUCTION

Databases will have a dominant role in telecommunication services based on the Intelligent Network architecture. The databases hold the information needed in operations and management of telecommunication services. The performance, reliability, and availability requirements of data access operations are hard. Thousands of retrievals must be executed in a second. Allowed down-time is only a few seconds in a year.

The globalization of services implies that databases managed by different operators must interoperate. This requires that the systems are open, use standardized protocols, are protected against misuse and intruders, and have a common logical view of information.

In this paper we review ITU-T Recommendations in Q.1200 (Intelligent Network), X.500 (Directory), and X.700 (OSI Management) Series. Our objective isto derive the requirements for database access in the IN architecture. The paper is organized as follows. In Section 2 we examine the IN long-term architecture framework. In Section 3 we examine the current Recommendations in Q.1200 Series. We also discuss the next Refinements based on the X.500 Directory. The X.500 Series of Recommendations are summarized in Section 4 and X.700 Series in Section 5.

2. PERSPECTIVE

The *Intelligent Network long-term architecture framework* is the structure that allows the integration of technologies developed in other standards activities into the *IN architecture*. Of these activities the most important are *Open Distributed Processing* (ODP) and *Telecommunications Management Network* (TMN). The IN framework will provide an open architecture that is achieved through the integration of computing information and telephony technologies. The architecture will be enhanced by emerging technologies including broadband capabilities, distributed processing, Open Systems Interconnection, object-oriented modelling, information technology, cooperative processing, distribution control, management of services and networks, verification/validation, and artificial intelligent.

It is intended that the *IN conceptual model* as described in Recommendation Q.1201 *Principles of Intelligent Network Architecture* remains consistent throughout the evolution of the IN architecture. In Q.1201 there is a list of attributes that intelligent networks will have. The list includes integrated services, integrated/shareable control, programmability, adaptability facilitated by modularized hardware and software, interoperability of networks and systems, and an OSI-aligned protocol architecture for all interfaces that facilitate communication between entities. For the database access the lastthree items are the the most important ones.

The need for interoperability calls for openness, particularly for open distributed processing. From the computational viewpoint[1],the services and the computing platforms can be represented asan aggregation of collections of objects.This object-oriented modelling is currently considered as the easiest way to provide a framework for mechanisms that implementthe necessary transparencies including access, concurrency, failure, location, and replication transparency.

The *location transparency* hides the location of an object.The *access transparency* gives a similar accessto the methods of local and remote objects.The *concurrency transparency* hides the existence of other concurrent users of an object. *Replication transparency* hides the effects of having multiple copies of an object while *failure transparency* provides fault tolerance. In brief, the transparencies enables greater freedom and independencein the design of applications and services.

It is intended that the IN long-term architecture takes advantage ofnew and emerging technologies as appropriate.Several of these technologies were mentioned at the beginning of this section. Among them were distributed processing, Open Systems Interconnection, object-oriented modelling, and management of services and networks.

Distributed processing is the mechanism for maintainingan environment which is composed of a variety of applications, protocols, and platforms.Important issues related to distributed processing include scalability, portability, and performance.Scalability and portability are the ways to achieve seamless and cost-efficient evolution of the computing platform.Data communication services that function as data transport servicesare an essential area in distributed processing, for its performance and fault tolerance.The concepts of database

---

[1]Viewpoints are pragmatic tools defined in the ISO *Reference Model of Open Distributed Processing* (RM-ODP). Each viewpoint - enterprise, information, computational, engineering, and technology - leads to a representation of the system with emphasis on a specific concern.Within IN the notion of viewpoints has been adopted in the conceptsof planes in the IN Conceptual Model. It is important to realize that multiple viewpoints must be considered to ensure portability.A recent introduction to RM-ODP can be found in Raymond [1993].

management must also be applied ina distributed environment.Database services are need to access and manage structured elements through the management of information processing and the database infrastructure.

The **OSI Reference Model** permits interworking betweendifferent systems.OSI is concerned not only with the transfer of information betweensystems, but also with their capability to interwork in order to achieve a common distributed task.For the database access, the higher protocol layers, particularly the application layer together with the OSI-applications - such as Directory (X.500), Security (X.800), and Transaction Processing (X.860) - is of fundamental importance.In addition, we cannot forget the management aspects of theOSI architecture (X.700).

Object-oriented modelling is pointed out bythe following quotation in Q.1201: "*The use of object modelling could satisfy the modelling needsof IN long-term architecture by the use of abstract object modelling concept.*" Our claim is even stronger than that.We are convinced that today object modelling is the easiest way to satisfy those modelling needs.

The object-oriented methodology allows strict modular system specifications. The strong encapsulation supported by object-oriented design isa prerequisite for evolutionary changes within a specificationso that the system can be regarded as an open system.The encapsulation hides all changes in the implementation of objects aslong as the new interfaces to the methods remain compatible with the old ones.

In order to ensure that large distributed systems are tractable,the systems must be designed in a way that minimizes the interdependencies between the components in the system.Object-oriented modelling techniques are particularlywell suited to these purposes because they providethe benefits of abstraction, encapsulation, and modularity. Abstraction is a tool that simplifies the description of systems through describing only those characteristics that are meaningful on the current description level.Encapsulation is a technique for only exposing the observable behaviour of an object's services and providing clients with information how to invoke these services.On the other hand, encapsulation hides the details of the object's implementation.Modularity is achieved because a system can be specified as an aggregation of collections of objects.The fact that subsystems can be treated as independent objects greatly simplifies system descriptions.

Object modelling promotes modularity by enabling the structuring of specifications into smaller parts. By constraining all interactions between objectsto take place at well defined interfaces,the interdependence of objects is minimized. By isolating and explicitly describing all interfaces between objects, the dynamic and evolutionary nature of distributed systems can be more easily modelled.

For the database access, the ongoing standardization in the area of object database management is seminal. A consortium has developed one standard called ODMG-93 Standard [Cattell, 1994]. Other standardization activities include the Object Data Model by the Object Management Group (OMG); see e.g. OMG [1992] and Soley [1992]. Unfortunately, the field of object-oriented database management is not yet mature.However, the group developing the RM-ODP is strongly committed to be compatible with the OMG specifications.

Management of services and network requiresthat the IN architecture and TMN concepts are integrated. In IN, managed object are very diverse.In order to realize the management of all the different kinds of managed objects in practice, the Management Information Base (MIB) must have an efficient implementation.This challenges the database architecture by implying a twofold functionality.Firstly, the database and its elements are managed objects.Secondly, the database could be the tool to implement the MIB.

To summarize, we have reviewed and briefly commented theRecommendation Q.1201.Our goal was to introduce the premises and requirements fora database architecture that can be deployed as the Service Data Function. The key requirements are 1) an OSI-aligned protocol architecturefor all interfaces, 2) compatibility with the OSI Reference Model ofOpen Distributed Processing (RM-ODP), and 3)adapting ITU-T Recommendations for Telecommunications Management Network (TMN).In particular, the two last requirements call for an object-oriented approach.

3. GUIDELINES FOR IMPLEMENTING DATABASE SERVICES IN CURRENT Q.1200 SERIES RECOMMENDATIONS

3.1 Functionality

The *distributed functional plane* (DFP) in the INCM consists of *functional entities* (FEs) and of the relationships between the FEs. Database services are provided by the FE called *Service Data Function* (SDF). Recommendation Q.1204 *Intelligent Network Distributed Functional Plane Architecture* specifies that the SDF contains customer and network data for real time access by the FE called *Service Control Function* (SCF) in the execution of an IN provided service. The SDF interfaces and interacts with SCFs and other SDFs. The SDF is managed, updated, and otherwise administered by an FE called *Service Management Function* (SMF).

It should be noted that the SDF contains data which are directly related to the provision of IN provided services. Therefore, the SDF does not necessarily encompass data provided by a third party but may provide access to these data. Examples of service data processing functions accessible to the SCF from the SDF include functions to access service information (e.g. subscription data parameters) and to update service information (e.g. sum of charging).

Capabilities in the Capability Set 1 (CS-1) are intended to support services[2]and service features[3] that fall into the category of "single ended", "single point of control" services. Such services are referred to as Type A services. All other services are placed in a category called Type B. Due to operational, implementation, and control complexity CS-1 standards do not encompass Type B services.

Recommendation Q.1211 *Introduction to Intelligent Network Capability Set 1* augments the functional specification of the SDF through requiring that SDF provides consistency checks on data. This implies that SDF should provide the transaction processing capabilities. Q.1211 also specifies that SDF hides the real data implementation and provides a logical data view to the SCF.

The service management aspects primarily address the network operator's interaction with the SCF, SDF, an FE called *Service Switching Function* (SSF), and an FE called *Specialized Resource Function* (SRF). Since this interaction normally takes place outside the context of a

---

[2]A service is a stand-alone commercial offering, characterized by one or more service features, and can be optionally enchanced by other service features

[3]A service feature is a specific aspect of a service that can also be used in conjunction with other services/service features as part of a commercial offering. A service feature is either a core part of a service or an optional part offered as an enhancement to a service.

particular call or service invocation, the treatment of management aspects is only cursory in the CS-1 Recommendations. However, Q.1211 requires that CS-1 must neither exclude nor constrain the capability of service customers to interact directly with customer-specific service management information. A personal service profile is an example of such information. Moreover, the following two points may be relevant to the CS-1 timeframe. Firstly, the SMF, an FE called *Service Creation Environment Function* (SCEF), an FE called *Service Management Access Function* (SMAF) may be used to add, change, or delete CS-1 based service related information or resources in the SSF, SCF, SDF, and SRF. Such changes should not interface with CS-1 based service invocations or calls that are already in progress. Secondly, the network operator may, at its discretion, give the service customer the ability to add, change, or delete appropriate customer-specific information. The mechanisms and safeguards that are put into place by the network operator for this interaction may take advantage of CS-1 functions and capabilities.

3.2 Relationships and Information Flows

In the DFP each interaction between a communication pair of FEs is termed an *Information Flow* (IF). The relationship between any communicating pair of FEs is defined by a set of information flows. If a communicating pair of FEs is located in physically separate entities, the relationship between them defines the information transfer requirements for a protocol between the physical entities. In order for one FE to invoke the capabilities provided by another FE, a relationship must be established between the two FEs concerned. This implies that the SDF is a server and that SCFs and SMFs are its clients. It should be noted that a relationship can only be established by the client FE.

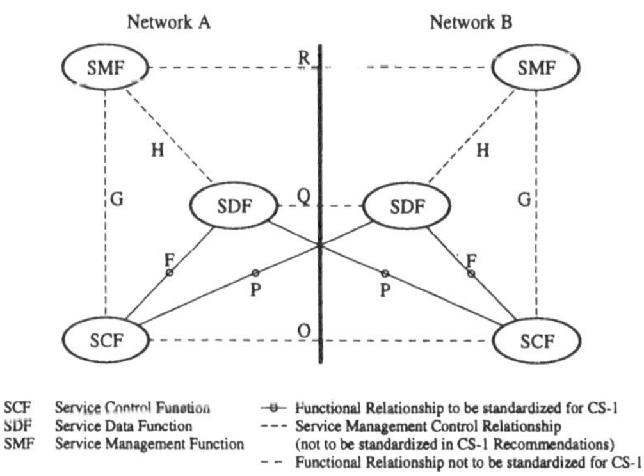

Figure 1. Functional Relationships and Reference Points for CS-1

The relationships recognized in Q.1204 include SCF-SDF, SDF-SDF, and SMF-SDF. Furthermore, Q.1211 identifies 18 distinct functional relationships, of which five are related to network interworking. Q.1211 identifies the relationships as reference points and assigns a unique one-letter identifier to each reference point. CS-1 divides the relationships into four control classes: 1) connection-control capabilities, 2) call-control (Non-IN service-control) capabilities, 3) IN service-control capabilities, and 4) management-related control capabilities. The difference between classes 2 and 3 is that the capabilities in class 3 involve the structured separations of the SSF from the SCF whereas the capabilities in class 2 do not involve that separation.

Figure 1 illustrates the relationships between SCFs, SDFs, and SMFs in two different networks. The functional relationship at a reference point may provide for one or more control classes. Each combinations of a functional relationship and a control class is referred to as a control relationship and is identified by an **<alpha>.<number>** string, where **<alpha>** identifies the functional relationship and **<number>** identifies the control class.

The physical aspects of the realization of each functional relationship do not imply a direct physical interface between the involved network functions. The CS-1 standardization defines the IN *application service elements* (ASEs) independent of the underlying protocol stack. However, it is recommended that the IN ASEs should be used with existing standardized protocol stacks. Control relationships involving the SDF are F.3, H.3, P.3, and Q.3, of which F.3 and P.3 are within the scope of CS-1. Q.1211 recommends SS No. 7/TCAP and DSS 1/Q.932 to be used for F.3. In addition, the final sentence in Clause 7.7 of Recommendation Q.1211 is noteworthy: *Wherever possible, the CS-1 Recommendations identify alternative interfaces (e.g. SCCP-GTT, X.500, or CMISE) for these domains.*

The ASEs at different reference points should be defined separately within a common structure. This helps to develop a modular and flexible *IN application protocol* (INAP). The INAP in turn, facilitates flexible packaging of the functional elements in the DFP into a variety of different *physical elements* (PEs) in the *physical plane* of the INCM. The decomposition of standardized *service independent building blocks* (SIBs) in Recommendation Q.1213 *Global Functional Plane for Intelligent Network CS-1* have been the basis for determining the number, nature, and content of the ASEs. In addition to the Basic Call Processing SIB, Q.1213 specifies 14 SIBs. The execution of the following five SIBs need the support of SDF: Log Call Information (LCI), Screen, Service Data Management (SDM), Status Notification, and Translate.

The definition of INAP ASEs reflects the capabilities that can be differentially applied to the three separate classes of services. **Class 1** includes the services that benefit from IN service control in call set-up and tear-down phases. **Class 2** includes the services that require mid-call control. **Class 3** includes the services that require topology manipulation. The focus in CS-1 has been on Class 1. CS-1 can only support a limited use of mid-call and topology manipulation capabilities.

Recommendation Q.1214 Distributed Functional Plane for Intelligent Network CS-1 divides the information flows into nine categories. Six of these categories fall into Class 1. Category r4 contains the information flows (IFs) between the SCF and SDF. The IFs in category r4 are: Query, Query Result, SDF Response, Update Confirmation, and Update Data.

Query is the IF that is used by the SCF to retrieve an item of data held in the SDF. It has three *information elements* (IEs): Database ID, Requested Info type, and Information key. The reply to Query by the SDF is the **Query Result** IF which has the IE Requested Info.

Update Data IF when requested by an SCF will entail an atomic execution of the update. However, problems such as concurrent access to the data are not solved by the IFs. This IF has four IEs: Function type, Database ID, Updated Info, and Information key.

Update Confirmation IF is the response to the Update data IF with the IE Outcome. The SDF sends this IF to an SCF to provide the result of writing to a specified service data object.

The IE Outcome describes the result of the request operation, either success or failure with a specific reason.

The SDF may issue the information flow **SDF Response** to the SCF as an interim response to the Query or Update Data IF. This IF is an indication that the request has been received but may take some time to execute. The IF will subsequently be followed by either a Query Result IF or an Update Confirmation IF.

The IE Database ID identifies the logical location of the database in which the requested information resides. It does not refer to a particular service, but rather to some specific data. The IE Requested Info type identifies the information whose value is requested. The structure of this IE is to be defined within each specific database.

Information key IE is used to locate the requested information fields. IEs in the initial DP IF[4] are all candidates for information key. The precise structure and possible values for the IEs will be service specific.

The information elements Function Type in Update data IF is used to indicate the action to be carried out on the particular data. Possibles values are replace, increment, and decrement. Update Info IE gives the new value of the data to be modified if the function type is replace. It gives the value by which the data is incremented (decremented) if the function type is increment (decrement).

3.3 Service Processing Models

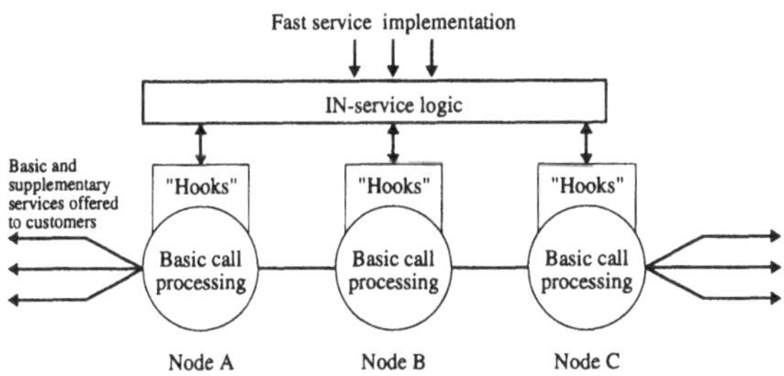

Figure 2. IN service processing model (Figure 16/Q.1201)

A high level overview of a desirable IN service processing model is illustrated in Figure 2. The three main elements of this model are: 1) the basic call processes, 2) the "hooks" that allow the basic call processes to interact with IN service logic, and 3) IN service logic that can be "programmed" to implement new supplementary services. Q.1211 gives a more detailed description: SCF receives and decodes the query, and interprets it in the context of a CS-1 supported service. It formulates, encodes, and sends a standardized response to the SSF. The

[4]The information flow Initial DP is issued from an SSF to the SCF. It is generated when the SSF detects a trigger at any *detection point* (DP) in the *Basic Call State Model* (BCSM) in order to request instructions from the SCF. The IEs of Initial DP are: Call ID, Service key, Call gapping encountered, Dialled digits, Called party number, Calling line identity, Calling party category, SSF/SRF capabilities, SRF available, Misc call info, Terminal type, Service profile identifier, Location number, Calling party business group ID, Calling party sub address, Original called party ID.

formulation of the response may involve complex service logic leading to a query to a separate SDF.

In Q.1214 we can find further refinements. An intelligent network has two realms related to call/service processing: 1) basic call processing and 2) service control. Service control resides in the SCF entity. It interacts with basic call processing via an SSF entity associated with the FE called *Call Control Function* (CCF).

A relationship is established between the SCF and SDF at the request of the SCF when the SCF requires to receive or modify some data contained within the SDF. The relationship is terminated by the SDF. Information flows related to the SDF may be associated with some degree of processing that depends on the supported service. This processing is related to data manipulation but not to call processing. Only logical view of data is known to the SCF. The IFs do not imply any physical organization of data or how they are stored. In particular, the fact that data are replicated is not known to the SCF.

To realized the functionality needed in the SCF, Q.1214 provides an SCF model shown in Figure 3. It should be noted that the model is conceptual and that it is not intended to imply an actual implementation of the SCF. However, the model provides a useful framework for a

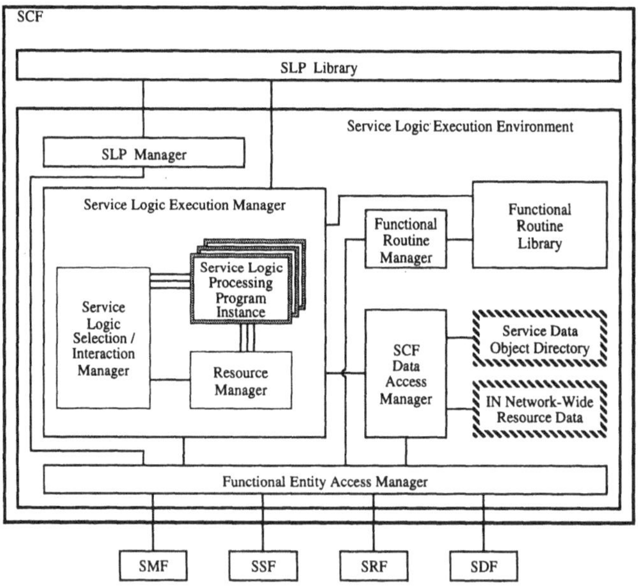

Figure 3. Service Control Function Model (Figure 4-19/Q.1214)

client of database services.

Let us briefly examine the SCF model. The *Service Logic Execution Manager* (SLEM) is the functionality that handles and controls the total service logic execution action. The SLEM interacts with *SCF Data Access Manager* and *Functional Entity Access Manager* to support the execution of a *Service Logic Processing program Instance* (SLPI). In particular, the SLEM needs functionality to manage SLPI access to SCF and SDF data via the SCF Data Access Manager.

The SCF Data Access Manager provides the functionality needed to provide for the storage, management, and access of shared and persistent information in the SCF, that is the information persisting beyond the lifetime of a SLPI. The SCF Data Access Manager also provides the functionality needed to access remote information in SDFs. The SCF Data Access Manager interacts with the SLEM to provide these functionalities to SLPIs.

The SCF data is contained in the *Service Data Object Directory* and in the *IN Network Wide Resource data*. The SDF Data Access Manager uses the Service Data Object Directory to locate service data objects in the network in a manner transparent to the SLEM and its SLPI. As such, the SLEM and its SLPIs have a global and uniform view of service data objects in the network.

The *Functional Entity Access Manager* (FEAM) provides the functionality needed by the SLEM to exchange information with other functional entities via messages. This message handling functionality should:

1) be transparent to SLPIs,
2) provide reliable message transfer,
3) ensure sequential message delivery,
4) allow message request/response pairs to be correlated,
5) allow multiple messages to be associated with each other, and
6) comply with OSI structures and principles.

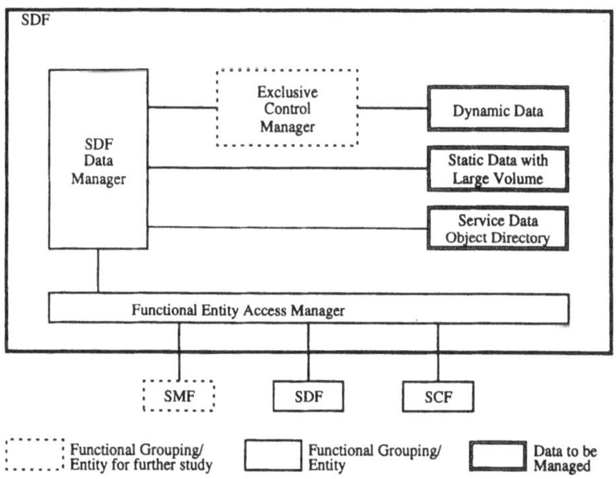

Figure 4. Service Data Function Model (Figure 4-20/Q.1214)

In addition to the SCF Model, Q.1214 provides a conceptual model of the SDF, which is shown in Figure 4. The SDF contains and manages the data which are related to Service Logic Processing programs (SLPs). The data are accessed in the execution of the SLP instances.

Therefore, data such as SLP selection data and SCF directory, which are accessed before the execution of an SLPI, are not included in the SDF handling data.

The functional entities in the SDF model are *SDF Data Manager, Exclusive Control Manager*, and *Functional Entity Access Manager*. The SDF Data Manager provides the functionality needed for storing, managing and accessing information in the SDF.

The Functional Entity Access Manager (FEAM) provides the functionality needed by the SDF Data Manager to exchange information with other functional entities (i.e. SCF, SDF, and SMF) via messages. This message handling functionality should:

1) provide reliable message transfer,
2) ensure sequential message delivery,
3) allow message request/response pairs to be correlated,
4) allow multiple messages to be associated with each other, and
5) comply with OSI structures and principles.

The FEAM may access other SDFs, because the data distribution in the network can completely be transparent to the SCF. However, this point as well as the functional relationship with the SMF is outside the scope of CS-1.

The Exclusive Control Manager provides the functionality needed to provide exclusive control, for example lock-unlock control, to ensure data integrity. However, this manager and its action method are identified as items for further study.

The data handled by the SDF is classified into types of "static" and "dynamic". The data that are "read-only" as far as SLPs are concerned are called static. The data that can be changed by SLPs are called dynamic. It should be noted that these definitions of "static" and "dynamic" are different from the definitions given in Recommendation Q.1290 *Vocabulary of Terms Used in the Definition of Intelligent Networks*.

The types of data in Q.1214 are further subdivided into six types.

Type 1 data is dynamic data that are local to an SLPI, e.g. call instance data parameters like the dialled number.

Type 2 data is static data that are feature-specificand are shared by SLPIs, e.g. subscription data parameters like day of week or time of day screening.

Type 3 data is dynamic data that are feature-specificand are shared by SLPIs, e.g. sum of charging or a counter for a call number limiting service.

Type 4 data is static data that belong to multipleservice features and are shared by SLPIs, e.g. a subscriber's phone number list to connect.

Type 5 data is dynamic data that belong to multipleservice features and are shared by SLPIs, e.g. subscriber's location data used by a service such as UPT.

Type 6 data is data in the Service Data Object Directory.

It is assumed that an SLP includes type 1 data. Besides the locally available service data objects, additional data is used to locate service data objects in other SDFs in the network. The additional data is used in a manner which is transparent to the SLEM and its SLPI in the SCF requesting the locally unavailable data.

Upon a data object retrieval request by the SCF, the SDF Data Manager will try to locate the data object locally. When the requested data object is not available, it will try to retrieve a reference to another SDF from the Service Data Object Directory. If the reference is available, the SDF will either refer this back to the requesting SCF, or try to retrieve the requested data directly from the referenced SDF. However, the latter mechanism is outside the scope of CS-1. If a reference is not available, the SDF Data Manager will return a failure to the requesting SCF.

3.4 Intelligent Network Application Protocol

The intelligent network application protocol (INAP) that is required for support of CS-1 is defined in Recommendation Q.1218 *Interface Recommendation for Intelligent Network CS-1.* The definition of INAP can be split into three sections: 1) the definition of *Single/Multiple Association Control Function* (SACF/MACF) rules for the protocol, 2) the definition of the operations transferred between entities, and 3) the definition of the actions taken at each entity.

The INAP is a ROSE user protocol. The ROSE protocol is contained within the component sublayer of TCAP (Recommendations Q.771 to 775) and DSS 1 (Recommendation Q.932). The ROSE *Application Protocol Data Units* (APDUs) are conveyed in transaction sublayer messages in SS No. 7 and in the Q.931 REGISTER, FACILITY, and call control messages in DSS 1. The INAP (as a ROSE user) and the ROSE protocol have been specified using the Abstract Syntax Notation One (ASN.1). At present, the only standardized way to encode the resulting PDUs is the Basic Encoding Rules (BERs).

The INAP will support any mapping of functional entities (FEs) to physical entities (PEs). Therefore the protocol is defined assuming maximum distribution, that is one PE per FE. The physical interface between an SCF locating in a *Service Control Point* (SCP) and an SDF locating in a *Service Data Point* (SDP) is illustrated in Figure 5. The interface will be INAP using TCAP which, in turn, uses the services of the connectionless SCCP and MTP. The SDF is responsible for any interworking to other protocols to access other types of networks.

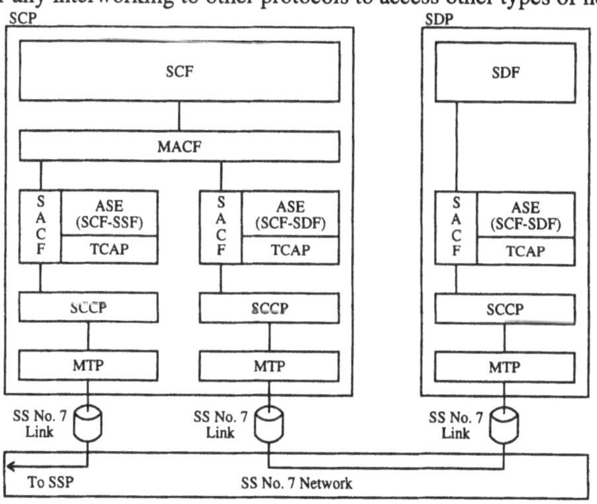

Figure 5. Physical Interface Between SCP and SDP (Figure 1/Q.1218)

The INAP is the collection of all specifications in all ASEs. Each ASE supports one or more operations. Each operation is tied with the action of corresponding functional entity. The use of the application context negation mechanism as defines in the Q.770-Series (*Transaction*

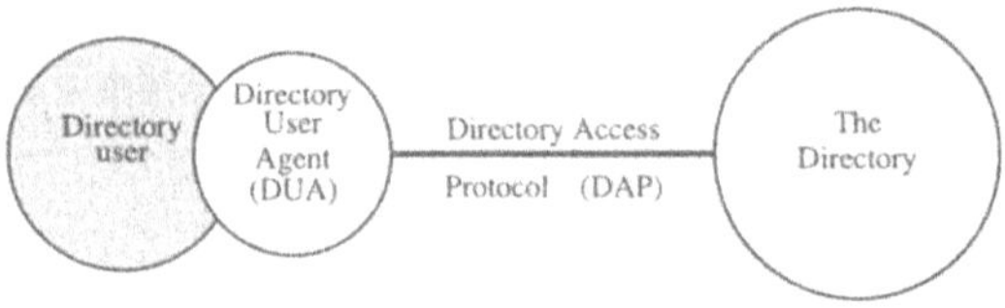

Figure 6. Functional Model of SDF Application Entity (Figure 41/Q.1218)

capabilities application part) allows the two communicating entities to identify exactly what their capabilities are and what the capabilities required on the interface should be. If the indication of a specific application context is not supported by a pair of communicating FEs, some mechanism to pre-arrange the context must be supported.

TCAP Application Context (AC) negotiation rules require that the proposed AC, if acceptable, is reflected in the first backwards message. If the AC is not acceptable, and the TC-User does not wish to continue the dialogue, it may provide an alternate AC to the initiator which can be used to start a new dialogue.

The *SDF Application Entity* (AE-SDF) includes TCAP and one or more ASEs called TC-users. The functional model of the AE-SDF is shown in Figure 6. The shaded area in the figure indicates the scope of Recommendation Q.1218. The ASEs interface to TCAP to communicate with the SCF. They also interface to maintenance functions. The interfaces use the TC-user ASE primitives specified in Recommendation Q.771 and N-Primitives specified in Recommendation Q.711. The operations of INAP are Query, UpdateData, and SdfResponse. They correspond to the information flows Query, Update Data, and SDF Response. Observe that the IFs Query Result and Update Confirmation are mapped on to operation Return Result from Query and UpdateData, respectively.

As far as CS-1 is concerned, the function of SDF is to synchronously respond to every request from the SCF. Therefore, the respective *Finite State Model* (FSM) shown in Figure 7 is trivial. In any state, if there is an error in a received operation, the maintenance functions are informed and the SDF FSM remains in the idle state. In addition, the error can be reported to the SCF using the appropriate component (see Recommendation Q.774). In any state, if the

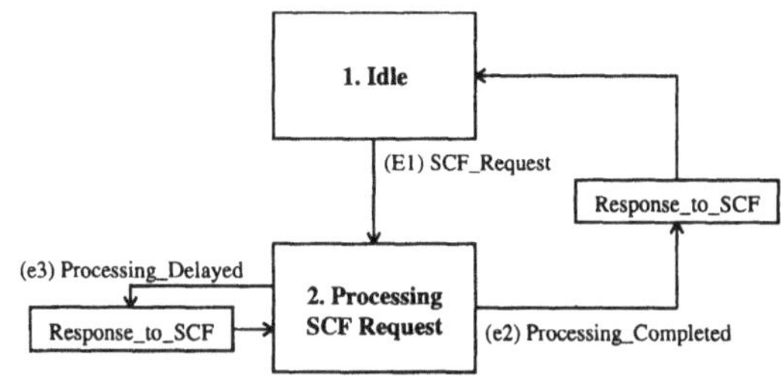

Figure 7. The SDF Finite State Model (Figure 42/Q.1218)

dialogue with the SCF is terminated, then the SDF FSM returns to idle state after ensuring that any resources allocated to the call have been deallocated.

The only event accepted in state "Idle" is **SCF_Request** caused by the reception of Query or UpdateData operation from the SCF. This event causes a transition to state "Processing SCF request".

The events accepted in state "Processing SCF request" are **Processing_Completed** and **Processing_Delayed**. The event **Processing_Completed** occurs, when the SCF request completes. This event causes the response to the Query or UpdateData operation to be sent to the SCF. In addition, the FSM transites into state "Idle". The event **Processing_Delayed** is an internal event caused by the recognition that the SCF request may be delayed. In order to inform the SCF about this situation, the SDF FSM sends the SdfResponse operation to the SCF.

It is important to keep in mind that for the network interworking purposes, the SDF may return to the SCF a reference to another physical entity containing the requested information.

3.5 Identified problems and proposed solutions

The CS-1 specifications have an ASE called *User Data Manipulation ASE* that provides the necessary communication capabilities to access an SDF from an SCF. The ASE consists of two operations: **Query** and **UpdateData**. In brief, the User Data Manipulation ASE is very simple, obviously oversimplified. Attempts to use the operations included in the ASE to support advanced services such as Universal Personal Telecommunication (UPT), Universal Mobile Telecommunication System (UMTS), or Virtual Private Network (VPN) have, according to Chatras and Gallant [1994], shown several shortcomings.

The two main concerns are the service-dependent semantics of the parameters in the operations and the authentication. The service-dependent semantics of the parameters imply that the actions performed by the SDF are also service-dependent. This means that the SDF must be separately reimplemented or tailored for each service. Therefore, the current specification of User Data Manipulation ASE contradicts the principles of IN long-term architecture, since the SDF behaviour must be service-independent.

The second main concern is the authentication. The globalization of services implies that SDFs will be accessed by foreign SCFs. Therefore, the SDF must perform the authentication of users and provide the access control. In the current specification the only way to carry out authentication is to use ordinary attribute value comparison. Therefore, the SDF does not know when authentication takes place.

To overcome the problems in the SDF access ETSI and later ITU-T SG11 have proposed that a new SCF-SDF interface is needed. This interface, which will be standardized in the near future, is based on the *Directory Access Protocol* (DAP) specified in X.500 Series of Recommendations.

When we look for the reasons why the database access in CS-1 Recommendations came to a dead-end, the explanation given most often is that the SDF operations were defined quite lately, on the fly, without any deep analysis of their use [Chatras and Gallant, 1994]. However, we believe that the fundamental reason was the current commercial database products based on the relational data model. The relational data model, which is inherently flat, has simple operations. However, the information needed in IN-services is strongly structured, more like a hierarchical tree than a table. Therefore, the lack of an explicit information model for CS-1 Recommendations is the main source of problems leading to the dead-end.

4. X.500 DIRECTORY

The next refinement of the SCF-SDF interface will be based on a subset of the *Directory Access Protocol* (DAP). According to Chatras and Gallant [1994] the time constraint was a determinant factor for the choice of DAP in CS-1 Refinements. The DAP solution was preferred to CMIS (X.700) for the following two reasons: 1) unambiguous security and authentication model and 2) distribution and replication procedures can be easily introduced.

In this section we give a brief introduction to the X.500 Directory. We emphasize the features and properties that are significant for our theme. Comprehensive introductions to X.500 can be found, for example, in Sykas and Lyberopoulos [1991] or in Bumbulis et al. [1993].

4.1 Information Model

The Directory can be viewed as a world-wide logical entity that contains all pieces of relevant information. In order to keep the Directory manageable and fault-tolerant, the Directory is distributed and replicated.

The information contained in the Directory is stored in *entries* and is organized in an information tree called *Directory Information Tree* (DIT). Collectively, the entries in the Directory are referred to as the *Directory Information Base* (DIB). Each entry in the DIB holds information about a single object. These pieces of information are organized as a set of *attributes*. Each attribute holds a certain piece of information about a single facet of the object. An attribute has the *attribute type* that indicates how the associated *attribute value* is to be interpreted. An attribute may be multi-valued. Each attribute type has an associated *attribute syntax* and *matching rule*. The syntax describes how the information of this type is represented in requests and replies. The matching rule determines how values of this type are to be compared. Typically, an attribute type is specified as a *subtype* (refinement) of some more general type. A subset of each entry's attribute values are distinguished to form the *relative distinguished name* (RDN) which is the unique (with the respect to all siblings of a common parent) name of this entry.

The class of a Directory entry is determined through the type of object that it represents. Entries of a particular class must have certain *mandatory attributes*. They may also have other *optional attributes*. Object classes are organized into a hierarchy according to "is-a" relationship. The class hierarchy makes it easy to create new classes that are extensions of existing classes. A special entry class **Alias** is defined to refer to an object by more than one name. The aliases imply the actual structure of DIT is directed graph.

To summarize, the Directory provides an object-oriented information model. X.500 information model is quite flexible but not as general-purpose as the information model in ODMG-93 [Cattell 1994] and in OMG [OMG 1992]. The information model permits generic operations which are parametrized by the data carried in the operations. The model is a formal description of the data that is needed to support the services. In addition to the formal description, the information model of X.500 provides a formal organization of the data.

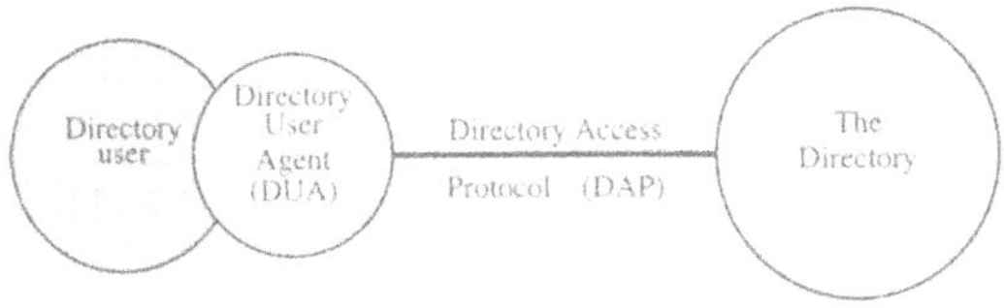

Figure 8. Directory Access Model

4.2 Directory Access

Directory users (persons or application processes) must access the Directory through *Directory User Agent* (DUA) which converses with the Directory using the *Directory Access Protocol* (DAP). When we relate the Directory Access Model in Figure 8 to the SCF Model in Figure 3, it is easy to see that the SCF Data Access Manager is the DUA and the Service Logic Processing Program Instances are the Directory users.

The DAP allows the DUA to issue requests without waiting for the completion of previous requests. Before an DUA can issue any requests, it must establish an association to the Directory. It should be noted that it is not specified in X.500 Series of Recommendations how the DUA obtains a presentations address for the Directory. During the establishment the authentication takes place. After the Directory has validated the user's identity, the DUA may issue requests; *queries* to receive information in the DIB and *updates* to modify either the structure or contents of the DIB.

The Directory provides four types of query operations: **Read**, **Compare**, **List**, and **Search**. **Search** operation will be in the subset of X.500 that is adapted in the next CS-1 Refinement. **Search** operation has four parameters: **base_object**, **subset**, **filter**, and **selection**. The **base_object** specifies the object from which the search begins. The **subset** gives the scope of the search in terms of entries. The **filter** gives the criteria that is used to eliminate entries. The **selection** specifies the attributes of the filtered entries which are returned as the result of the operation.

There are four types of update operations: **AddEntry**, **RemoveEntry**, **ModifyEntry**, and **ModifyDN**. The **ModifyEntry** will be in the next CS-1 Refinement, whereas the use of **AddEntry** and **RemoveEntry** operations is still under further study. The **ModifyEntry** has two arguments: **object** and **changes**. The **object** gives the name of the entry that is modified. The **changes** gives a list of changes: add attribute or value, remove attribute or value.

4.3 Structure of the Directory

When the IN services are globalized, the data hold in the the Directory can be viewed as a world-wide logical entity, into which many different organizations may provide entries. The functional model of the Directory is shown in Figure 9. The Directory consists of processes called *Directory System Agents* (DSAs) which communicate with one another using the *Directory System Protocol* (DSP). When we relate the Functional Model of the Directory in Figure 9 and the functional entities on the distributed functional plane of the IN conceptual model given in Figure 1, we see that an SDF is a DSA. This resolves the unspecified interface between the SDFs.

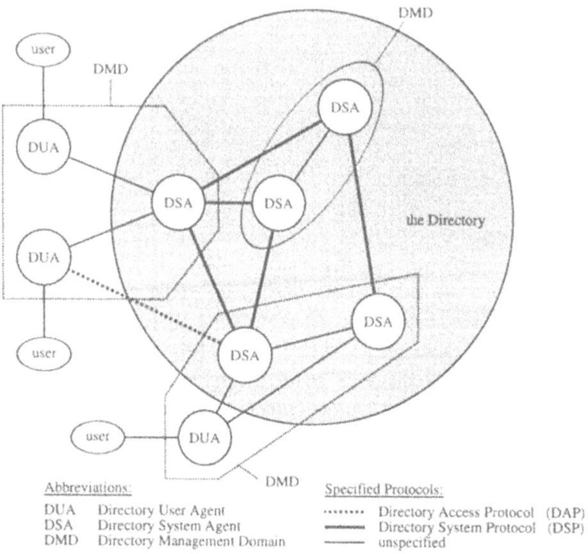

Figure 9. Functional Model of the Directory

Suppose that a DSA has received a request that it cannot complete. The DSA has now two options. Firstly, it can communicate with other DSAs on the behalf of the DUA to complete the request. Secondly, it can simply *refer* the DUA to a DSA that it believes will be better able to complete the request. The refer option implies that each DSA must maintain a set of *knowledge references*. Since each DSA contains a part of the DIT, the DSA must know the presentation addresses of those DSAs which contain the immediate parents and children of its entries. The X.500 Recommendations specify the protocols that are used to create, maintain, and extend the knowledge held by each DSA. These protocols are not in the scope of the CS-1 Refinements but they should be studied for the management aspects of the future IN Capability Sets.

In the Directory exactly one DSA is responsible for the creation and maintenance of an entry. This DSA is called the *master* DSA for this piece of information. The Directory allows that other DSAs may have copies of some pieces of information. To simplify the implementation of the DSAs, the updates to the master information need not be synchronized with updates to its copies. As a result, the DIB may, at times, be in an inconsistent state. This is not a serious deficiency, since the users have always the option of requesting access to the master information.

The 1992 Amendments to X.500 present two different mechanisms for replication information: *caching* and *shadowing*. Caching occurs when a DSA stores locally a copy of an entry used to fulfill a request. This may result in a substantial performance improvement if the same entry is rerequested. Shadowing occurs when one DSA mirrors some information held by another DSA. The shadowed information is brought into agreement with the master information on some predetermined schedule.

The replication mechanisms of X.500 need further study. We need to examine the semantics of correctness in transaction processing. Do we need strict serializability or can we relax it? Another topic for further study is how the shadowed information should be updated. Finally, we are interested in extending the caching. In some situations, the nearest DSA/SDF is still too far away from the DUA/SCF. We want to learn if it is feasible that the DUA/SCF caches some very frequently requested entries.

4.4 Administration and Security

In addition to attributes that hold information about single facets of an object, there are attributes that hold administrative information, e.g., knowledge references and access control items. Such attributes are known as *operational attributes* that are normally invisible to ordinary users. Operational attributes that apply to more than one entry are stored in a special kind of entry known as a *subentry*[5]. Each subentry has a **subtreespecification** attribute that specifies the set a entries to which the remaining attributes apply.

Access control to the information in the DIB can be based on an entry, on an attribute, or on an attribute value. Two methods for verifying user's identity have been specified: *simple* and *strong authentication*. Simple authentication is based on passwords while strong authentication is based on public key cryptography; for details, see e.g. Bumbulis et al. [1993], Burrows et al. [1990], Fumy and Leclerc [1993].

The needs for authentication in IN services should be carefully analyzed and evaluated. Intuitively, we believe that simple authentication is sufficient for most of queries that need authentication. Many queries may be processed without any authentication. However, some queries, most updates and all access to operational attributes may require strong authentication.

5. MANAGEMENT ASPECTS

The *Telecommunication Management Network* (TMN) is a generic architecture to be used for all kinds of management services. It is based on the principles of the OSI Management (X.700 Series of Recommendations) and is standardized in M.3000 Series of Recommendations. The integration of IN and TMN architectures is one of the most important goals in the IN long-term architecture. The reason is that it will be impractical to support two independent architectures (TMN and IN) when applications on both architectures must interoperate.

In this section we briefly summarize IN Management, TMN and OSI Management. We concentrate on the aspects that are important in the database access. Other aspects can, for example, be found in Appeldorn et al. [1993] and in Yemini [1993]. As we have already noticed in the Perspective, the database has a twofold functionality. Firstly, the database and its elements are managed objects. Secondly, the database may be used to implement the *Management Information Tree* (MIT) which is a focal point in TMN (and in OSI Management).

[5]Besides storing administrative information, subentries can also be used to store attributes shared by more one entry.

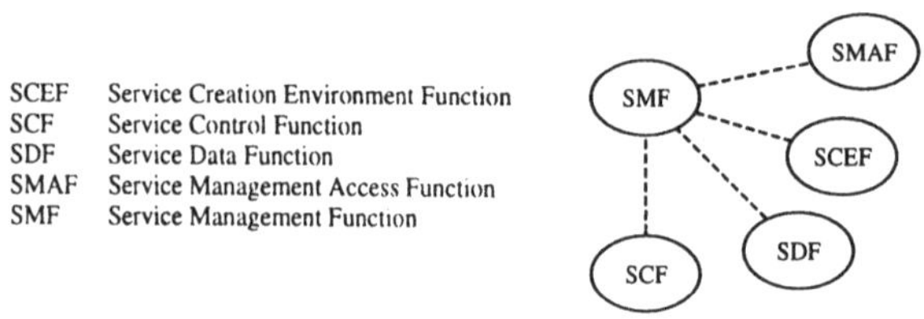

Figure 10. Some Functional Entities and Management Control Relationships

5.1 IN Management and TMN

The management aspects of IN are primarily concentrated on service management. On the distributed functional plane of the IN conceptual model there are three functional entities related to management. These are the *Service Management Access Function* (SMAF), the *Service Management Function* (SMF), and the *Service Creation Environment Function* (SCEF). Figure 10 shows the management control relationships between these FEs, the SCF, and the SDF. As the figure indicates, the SCF and SDF only need a management interface to the SMF. It should be noticed that the management interfaces have not yet been specified. However, the integration of IN and TMN would lead to the usage of a TMN-type stack of protocols including ACSE, ROSE, and CMISE (*Common Management Information Service Element*).

The SMF handles five groups of activities: service deployment, service provisioning, service control, billing, and service monitoring. In each group there are activities that are related to data manipulation. Service deployment includes allocation of service generic data. Service provisioning includes introduction and allocation of customer specific data. Service control contains updating service generic and customer specific data. Billing includes generating and storing charging records, collecting charging records, and modifying tariffs. Service monitoring includes collecting measurement data.

TMN has identified several *TMN Management Services* that are built on a set of *TMN Management Service Components*. The components, in their turn, are based on *TMN Management Functions*. The TMN functional architecture is described through *function blocks* which are similar to IN functional entities. The function blocks identified in TMN are the *Network Element Function* (NEF), the *Operations System Function* (OSF), and the *Work Station Function* (WSF).

The NEFs model all entities that form the network to be managed. Each NEF should physically locate on the network element it manages. The WSFs represents the functionalities and information that relate to the man-machine communication in TMN. The OSFs include the functions for processing, storing, and retrieving management information. TMN has identifies four different OSFs that correspond to the business, service, network, and element management layers. These OSFs are known as **B-OSF**, **S-OSF**, **N-OSF**, and **Ne-OSF**.

5.2 OSI Management

The fundamental idea in the OSI Management is that the knowledge representing the information used for management is separated from the functional modules performing the

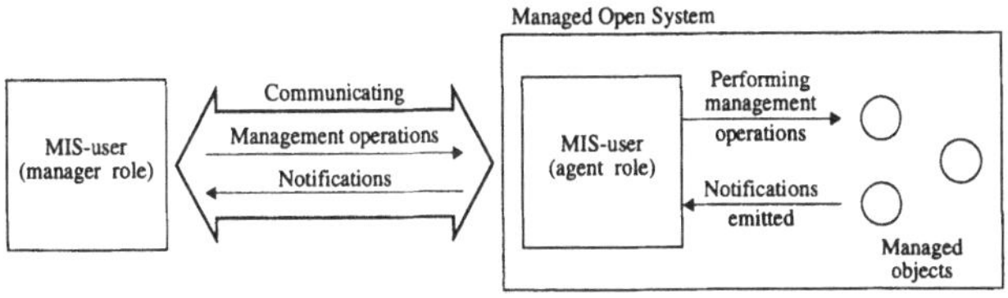

Figure 11. Systems Management Interactions (Figure 1/X.701)

management actions. This can be seen in the information and functional aspects of the systems management model. Another important feature of OSI Management is distributed processing. Since the environment to be managed is distributed, the individual components of the management activities are themselves distributed.

As shown in Figure 11 the interactions that take place are abstracted in terms of *management operations* and *notifications*. Management activities are effected through the manipulation of *managed objects* (MOs). An MO is the abstraction of a resource that represents its properties as seen by management. An essential part of the definition of a managed object is the relationship between these properties and the operational behaviour of the resource.

Systems management application entities (MIS-users) can take either manager role or agent role. An agent manages the MOs within its local system environment. It performs management operations on MOs as a consequence of management operations issued by a manager. An agent may also forward notifications emitted by MOs to a manager. The agent maintains a part of the *Management Information Tree* (MIT), which is a dynamic database. The MIT contains instances of MOs organized on a hierarchical database tree, similar to the DIT in X.500 Directory.

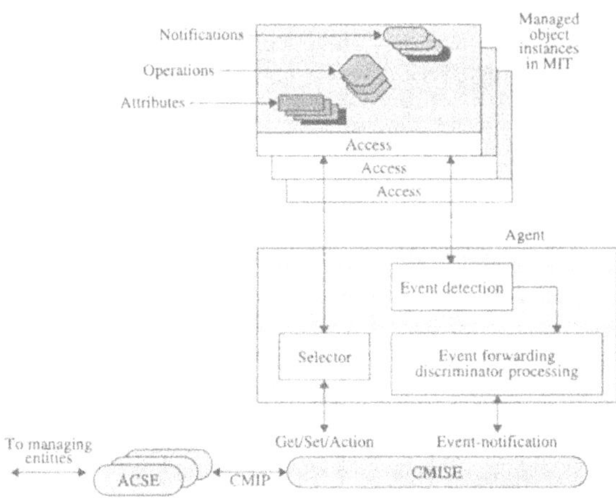

Figure 12. Agent Communication Architecture in OSI Management

Agents and managers are viewed as peer applications. They exchange information using the services of CMISE. CMISE uses the *Common Management Information Protocol* (CMIP) and utilizes the services of ACSE and ROSE. Figure 12 exemplifies the structure of an agent communication architecture.

When we relate the OSI Management to the SDF and SMF, the agent corresponds to the Maintenance Functions in the Functional Model of SDF Application Entity given in Figure 6. The SMF corresponds to the manager. This is for the SDF management. Another part of the SDF-SMF interactions is the data manipulation needs of the SMF. In billing the SMF stores charging records. Now the SDF can provide the database for those records. The MIT is not necessarily the best way of organizing the charging database. A traditional database based on the relational model can be better suited to such data handling. However, this increases the complexity of SDF.

6. SUMMARY

We have examined current ITU-T Recommendations in Q.1200, X.500, and X.700 Series which are relevant to database access in the IN architecture. The database must implement the Directory Information Tree and the Directory System Agent. In addition, the database must provide User Data Manipulation ASE with Query and UpdateData operations in order to be compatible with the CS-1 INAP. Moreover, the maintenance functions of the SDF must implement the OSI Management Agent (or NEF function block of TMN). This implies that the SDF must include CMISE. It must also maintain the dynamic Management Information Tree database. The future IN Capability Sets, perhaps CS-3, may require additional database capabilities supporting SMF data handling needs.

In addition to the requirements above, the implementation of the database system to be used in IN calls for further study in various research areas. The research topics include: real-time transaction processing, distributed main memory databases, object-oriented data models, relaxed serializability for correctness of transaction, performance and reliability evaluation methodologies for real-time databases. These questions are examined in the Darfin[6]-project. During the academic year 1994/5 the project is funded by Telecom Finland.

REFERENCES

Appeldorn, M.; Kung, R.; and Saraccor, R. (1993) "TMN + IN = TINA". IEEE Communications Magazine 31, 3 (Mar.) 78 -85.

Bumbulis, P. J.; Cowan, D. D.; Durance, C. M.; and Stepien, T. M. (1993) "An Introduction to the OSI Directory Services". Computer Networks and ISDN Systems 26, 2 (Oct.) 239 -249.

Burrows, M.; abadi, M.; and Needham, R. (1990) "A Logic of Authentication".ACM Transactions on Computer Systems 8, 1 (Feb.) 18 -36.

---

[6]Database ARchitecture For Intelligent Networks

Cattell, R. G. G.; ed. (1994)The Object Database Standard: ODMG-93.Morgan Kaufmann, San Mateo, Calif.

Chatras, B. and Gallant, F. (1994) "Protocols for Remote Data Management in Intelligent Networks CS1", In Workshop Record of IEEE Intelligent Network '94 Workshop,Volume 1. (Heidelberg, Germany; May 24 -26, 1994).

Fumy, W. and Leclerc, M. (1993) "Placement of Cryptographic Key Distribution Within OSI: Design Alternatives and Assesment".Computer Networks and ISDN Systems 26, 2 (Oct.) 217 -225.

OMG Object Model (Draft), May, 1992.OMG TC Document 92.5.1.

Raymond, K. A. (1993)Reference Model of Open Distributed Processing: A Tutotial.In Procedings of the Intenational Conference onOpen Distributed Processing (Berlin, Germany; Sep. 13 -16, 1993).

Soley, R.; Ed. (1993)Object Management Architecture Guide, Second Revision.Object Management Group, Framingham, Mass.

Sykas, E. D. and Lyberopoulos, G. L. (1991) "Overview of the CCITT X.500 Recommendations Series". Computer Communications 15, 9 (Nov.) 545 -556.

Yemini, Y. (1993) "The OSI Network Management Model".IEEE Communications Magazine 31, 5 (May) 20 -29.

14

Problem classes in intelligent network database design

Juha Taina
University of Helsinki, Department of Computer Science
P.O. Box 26 (Teollisuuskatu 23)
FIN-00014 University of Helsinki, Finland
Phone: int + 358 0 708 4247
Fax: int + 358 0 708 4441
Email: taina@cs.helsinki.fi

Abstract

The demands to an IN database make it a distributed real-time heterogeneous database, where each of the nodes can be a parallel database. We will discuss the problems that the IN structure and its demands create, and how the current database research can fulfill them. Our interests are in general solution level, logical data model, query languages, speed, and transaction handling and recovery.

1. PREFACE

In time of faster service creation and high competition, it is necessary to have open and compatible network systems [AiPW91]. The telephone network solution to the dilemma is both Intelligent Network (IN), that creates the background for new services, and Telecommunication Management Network (TMN), that creates the general telephone network model. Together they form the needed environment [ApKS93]. Our interests are in IN, and especially in the database that supports the information exchange.

In IN, the database must be able to handle 95 % of the requests in less than 15 ms, and the total unreachable time in a year must not exceed 10 seconds. If these demands are not met, customers will notice delays in the telephone interface. In this article we will examine the demands, and what solutions we think the current database research can offer.

The rest of the paper is divided into seven chapters. Chapter 2 is a short introduction to IN. Chapter 3 deals the database theory in general level. Chapter 4 is about possible logical data models of an IN database. Chapter 5 is about query languages, both on maintenance and on user level. Chapter 6 is about speed demands and how they can be met. Chapter 7 is about transaction processing and recovery from failure conditions. Chapter 8 is summary.

2 INTRODUCTION TO IN

An Intelligent Network (IN) is able to separate the specification, creation, and control of telephony services from physical switching networks [HoHa92]. IN and its standardization is also the first step to create an open standardized telephone network where new services can be intro duced in more rapid speed and systems are compatible with each others. In the future the trend will be more to open systems, where calls don't have to come from the same network ([Hade90],[LoKM90]). This cannot be handled without clear standardization and well defined protocols.

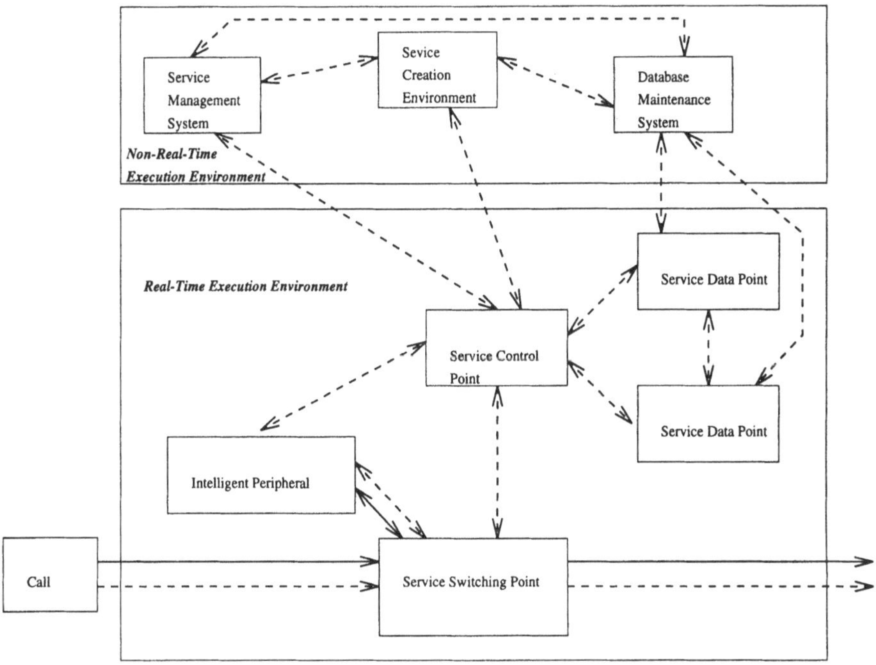

Figure 1. Intelligent Network overview. Dotted lines show signaling, solid lines show transport (original picture without Service Data Point [SDP]-objects is in [HoHa92]).

An Intelligent Network consist of the following parts (figure 1): Service Switching Points (SSP), Intelligent Peripherals (IP), Service Control Points (SCP), and Service Data Points (SDP). A SSP accepts a phone call. It requests the caller profile from the SDP database, and then waits for the caller to dial a number. If the number needs IN abilities it triggers a SCP. A SCP executes service creation blocks that are needed for the triggered services. It might or might not consult one or more SDPs for getting the information. The results are returned to the SSP. It is also possible that the SSP triggers an IP, that is responsible of tasks like customized and concatenated voice announcements, voice recognition, and Dual Tone Multi-Frequencies digit collection [MoMa94].

Our focus is in SDPs and how they interact with the other IN objects. A SDP is responsible of keeping and maintaining data that IN needs. Together, all SDPs form a complete distributed IN database.

The requirements that the distributed database must fulfill are not easy. The call profile can vary a lot, and yet it should look transparent to the customer. Let's suppose that Mr. Virtanen wants to make a phone call. He picks the telephone, dials the numbers and waits for the dial tone. He expects the procedure to happen without noticeable delays. If the dial tone is immediate he thinks that he made an error somewhere. If the dial tone is delayed for more than a second or two he gets frustrated to waiting. However, the call might trigger very different operations in the telephone network. In the simplest case the call doesn't need IN abilities, or doesn't need to make queries to a service data point (SDP). In a more complex case the call needs to make one or more SDP queries that might trigger more queries to other SDPs (figure 2). In the worst case it is possible that the information have to be queried from several database nodes. In all cases the response time should be within reasonable limits.

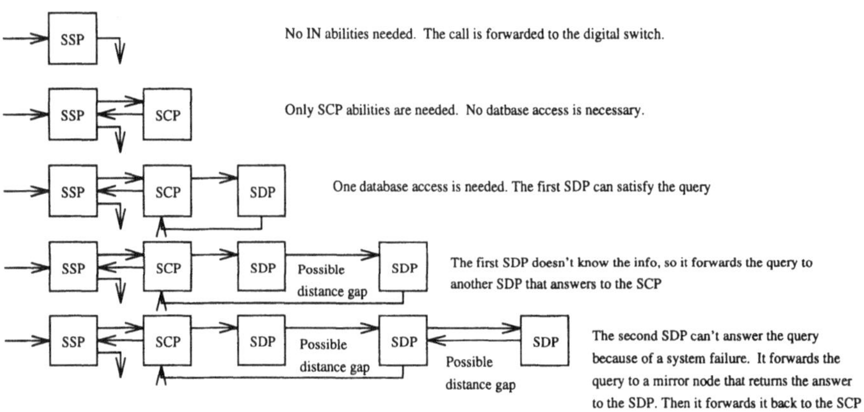

Figure 2. Different examples of call and signal flows in different calls.

3 HIGH-LEVEL DATABASE MODEL

A possible database solution can be distributed to several nodes on two levels: visible and node level. In the visible level a number of possibly heterogeneous databases work together to create the whole distributed IN database. In the node level each visible node consists of several distinct subnodes. The division between the two levels helps to keep the whole database reliable and to reduce recovery times. With only one level, either the information in the databases would be unreachable during a node recovery, or the information in every node would be replicated to the others.

A typical undistributed database is already quite a complex system (figure 3). It has several interfaces to different types of users that work on different level, several processes, and at least two data stores.

The interfaces vary from system maintenance command interface to a simple query interface. Typically not all users from all interfaces have access to all data. In an IN database the most important interfaces are the maintenance interfaces and application program interfaces. The latter is especially important, as it is the main interface for IN queries.

The database processes handle query processing and database access. The system is hierarchical the way that the lowest level handles direct access to physical data, while the higher level processes use the services that the lower level offers. The most important process is the run time database processor that handles processed queries and updates. Another important process is the DDL (Data Description Language) translator that translates system maintenance commands to the database language.

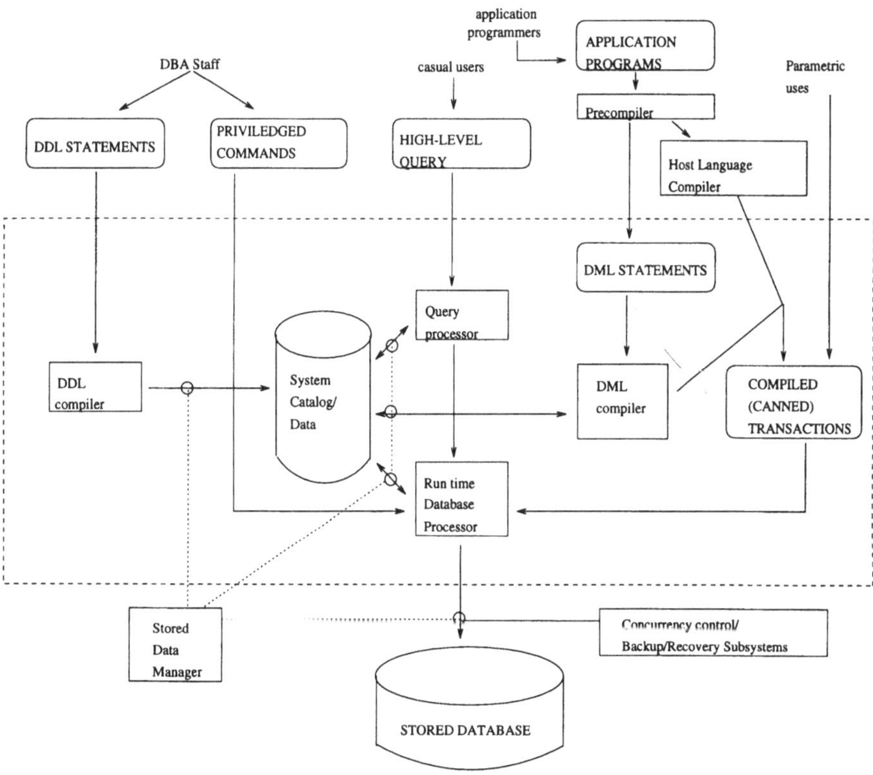

Figure 3. Components of a DBMS (DataBase Management System). Dotted lines show access under the control of the stored data manager [ElNa89].

The two necessary data stores are the physical data store and the system catalog. The former has the data itself, the latter has the interpretation of data. The separation between physical data and its interpretation is important as it allows different data to be stored in uniform way. Only the system catalog contents have to be changed. Both the system catalog and the stored database are controlled by the stored data manager, which can also control disks.

The stored data manager is not necessarily a database process. It can handle requests from other programs as well. Other low level subsystems that are outside the database management processes are the concurrency control and backup/recovery subsystems. The former is responsible of keeping simultaneous users unaware of each other. The latter is responsible of keeping stored data available after a system crash. Sometimes the recovery subsystem is also responsible of keeping as much of the database available during the recovery.

The picture becomes more complex when several database nodes interact together. Such systems are called distributed databases. The idea behind a distributed database is simple. Every node is a complete database itself, and they communicate with each other. Thus the figure 3 is valid to a single node, although there is a new component that takes care of the distribution algorithms. Alternatively it is possible to divide the tasks into different components, and only add a low level component that takes care of communication to other nodes.

In IN the database architecture is basically a distributed real-time database where each node is a parallel database. The database nodes don't have to be homogeneous. Thus the total picture is a real-time distributed heterogeneous database where nodes can be parallel databases.

There are at least three different levels we can design and analyze the system. First there is the total database level, where the group of SDPs creates the total database. Then there is the node level, where each of the SDPs creates a database. In such a level the databases are usually parallel databases with enough replication. And finally, there is the parallel database node level, where each of the parallel nodes is a regular database, like in figure 3.

In a parallel database data and processing are divided into several processors that interact with each other. The parallel nature of a database node creates new processes or alters the current ones. There are three different kind of parallel databases, depending on the level of resource sharing: shared nothing systems, shared disk systems and shared all systems (figure 4).

In a shared nothing system each of the processors have private memory and disk. The whole system consists of distinct computers that communicate over a fast network. The connected computers are called parallel database nodes.

Usually one of the nodes is a leader that shares the tasks to the other nodes. Not all the nodes need to be connected to each other, but each of them can be reached from the others.

The advantage of a shared nothing system is that the parallel computing makes complex queries easier to handle, as long as the problems are divisible to smaller subtasks [Seli90]. It is also a working solution if queries can be scheduled to different nodes without overloading any of them. A problem is that queries don't usually access data evenly, and thus some of the nodes are overworked. Also sometimes parallel power is lost due to the overhead that is needed to node communication and task sharing.

In a shared disk system the nodes have private memories, but disks are shared. The communication between different processes is simpler than in a shared nothing system because disks can be used to message sharing. However the shared disks themselves can become a bottleneck.

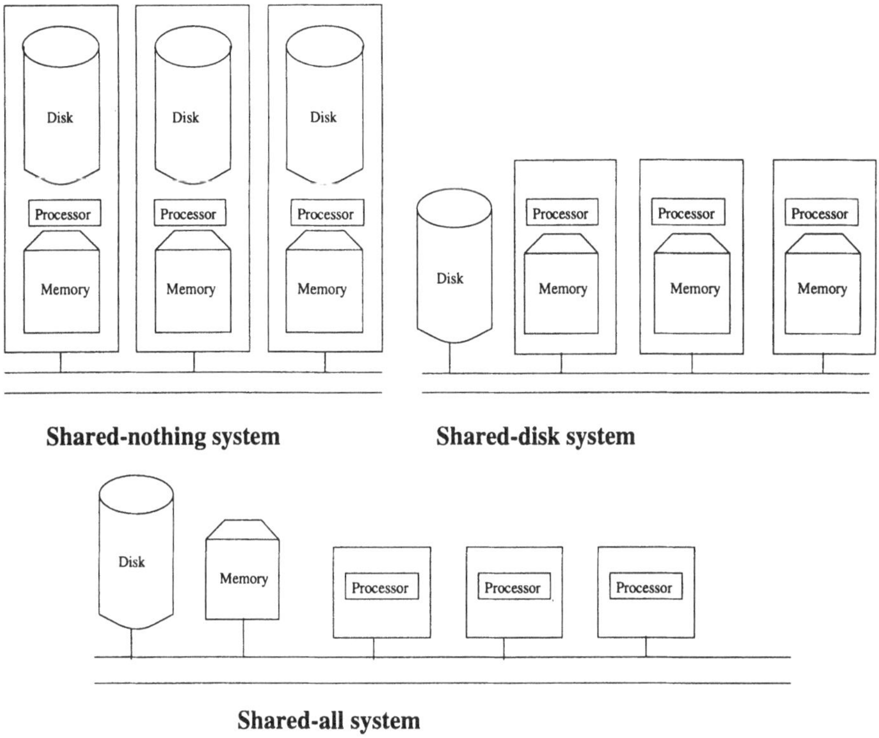

Figure 4. Different parallel architectures.

In a shared all system both the main memory and the disks are shared. The communication channels can be implemented to the main memory, but then both the disks and the memory can become a bottleneck.

An IN database node fills the requirements of a shared nothing database. The IN queries are relatively short and in principle it should be possible to cluster data between subnodes the way that queries can be distributed relatively evenly. The number of nodes in a parallel database can grow higher than in shared disk and shared nothing systems. Also the parallel nodes can be regular computers, which makes the implementation cheaper and better suited to an open environment.

4 LOGICAL DATA MODEL

The logical data model must be flexible enough to handle the demands of the network protocols and the information models. Information modeling will be object oriented [CCIT92], so a natural decision for the logical data model is also object-oriented. Another possible solution is a relational database that has an object-oriented interface. The former is more flexible. The latter is easier to implement, but it can become limiting in the long run.

Building an object-oriented front-end to a relational database is a relatively simple task. The immediate advantage of the decision is that the theory behind relational databases is well known, and there are already implementations of fast relational databases. The disadvantage is flatness that relational database automatically creates to the system. In a system where objects are simple, it is possible to do simple mappings between objects and relations. The more complex the system becomes the harder the mapping becomes. Eventually the mapping becomes so complex and hard to maintain, that the system slows down to an unacceptable level. Currently the IN database mostly consists of simple structured data, although the amount of it can make some processing complex. However, we don't know what kind of objects are needed in the future. While we might get immediate advantage in choosing a front-end to a relational database, it could mean that later we would have to design and implement a totally new object-oriented database management system.

An ability to handle complex objects doesn't mean that the system is an object-oriented database. An object-oriented database system must satisfy two criteria: it should be a database management system and it should be object-oriented system. The authors in [Atki92] list 13 golden rules that any object oriented system should fulfill:

1. Complex Objects

An object-oriented database must support objects that are built from the simple undivisible objects, such as strings and integers. Possible complex objects are tuples or structures, lists, and sets. A tuple is the simplest form, where a group of objects are joined under a common name. A tuple is also the basic building block of relational databases. A list and a set consist of a group of similar objects, that can be complex. A list is ordered, while a set is not.

2. Object Identity

Object identity means that every object is an unique entity. If two objects have similar attributes in relational database they are considered a single object. In an object-oriented database they are two different objects. Object identity comes to the picture in object sharing. In complex objects, one part of the object can be an unique object alone. If it is shared between two objects, then in object model we know that the object is the same. In value-based models (such as relational model) we don't, because the values can be the same.

3. Encapsulation

Encapsulation means that the specification and the implementation of an object are distinct. An object might show out different than it actually is. Some fields can be hidden, or a field can actually be a function that is counted every time it is referenced. When the object is stored to the database, both data and operations are stored.

4. Types and Classes

A type in an object-oriented system tells what the abstract model of the object is like. It has two parts, the interface and the implementation. Together they define how the object looks like and how it behaves when referenced. Thus the type is a static definition of an object and its meta-information. The object itself can vary in the limits of its type. A class is a similar term than type, but it has a more active role. A class contains two aspects: an object factory and an object warehouse. Class-based objects are created and maintained in runtime, and their role is free in most aspects. While the extra freedom gives more possible implementations, it also means that all compatibility checking must be done in real time.

5. *Class or Type Hierarchies*

A class or type hierarchy means that an object can inherit some of its subobjects (fields) from another object type. Object inheritance is a very powerful modeling tool because it gives a chance to describe the world in more detail. It also helps implementations where the object inheritance is needed. In relational databases the inheritance model must be wired to the application programs. Such a solution is more accident prone and harder to maintain.

6. *Overriding, Overloading and Late Binding*

Sometimes it is useful to have the same name to several different operations or fields. This becomes handy when it is combined with object hierarchies. For instance, let's suppose that there is a display function for all objects. According to the object type the function performs different operations. Some objects should be printed to a line printer, while others can be shown in interactive diagrams. In object-oriented system the high-level object type has the default display-function, while its subobjects have their own display functions. When referring to a subobject, its own display function is used. This is called overriding. It also means that the name display means several different functions that can be distinguished only from the object that owns the function. This is called overloading. Late binding means that the real operation used is decided at runtime, not at object creation time.

7. *Computational Completeness*

Computational completeness means that the database management language (DML) can use any computable function. This is obvious in programming languages, where each language is as expressive. It is not obvious on database languages. For instance, SQL in relational databases is not computationally complete.

8. *Extensibility*

Extensibility means that it is possible to define and create new types from old ones. There shouldn't be a distinction between old and new types, nor should a user notice when an old type is used instead of a new one.

The previous items described the necessary attributes that distinguish an object-oriented database from other logical model databases. The next ones distinguish an object-oriented database from an object-oriented programming language. As the IN database is a database, these items automatically belong to the definition.

9. *Persistence*

Persistence means that data will stay between different executions. It is obvious in databases, but not in programming languages.

10. Secondary Storage Management

One typical thing to databases is that the amount of data is high and it often grows during the database life span. The main memory available is not big and reliable enough to keep the data. Thus the database management system must be able to deal with secondary storage, such as disks. We will discuss more of this in chapter 6.

11.Concurrency

Concurrency means that the system should be able to let several simultaneous users to use the database without them interfering with each other. It is important in IN, where the response times must be kept low. Without concurrency one difficult query could block the whole system. We will discuss more of this in chapter 7.

12.Recovery

The database should not loose any of the stored information in case of a failure. In IN database, the recovery level must exceed this; it is not allowed to let the users notice the failure at all. We will discuss more of this in chapter 7.

13.Ad Hoc Query Facility

Ad Hoc Query Facility means that a user must be able to do simple interactive database queries. This is true to IN databases as well, although most queries will be done by a SCP and thus the language doesn't have to be that informative. We will discuss more of this in chapter 5.

5 QUERY LANGUAGES

The IN database can be accessed either from a SCP or from a database maintenance system. In the former case the language can be in a low level, as long as it supports the queries and updates that are needed in the IN services, and it is extensible. It is important to ensure that query processing doesn't slow down the system. The fastest solution would be to do query optimization before the query is sent to the database.

The demand to query optimization before database access sounds strong, but actually it is not. A typical query from SCP to SDP is simple, and thus the query doesn't need much optimization. The updates from SCP to SDP are rare and simple. So in general the database access language doesn't need much resources to handle optimizing the queries and updates.

The database maintenance system is responsible of statistical queries and general database updates. The queries are made by the maintenance staff, and so a high level query language is needed. However, speed is not a critical issue in the language, because the maintenance systems are not usually time-critical.

Usually the chosen language to database management systems is SQL or some of its variations. In ODMG - 93 the language is called Object Query Language (OQL) [Catt94]. The authors in [Catt94] suggest the following principles and assumptions that also fit an IN high-level query language:

- OQL provides an easy access to the database. A simple interface is more important than computational completeness item OQL provides declarative access to the database
- OQL is object-oriented

- OQL has an abstract syntax
- the formal semantics of OQL can easily be defined
- OQL has SQL-like syntax, but it is possible to use other syntaxes by merging OQL with a high level programming language such as C++
- OQL provides high level access to sets of objects, and primitives to deal with structures and lists
- OQL does not provide explicit update operators. It relies on operations that are defined on objects. thus every object class must have update operations, and
- OQL can be easily optimized.

We believe that the same principles work in an IN database query language. It is probable that the language is mostly used in embedded form, where it is part of a high level programming language. The interactive language interface itself is not among the most important subjects in the database design, because it is not used as often as in a regular database. Most queries are done through the SCP interface. The database maintenance staff needs powerful tools to access the database, and such tools can be built with the embedded query language.

6 SPEED

Speed is the most critical aspect of the database. Although a good hardware configuration will lower response times, it is not enough to solve the real time response times.

The fastest way to reduce response times is to reduce disk access. Every disk access needs a magnitude more time than a memory access. Thus if data can be kept in main memory, it will lower the response times noticeable. Such a database is called a main memory database.

The idea of a main memory database has been introduced in [DeWi84] and [AmHK85]. Their idea is to move all data to the main memory and thus cut the disk access off completely. However, this ideal case is not realistic, as the amount of data usually exceeds the size of the main memory. Priorities are needed in order to decide what data is kept in main memory and what needs disk operations. Also a copy of the data in main memory must be kept in a disk to prevent data loss in case of a system failure. Usually an update is done both to the main memory and to some nonvolatile memory, where it can be moved to a backup disk with a lower priority process.

The physical location of the data doesn't necessarily affect the database management system architecture. In an ideal case only the stored data manager and the backup/recovery subsystem have to be changed to handle different data locations. However, optimizing the data locations, whilekeeping it invisible to the system itself, will make the subsystems more complex.

A main memory database should not use virtual memory, because the virtual memory algorithms are not optimized to heavy database access. The result can be disastrous to the speed. This lowers the amount of data that can be put to the main memory, as also the database software must be kept there. If virtual memory is used, totally new solutions are also needed. Such solutions have been introduced in literature, for instance in [ChSi92] and [KLVA93], but they need special virtual memory hardware that doesn't exist yet.

In IN databases the best solution is to keep the most time critical data in the main memory. It is not possible to keep all data in main memory, because history data alone exceeds the limits. One solution is to use priorities for data and queries. The lower priority data can reside on a disk, as long as it is reachable in few disk accesses. It is possible to use a lot of small parallel nodes to gain more main memory, but it is expensive. Thus the disks can't totally be replaced.

7 TRANSACTIONS AND RECOVERY

A database communication is based on undivisible execution units called transactions. A transaction may have several database access operations that are either all accepted or none accepted. The database is consistent both before and after the transaction. Usually all resources that are needed to fulfill the read and write operations are locked before the transaction and released after it. If a resource can't be locked, the transaction is blocked until some other transaction releases the needed resource. This is called a two-phase commit protocol, and it was introduced in [EGLT76].

Transactions are needed for two reasons: They keep concurrent users from interfering with each other, and they keep the database consistent in case of a failure. A transaction manager, which is part of the concurrency control/recovery subsystem, is responsible of keeping the transactions in order. It also locks and releases resources.

The number of different operations in a single transaction, and the size of locked resources depend on the application in question. A rule of thumb is that transactions should be as short as possible, and yet keep logically undivisible units together. The size of a locked resource affects the concurrency level in a database. For instance, if every transaction locks the whole database then only one transaction can be active in a given time. If only an object is locked at a time then concurrent transactions can execute at the same time as long as they don't want to lock the same object. The drawback of this is extra overhead and extra data, as every object must have a lock attached to it.

There are at least three different types of transactions in an IN database:

1. Transactions that need only read main memory data, such as regular SCP-based transactions
2. Transactions that need to read disk data, or write main memory data, such as updating SCP-based transactions, SCP-based transactions that need extra data, and some statistical data gathering maintenance-based transactions.
3. Transactions that need to read several objects of disk data, or write to disk, such as transactions that backup main memory data to a disk, and most maintenance-based transactions.

The first class logical transactions are usually very short, and the extra overhead for handling object-level locks might be a major bottleneck in the system. Thus it is possible to let such transactions run alone in the database, as long as they are finished fast. This is true to almost all SCP-based transactions, especially when they use only data that resides in main memory. It's also easy to calculate the maximum time needed for such a transaction, which can help scheduling between parallel nodes. The second and third class transactions should use normal locking procedures, but in the second class the overhead should be minimized in cost of concurrency. Any SCP-based transaction should be finished as soon as possible. However, if all classes locked the whole database, the low-level transactions would block the database for unacceptable long times.

All write operations are saved into a transaction log during a transaction. Each of the operations consists of the old value and the new value. In case of a transaction failure, the transaction is canceled and all changes are returned back to the original values. A finished transaction also ensures that the changes it had made have been successfully stored to the database. In IN databases that keep the accessed data in main memory, a finished transaction can either mean that the data is written both to a disk and a main memory, or only to a main memory. In the former case, the data can be restored from the disks, unless a disk failure had happened, but it also means slower updates. In the latter case the updates are very fast, but the data can still be lost if a system failure occurs after a finished transaction.

Recovery problems are some of the most difficult to solve. The algorithms and solutions depend on the data model and the amount of main memory data used. The following failure situations can occur in a distributed IN database:

Transaction failures

where a transaction can't be finished due to illegal operations in the transaction itself. The transaction is canceled and the database is returned to the original state. If the transaction would update several SDP-nodes, then it is better to write all updates to a secondary storage at first, and do the real updates to the database nodes after the transaction is validated. That way it is not necessary to send cancel commands over the network.

Node failures

where a database node or part of it crashes. The simple solution is to block off the recovering node from the net and then reload all data. However, this might cause unwanted delay to the whole throughput. The other choice is to load data to the main memory when it is needed and index everything on the run. That way the whole recovery will take longer but it won't block the whole node. In any case, there has to be a backup system for avoiding data losses.

Secondary storage failures

where a disk or other secondary storage system crashes. The data in the disk is lost, but in a fault tolerant system the system itself can function. If the database node can be shut down without affecting the general database throughput too much, then it is possible to restore the data from other database nodes and backup systems. This means that all data must be replicated to several nodes. If the node can't be shut down, then the replication must be made in the node itself. This can be done either in a subnode level in a parallel database, or in a disk level. In the first case, the other subnodes take care of the database activity while the disk is restored. In the second case, the other disks offer the same data, or the data can be restored from the other disks. This is often done in RAID-disks ([PaGK88],[Chen93]). A RAID-disk consists of several small disks that cooperate with each other. The data is stored the way that when any of the disks fails the data in it can be constructed from the others. The bottleneck of a RAID-system is its controller, which can also fail. The system performance can also lower drastically if one of the disks fails.

Network failures
where a database node can't be reached due to a network failure. The recovery is left to the network maintenance system that is outside the database management system.

8 SUMMARY

In this paper we have examined and analyzed what the current database research can offer to IN databases. In principle it is already possible to build a database system that fulfills the demands. A distributed database where each of the nodes is a parallel main memory database, and where the database languages and recovery systems are optimized to fast queries is a possible system. However, there are still a lot of open questions in areas such as data distribution and replication between nodes, preferable recovery algorithms, and query languages.

REFERENCES

[AiPW91] Aiken J. A., Parker S. T. and Woodwell D. R., Achieving Interoperability with Distributed Relational Databases. IEEE Network Magazine 5,1 (1991), pp. 38 - 45

[AmHK85] Ammann A., Hanrahan M. and Krishnamurthy R., Design of a Memory Resident DBMS. pp. 54 - 57 in IEEE Spring COMPCON 85 Proceedings. IEEE, 1985

[ApKS93] Appeldorn M., Kung R. and Saracco R., TMN + IN = TINA. IEEE Communications Magazine 31,3 (1993), pp. 78 - 85

[Atki92] Atkinson M., Bancilhon F., DeWitt D., Dittrich K., Maier D. and Zdonik S., The Object-Oriented Database System Manifesto. pp. 3 - 20 in Building an Object-Oriented Database System - the Story of O2 (Bancilhon F., Delobel C. and Kannellakis P. eds). Morgan Kaufman Publishers, 1992

[Catt94] Cattel R. G. G., Object Database Standard: ODMG - 93. Morgan Kaufman Publishers, 1994

[CCIT92] Principles of Intelligent Network Architecture. Recommendation I.312 / Q.1201, CCITT, 1992

[Chen93] Chen P. M., Lee E. K., Gibson G. A., Katz R. H. and Patterson D. A., RAID: High-Performance, Reliable Secondary Storage. Technical report, University of California, Berkeley, 1993

[ChSi92] Chew K-M. and Silberschatz A., Toward Operating System Support for Recoverable-Persistent Main Memory Database Systems. Technical report, University of Texas at Austin, 1992

[DeWi84] DeWitt D. et al., Implementation Techniques for Main Memory Database Systems. pp. 1 - 8 in ACM SIGMOD Conference Proceedings. ACM Press, 1984

[EGLT76] Eswaran K. P., Gray J. N., Lorie R. A. and Traiger I. L., The Notions of Consistency and Predicate Locks in a Database System. Communications of the ACM 19,11 (1976), pp. 624 -633

[ElNa89] Elmasri R. and Navathe S. B., Fundamentals of Database Systems. The Benjamin/Cummings Publishing Company, Inc., 1989

[Hade90] Haderle D. J., Database Role in Information Systems: The Evolution of Database Technology and its Impact on Enterprise Information Systems. pp. 1 - 14 in Database Systems of the 90s (Blaser A. ed). Springer-Verlag, 1990

[HoHa92] Homa J. and Harris S., Intelligent Network Requirements for Personal communications Services. IEEE Communications Magazine 30,2 (1992), pp. 70 - 81

[KLVA93] Krueger K., Loftesness D., Vahdat A. and Anderson T., Tools for the Development of Application-Specific Virtual Memory Management. ACM SIGPLAN Notices 28,10 (1993), pp. 48 - 64

[LoKM90] Lockemann P. C., Kemper A. and Moerkotte G., Future Database Technology: Driving Forces and Directions. pp. 15 - 34 in Database Systems of the 90s (Blaser A. ed). Springer-Verlag, 1990

[MoMa94] Molin K. and Martikainen O., Intelligent Network Tutorial for the Second Winterschool on Telecommunications. Technical report, Lappeenranta University of Technology & Telecom Finland, 1994

[PaGK88] Patterson D. A., Gibson G. and Katz R. H., A case for Redundant Arrays of Inexpensive Disks (RAID). pp. 109 - 116 in ACM SIGMOD Conference Proceedings. ACM, 1988

[Seli90] Selinger P. G., The Impact of Hardware on Database Systems. pp. 316 - 334 in Database Systems of the 90s (Blaser A. ed). Springer-Verlag, 1990

15

Customer value creation in value added telecommunication services

[a]Janne Takala, [a]Olli Martikainen, [b]Jukka Ruusunen
[a]Telecom Finland, P.O. Box 145,
FIN-00511 Helsinki, Finland
Tel. +358 20401, Fax +358 2040 2320
[b]Systems Analysis Laboratory, Helsinki University of Technology,
FIN-02150 Espoo, Finland

Abstract

Advanced value-added telecommunication services provide high growth potential for telecommunication operators. Intelligent networks provide new opportunities for the development of differentiated value-added services. With increasing competition customer oriented service development becomes very important. In this paper we will present a framework for service development that includes both customer values and operations of the service provider. The framework of value network will be formed of customer preferences and the operations of the service provider. Our approach thus combines the theory of value creation in the company with consumer theory of how consumers perceive created values. We also give an example of the utilisation of the framework by analysing how customer value is created with the 9700-service of Telecom Finland.

1 INTRODUCTION

The demand for sophisticated telecommunication services and the rapid development of telecommunication techniques have created a basis for the development of Intelligent Networks (IN). IN support management and distribution of advanced telecommunication services. IN provides possibilities to integrate existing and new services, introduce new services rapidly, and differentiate services according to the needs of different customer segments. These new value-added services are created on top of existing telecommunications infrastructure and they utilise its services in value-added service applications. IN technology provides opportunities to offer value-added services to smaller and more profitable customer segments.

The demand for advanced services will grow in the future. Currently only 25 % of the revenue generated by telecommunication services is accounted for by advanced services. In the future the share will be higher, since the estimated growth rate of advanced services will be close to 20 %, as the estimated growth rate of basic services will be less than 5 %.[3]

The development of new value-added telecommunications services requires knowledge about markets, customer needs, and technological possibilities. With the knowledge it is possible to develop successfully differentiated services that satisfy the needs of different customer segments and also provide profit for the service provider. Telecommunications service development has so far been technology oriented. As competition increases, customers' needs become more important and service development has to be customer oriented. The objective is to create low-cost services that provide high value to customers. The problem is how to develop the wanted services for the right customer segments so that they are profitable.

This paper will concentrate on the customer value creation of the value-added services. Our objective is to discuss customer values and how they are related to service differentiation in the modern telecommunications environment. We will formulate a framework that describes how customer value of a modern service is created in the value chain of a firm. The approach can be conceived as a part of a service portfolio analysis. We will use the concept of value networks instead of a value chain to describe the firm's value creating activities. The framework developed here can be used in service development, pricing, customer segmentation, and operations analysis.

In chapter 2 we will discuss customer values and service differentiation. In chapter 3 we will present a modern product concept that is subsequently used in chapter 4 where we will formulate a model for customer value creation of a value-added telecommunication service. In chapter 5 we will present an example of value-added services and its customer value creation. In chapter 6 we discuss further research possibilities and state our conclusions.

2 CUSTOMER VALUES AND DIFFERENTIATION

In this chapter we will first discuss definitions and meanings of customer value, since the concept does not have a unique definition. Secondly, we will discuss factors that influence how a customer perceives a service and what makes it valuable. Thirdly, we will address the issue of differentiation and its relation to the customer value.

2.1 Customer values

The concept of customer value is diverse. Its definition varies depending on the approach to the subject. Researchers, manufacturing managers, marketing executives and sales representatives understand it in different ways. Most definitions resemble each other in several ways but emphasize different aspects of the concept. In most definitions customer value is created in the customer's value chain by satisfying customer needs and wants, for example, in forms of increased efficiency, cost savings, increased revenue, pleasure, or image.

The concept of customer value is used in microeconomics, consumer behavior, marketing, and strategic management. Porter defines customer value as the amount buyers are willing to pay for what a firm provides them [10]. Moreover, he defines a concept of actual value or cumulative value that is the actual impact of the product on customer value. In microeconomic theory the concepts of total value and marginal value are used. A necessary condition for all the definitions of customer value is that the customer should perceive it. From the point of view of our analysis, we find two slightly different definitions useful.

The first definition of **revealed customer value** is defined as a sum of money the customer reveals (by buying or telling) to be willing to pay for the service. If the value is revealed by buying, the paid price is the revealed customer value, and it is assumed to be close to what customer is ready to pay. If the value is not measured by actions but by intentions or willingeness to pay asked in questionnaires, it is assumed that the customer does not distort his or her true preferences. The service may offer more value to customers than what they reveal, but due to competitive situation, uncertainties, or a desire to get a lower price, they can express a lower value. The definition is easy to conceptualize to customers. It is also meaningful to the firm, since the revealed customer value is what the firm would get from the service. The total value created in the firm is the total revenue. It is also important when trying to estimate the value of a new service that is not available yet.

The second definition of **actual customer value** is based on the actual value the service creates to the customer in the course of time of its utilisation. This definition is close to the total value concept of Glahe and Lee [4]. Porter defines this concept as actual value or the cumulative customer value of differentiation [10]. It is the upper limit for the price the service provider can charge from the customer, and in this way the upper limit for the revealed customer value. The actual customer value is a valuable asset to the firm if it is able to estimate it and communicate it to the customer, since it can provide an competitive advantage over competitors.

The estimation of the actual customer value is based on the understanding of the customer's value chain and its links to the service. The customer research methods and questionnaires may be used to estimate the actual customer value, but not all customer's are able to conceive the definition or estimate it. Often customers are not able to perceive all of the actual customer value even after they have utilised the service for a while. This often happens with services that are intangible and have high credence qualities [7]. This complicates the estimation process. The second definition provides a way to estimate to what extent customer's needs are satisfied or how much quality is offered.

It is important to understand the distinction between the two definitions. In service development it is important to maximize the actual customer value for two reasons. Firstly, a higher actual customer value facilitates charging of higher prices. Secondly, if the actual customer value increases more than the price, the probability of purchase increases since the customer will "get more value for the same money". The revealed customer value is important also for two reasons. Firstly, it is needed in price determination. Secondly, it is needed in profitability evaluations.

2.2 Customer perceptions of services

This service perceived by the service provider is a combination of features created in the service design, production, marketing and sales processes. The presence of all these features is intentional. The point of view is functional and operational and differs from customers' perceptions of the same service.

Customers perceive the service in relation to their needs the service satisfies. The customer conceives the service as a bundle of various service attributes [7]. The attributes are closely related to the customer's needs the service satisfies. Some of these attributes are images of features, some are combinations of several features, and some are a creation of the customer's imagination and have no connection with the features.

Recent consumer models also classify the attributes according to their level of concreteness to the customer [9, 11]. From the consumer's point of view a service value chain consists of service attributes, psychological and social consequences of using the service, and basic values to which the service consumption relates. However, in the following model we include only service attributes.

Examples of telecommunications service attributes are accesibility, communication, reliability, detailed billing, value-added services, mobility, price, status, convinience, capacity, consistency, quality, and information security. The market can be segmented also according to the attributes that are demanded by different customer groups. Customer research methods utilise attribute evaluations in questionnaires [6]. The obtained data can be used in the service development process.

2.3 Service differentiation

The value-added services created with IN are essentially differentiated services, ie. they have some features that are not prevalent in the basic services. Differentiation is one of the three types of generic strategy to gain competitive advantage, and one of the two basic types of competitive advantage [10]. In service industries differentiation is an alternative solution to price competition.

The idea of differentiation is to provide something unique and different that has value to the customer, ie. to satisfy customers' needs in a new way better than competitors. Differentiation allows the firm to charge a premium price from its customers, to sell more of its service, or to gain intangible benefits like increased buyer loyalty [10]. The ideal situation would be when differentiation helps the firm to increase both the margin and the volume of sales while satisfying customers. A service can be differentiated by adding a unique feature to the set of existing service features. Another way to differentiate is to create the existing service features in a new and unique way.

Service differentiation is therefore succesful if the firm can offer something that satisfies the needs and wants of the customer that have not been satisfied yet. The objective is to provide quality solutions and in this way maximize the actual customer value. If the firm is able to communicate the increase in actual customer value, it can increase the sales volume, charge a premium price and keep customers satisfied.

Service differentiation is not an easy strategy for the firm. Customer needs may have been interpreted in a wrong way and the service fails to generate any sales. The service development may be too costly and the service is never profitable. The initial price may be too high and the predicted demand never actualizes. Even if the above mentioned pitfalls are avoided, competitors may soon imitate the service thus eroding the competitive advantage. In the long run the differentiated features may lose their importance to customers and become more generic.

The keyword to the issue is customer orientation: focusing on customer needs and wants, differentiating the service according to them and providing quality solutions. Customer orientation will be an important source for competitive advantage in telecommunications business in the future. Service providers will use this approach especially with business clients who are a major customer group for value-added telecommunications services.

With IN technology it can be possible to create differentiated value-added services cost-effectively, but several questions remain. How to create cost-effectively services that satisfy customer needs and create actual customer value? What is the correct price level considering the revealed customer value and competitors' pricing? The problem is how to identify different consumer needs, how to express the identified customer needs in monetary terms as customer value, and how the value is cost-effectively created in the firm's value chain.

Our objective is to create a framework that would help to identify how customer value of a differentiated value-added service is created in different parts of the value chain, i.e. to explain how resources and technological aspects are transformed to tangible and intangible attributes that are conceivable to customers in monetary terms. The subject is approached from the point of view that is applicable to a telecommunications environment.

In order to do that, we must define such concepts that can be used in value chain analysis and in customer research. Therefore, we must define the value chain of a company so that it is more applicable to the modern telecommunications environment. We also have to determine the connection between the attributes that are familiar to customers and the features created in the value chain. We will use a modern product concept that helps to approach the problem. With the product concept it is also practical to define production platforms that are used to produce and create the different elements of a modern service. The concepts are defined in the next chapter. Using these concepts we will formulate a model of customer value creation.

3 A MODERN PRODUCT[1] CONCEPT AND PLATFORMS

Lahti and Martikainen [8] define modern products as combinations of product elements provided by the firm. These product elements are defined as the presentations of the basic and

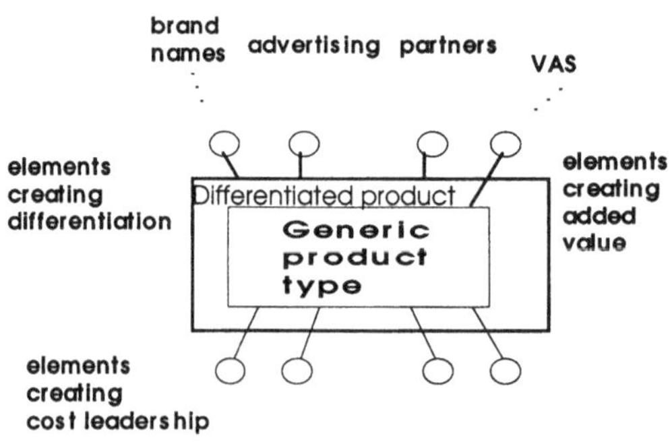

Figure 1. Elements of a product.

support activities of the firm in the products. Different products are different combinations of

[1] The product concepts and definitions apply also to services.

these elements. Some of these elements create cost leadership, value adding and differentiation characteristics of the product. One element can therefore create more than one of these characteristics. The elements, which are the same as product features, provide competences for the product. Hereafter when we discuss features we refer to product elements. In the definition of the modern product concept we emphasize the difference between the generic and the differentiated product. This is presented in figure 1.

Examples of telecommunications product features are access, transmission, switching, billing, database, maintenance & network supervision, administration, mobility, intelligent switching, and value-added service features. The features can be created by different companies. Companies can be telecommunication operators, telecommunication service providers, 3rd party service providers, or even customers.

The way these features are combined in the value chain is defined in product development. Basic questions are what features should be included in the products and in what extent, how these features are produced in the value chain, and how customer value is distributed among the features. If the cumulative cost of producing these features is higher than their cumulative value in the long run, it is difficult to justify the inclusion of the features in the product.

Lahti and Martikainen define the combination of synergic product features as platforms. Platforms can be regarded as a resource base that is dedicated to produce these synergic product features. The product features are created in the platforms. Platforms are determined based on the feature synergies. The synergies between the features are based on economies of scale and scope. Logistics, resource, and marketing synergies can also exist between two or more platforms. We can identify for example production platforms, channel platforms, and advertising platforms. The definition of platforms generally as a synergic concept leaves flexibility to the framework presentation.

Examples of telecommunication platforms are R&D, access network, local trunk network, trunk (national) network, IN, advertising, sales, customer service, accounting and invoicing, and administration. Telecom Finland uses the platform concept in its business analysis.

In order to present the framework focusing on the features, attributes, and synergies existing in the process we need a description of the firm's value activities that is different from the value chain concept as defined by Porter [10]. Instead we will use a value network concept that is defined in the next chapter. We use the concept to model a customer value network consisting of attributes and a service provider value network consisting of features and platforms. In these value networks it is possible to analyse the nature and extent of synergies inside the firm, between the attributes, and in the total framework.

4 CUSTOMER VALUE CREATION

In this chapter we discuss first the customer value network which consists of service attributes important to the customer and relations between them. Secondly, we analyse in the same approach the value network of platforms and features, and the cost accumulation in the value network. Thirdly, we present a connection between the networks to form the customer value framework. Fourthly, we discuss possibilities to proceed to analyse the customer values using the framework as a conceptual basis.

4.1 Attribute network

As described in section 2.2 the customer perceives the service as a set of attributes. These attributes include usually several product features. The customer is able to estimate the strength of preference based on the number and quality of potential features in the attribute. The attributes are also affected by other factors that are not features, like subjective beliefs about product performance. The overall value of an attribute is related to the needs it satisfies.

The attributes present in the service create the customer value. Each attribute has some value to the customer depending on the level of satisfaction it provides. A part of this value is independent of the other attributes in the service, and a part of the value depends on the value of the other attributes. The value that is dependent on the value of other attributes is synergic value. The attributes and their values form **a customer value network** that describes the overall customer value of the service and the synergic values of the attributes. The customer value network is described in figure 2.

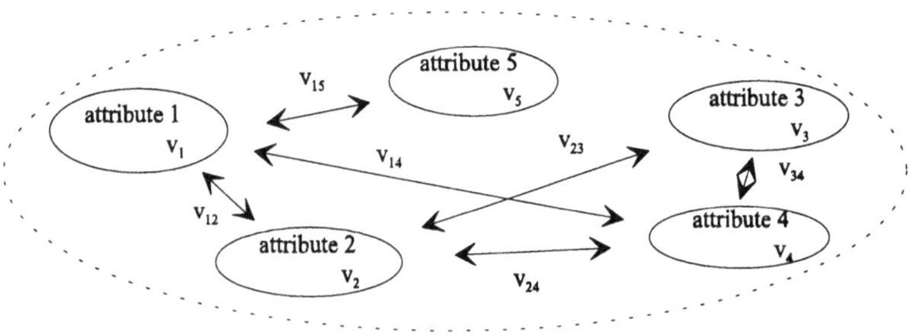

Figure 2. The customer value network.

The customer value network can be thought of as a part of a larger system, an attribute space $\mathcal{A}$. The attribute space $\mathcal{A}$ is set of all the possible attributes conceivable to customers. The services are synergic combinations of the attributes in $\mathcal{A}$. These combinations and their synergies can be described as value networks.

The customer value of the service that is formed in the attribute network can be described mathematically as a value function (see [12]). The value function includes as variables the attributes present in the service, all the features included in attributes and some other factors relevant to the decision making. The general form of the value function of the *i*th service in the service portfolio (*N* services) is

$$V_i = f\left(X_{ij}\left(y_{jk}, e_{jl}\right)\right), \quad i = 1 \ldots N,\ j = 1 \ldots M,\ k = 1 \ldots P,\ l = 1 \ldots R.$$

X_{ij} is the *j*th attribute in the *i*th service, y_{jk} is the *k*th feature in the *j*th attribute, and e_{il} is the *l*th non-feature element in the *i*th attribute.

Depending on the nature of interdependence between the attributes it is possible to derive a mathematical form for the value function. Examples are additive and multiplicative multi-attribute value functions. A necessary condition for multiplicative value function is mutual preferential independence between all the attributes. For an additive value function a stronger necessary condition of additivity independence is required [2,12]. An example of a multiplicative attribute model is presented in [6]. Additive functions do not include interdependencies between attributes, but since they are less complex than multiplicative models, they have been used much in empirical consumer research [11].

Previously customer research methods have been developed to predict customer preferences based on their beliefs of attributes and, in some models, also weights given to them [11]. Questionnaires provide possibilities to obtain information of customer preferences between products, services, attributes, and features. New and differentiated services, attributes, and features are compared with existing and more generic counterparts or with an "ideal service". The methods provide results that are measured both with qualitative measures and also on some quantitative scale like in monetary terms. However, the network approach in which also the interdependencies of different service attributes are studied, has not been very common so far.

Telecom Finland has studied customer preferences and evaluations of several of its services. Research has been performed also on the services utilising IN. The customer research of 9700-service will be used in the example in chapter 5. In the future the increase of competition in the telecommunication markets will increase the importance of customer research in general.

Based on empirical customer research such methods as conjoint analysis and multidimensional analysis have been used to get quantitative results about customer values [5,12]. Another similar method that provides qualitative results about customer requirements is Quality Function Deployment (QFD) process (Akao [1]). These methods are useful in customer segmentation, service development and pricing decisions.

4.2 Feature network

The features and platforms form **a service provider's value network** where the customer value is created in the service creation processes. The feature network describes actually the cost accumulation of the service creation process. There are costs incurred by the creation of each of these features and costs of combining the features to the service.

Platforms, which are synergic concepts and combinations of synergic service features, describe the basic and the support functions of the firm. They form a value network that describes the nature and extent of synergies in the firm. The value network composed of features describes the value creation of one particular service. The platform network creates all the features that are included in the services of a certain service portfolio. Since the features are elements of platforms, these networks are actually one value network that describes the cost accumulation and synergies in the firm. These synergies are results from economies of scale and scope in the platforms, and logistics, marketing, and resource synergies between the platforms.

We can describe the value network of features and platforms as a feature space $\mathcal{F}$. The feature space $\mathcal{F}$ is a set of all the features that the firm is able to create. The services are a union of the elements (features) present in the service. The platforms are unions of synergic features, and therefore subspaces of the feature space. The value network can be described as relations between the platforms when we are studying the service portfolio and as relations between some features when we are studying a single service.

Figure 3 illustrates the feature space $\mathcal{F}$. The solid black circles and ellipsoids describe the features. The platforms are described with ellipsoids drawn with dotted lines. The feature space is bounded by the large ellipsoid. For example, the grey circles are the transmission and switching features created in trunk, local, and access network platforms.

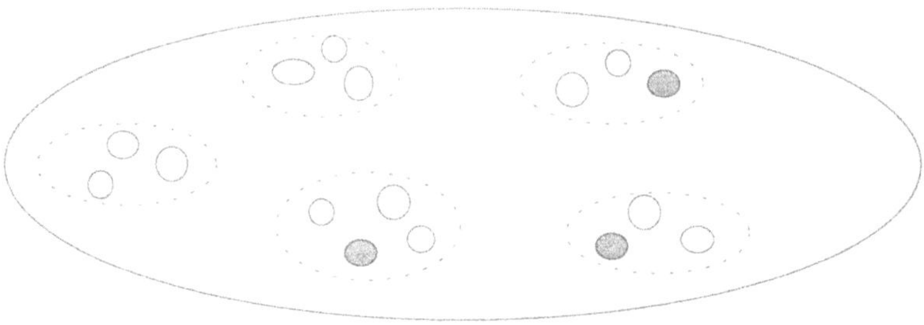

Figure 3. The feature space $\mathcal{F}$ and platforms.

In telecommunications industry the cost structure of platforms is different from, for example, the process industry. The platform can be described as a common resource for these services. In many platforms the majority of costs are not variable but fixed, and in most cases they are not direct but indirect costs. The features of one platform are included in several services, so the costs of the platform should be allocated to the services according to the amounts these features are used by each service. In consequence, changes in the volumes of one service affect the cost structure of all the services. The cost allocation is problematic, but using principles of activity based costing it is possible to allocate the costs in a way that describes the actual cost accumulation.

Telecom Finland has developed a cash flow model for the analysis of the value network. The model includes the services and the platforms of Telecom Finland, costs, investments, and revenues on a yearly basis for five years. The cash flow model can be used for many purposes: in intertemporal strategic analysis, service portfolio analysis, and scenario calculations.

4.3 The connection between the networks

The connection between the feature space and the service attribute space is a mapping $\mathcal{M}$ that interconnects the customer value of the service to the features included in the service, and maps the cost accumulation of features to the attributes. The mapping is described in figure 4.

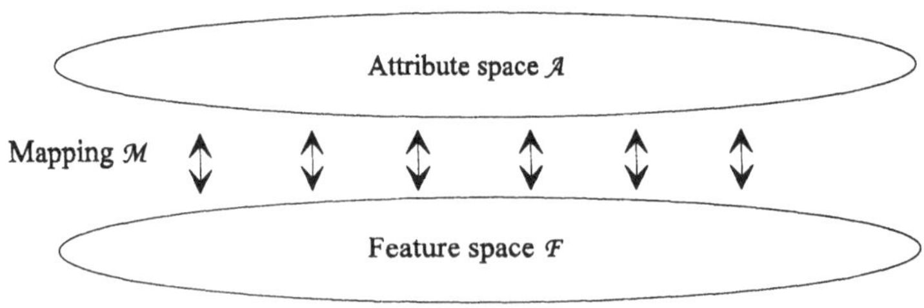

Figure 4. Mapping $\mathcal{M}$ between the feature space and the attribute space.

The mapping $\mathcal{M}$ should be able to describe the synergies between the attribute space and the feature space as well as the synergies between each attribute and feature. Such a mapping would be highly nonlinear and currently we cannot present any explicit way to present it. In general we have a triple that consists of the attribute space $\mathcal{A}$, the feature space $\mathcal{F}$, and the mapping between the spaces. The mapping $\mathcal{M}$ interconnects the customer values of all the services to the features and also costs of all the features to all the attributes. In this way it is possible to analyse what features and attributes are profitable to provide.

The mapping $\mathcal{M}$ defines how the customer value network and the service provider value network are connected. It describes how the information from one network is transferred to the other network. An example is the connection between a customer value function and the cash flow model of Telecom Finland that is illustrated in figure 5.

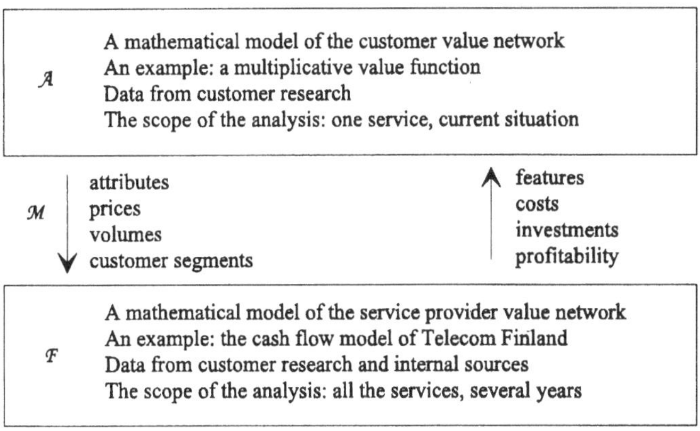

Figure 5. The connection of value models.

With some restrictive assumptions it is possible to derive a mathematical form for the triple. Such assumptions include that, for example, not too many features and attributes are present in the service, and mutual preferential independence exists between the attributes. In chapter 5 we present an example of value formation in Telemedia services.

4.4 Using the framework

Since the exact form of the general model is difficult to derive some other methods can be utilised to get service development information. It is possible to utilise customer research methods to get results for a service that is differentiated by adding only one or two new attributes. In such a case it is possible to compare the old, more generic service with the new, differentiated service and deduct the value of the additional differentiated attributes. Such questionnaires are quite possible to formulate using the customer research experience (see section 4.2). Using these methods it is possible to identify segments on the market, estimate the potential sales volumes in each segment, and to predict the suitable price level in each customer segment.

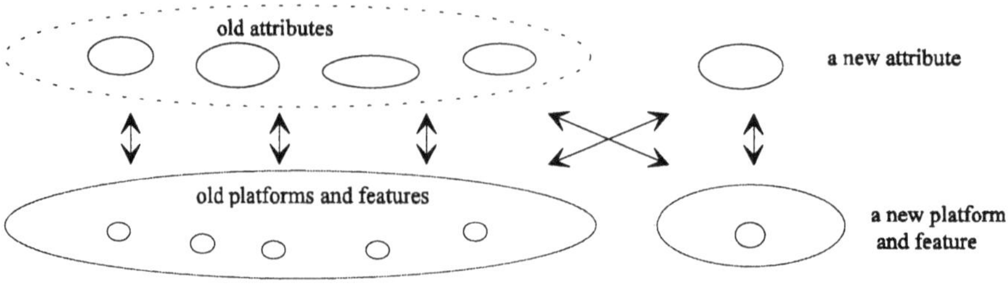

Figure 6. Customer value and cost evalution description of a new service.

With the same logic it is possible to calculate the cost accumulation in the firm with the old service portfolio and sales volumes. Using the customer research data it is possible to estimate product development costs, marketing costs of the new services, and the overall cost structure of the firm with the new services and the new sales volumes. After this it is possible to calculate scenarios for different market situations and estimate which service versions are profitable in which market segments.

Similar methods have been used in service and service development processes for a long time. What is new in our approach that it includes customer values (and higher degree of customer orientation), a systematic approach in which the value hierarchy from customer value attributes is matched with the value network of the company. It is an approach which also models the synergies between the elements in the hierarchy.

There are, however, many questions to be answered. The first is: Is it possible to get concrete results that are applicable in business operations? This has already been proved to some extent by some examples with a not-so-comprehensive approach to the subject [6]. Using customer research methods it is possible to estimate the strengths of preferences for the different attributes and whether the necessary conditions of additive or multiplicative functions hold. With the help of thorough insight to the firm's operations and activity based costing it is possible to model the feature and platform networks. Using expert estimates and research data it will be possible to determine the mapping, first with strong assumptions, and in the course of time with increasing insight, with weaker assumptions. The framework is therefore identifiable.

The second question is: Is it worth the trouble? Somewhat similar results can be obtained with ad hoc methods. The difference is that the theoretical basis increases the applicability of the results. Moreover, the process of studying value hierarchy can result in invaluable insights into the business. The analytical results can provide new information not only for service development and market segmentation, but also for inceasing the efficiency of existing platforms. A significant benefit is also that the clear concepts of the systematic approach facilitate communication in the organization. Misunderstandings can be avoided and common goals become evident to everybody.

5 EXAMPLE: TELEMEDIA-SERVICES

In this chapter we present an example of structuring the value hierarchy. The objective is to describe the logic of the framework and not so much to derive analytical results. The object of the example is a service in Telemedia service portfolio of the case firm Telecom Finland.

Telemedia services are a group of modern telecommunication services based on sophisticated technology to provide customer oriented applications. The major service groups in Telemedia include EDI and telematic services, intelligent telephony service, and video and telephone conferencing. The Telemedia services include 9700-Telemarket, 9800-Free Phone, electric mail, Telecard-services and voice center services. The core idea of the Telemedia service group is to provide services related to intelligent information transmission and delivery.

Instead of using all these services we focus on 9700 Telemarket services. There are several reasons for this. Firstly, the 9700-service generates a major part of the revenue of Telemedia. Secondly, the Telemedia services have different features so it would be difficult to analyse all of them together. Also the basic idea of our approach is to analyse customer values of a single service from a larger service portfolio. Thirdly, customer segments differ from one service to another, which means that service attributes vary much. This would also complicate our analysis.

The 9700-Telemarket service is similar to 900-numbers in the USA. Telecom Finland, the telecommunications operator, sells these services to companies. These companies, 3rd party service providers, create the information or the actual service and market the service to end users. The end users vary much from consumers to business customers. Telecom Finland provides the intelligent infrastructure on which the 3rd party service providers can create services that customers demand. Telecom Finland charges the end users in phone bills, deducts its own revenue and forwards the money to the 3rd party service providers. Telecom Finland also provides reports and consulting services to the service providers.

5.1 3rd party service providers as customers

We first analyse the service excluding the end users and focusing on the 3rd party service providers as customers. The inclusion of the end users in the analysis would be most

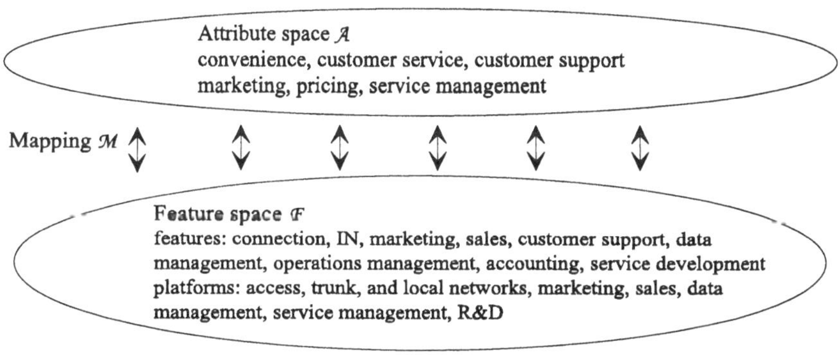

Figure 7. A model of 9700-service with 3rd party service providers as customers.

interesting but currently we do not have empirical data for that purpose. We will discuss this aspect in chapter 5.2. As for the customer value attributes, they are defined from the point of view of the 3rd party service providers. These companies have also graded the performance of Telecom Finland on each attribute and estimated the weights of the attributes. The features of the 9700-service have been defined by Telecom Finland. Activity based costing methods have been used to calculate the cost structure of platforms and features. The value hierarchy is presented in figure 7.

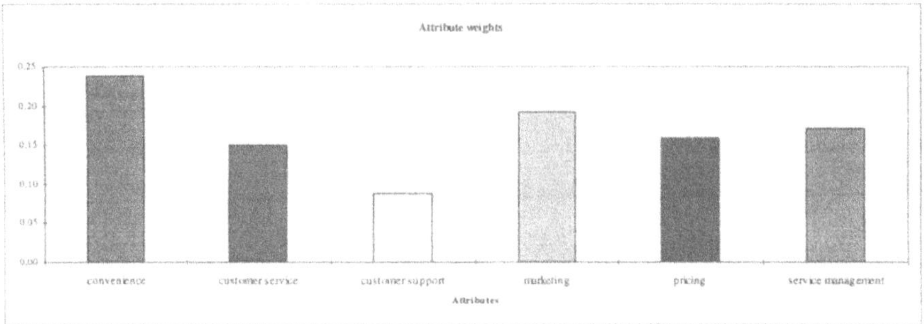

Figure 8. Attribute weights.

In customer value calculations we use the definition of revealed customer value, since we use the revenue as a measure of customer value. We make an assumption that the mutual preferential independence exists between all the attributes. Moreover, we also make a strong assumption of additivity independence between the elements. The available research data does not support these very strong assumptions, but since our objective is to provide an example of the logic of value creation, we keep the example simple.

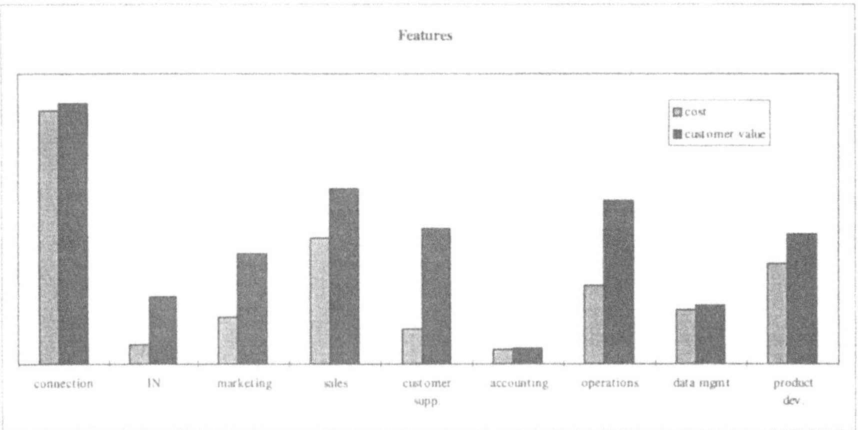

Figure 9. Feature cost and customer value structure.

The attribute weights are given in figure 8. They have been calculated using customer research data and based on the above mentioned assumptions. In the calculations both the relative importance of the attribute and the customer satisfaction in the attribute have been considered.

Using the attribute weights and estimated relations between the attributes and the features we have calculated the value structures of the features and the platforms. We assume that each feature is necessary in the service from the operational point of view, and due to this the customer values of the features exceed their costs. The cost and customer value structures of features and platforms are given in figures 9 and 10.

The service feature IN is in this case the intelligent switching and it is "created" in the service management platform. The customer support feature is the after-sales service that handles application support. The operations feature manages the actual service creation excluding the IN. The customer service platform in figure 10 is responsible for customer support and accounting. The service management is responsible for operations and IN.

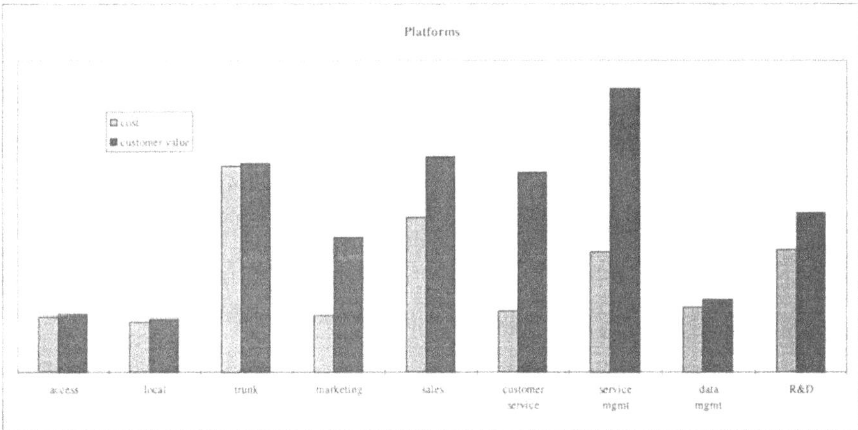

Figure 10. Platform cost and customer value structure.

The results provide information about customer attitudes and possible efficiency improvement targets. Although the results are just descriptive[2], the small difference between customer value and costs in the network platforms and the higher importance of more customer oriented features is not unexpectable.

5.2 3rd party service providers as platforms

A more comprehensive approach to the service concept includes the 3rd party service providers as platforms in the total value network, which includes infrastucture and end users. The approach is important because it includes the real source of customer value, namely the end users. Their needs and attributes should be considered in the service development. The customer value in the new approach is the amount paid by the end user.

---

[2] Original cost and revenue figures have been altered because of business secrecy.

In our framework a more comprehensive view including end users brings more information about the value network and the distribution of value among features, platforms, and service providers. The distribution of value between telecommunications operators and 3rd party service providers will be an important issue in the future, when the number of 3rd party service providers increases.

The more comprehensive approach would change the system of our 9700-service example. The 3rd party service provider's operations are included as additional features and platforms as described in figure 11.

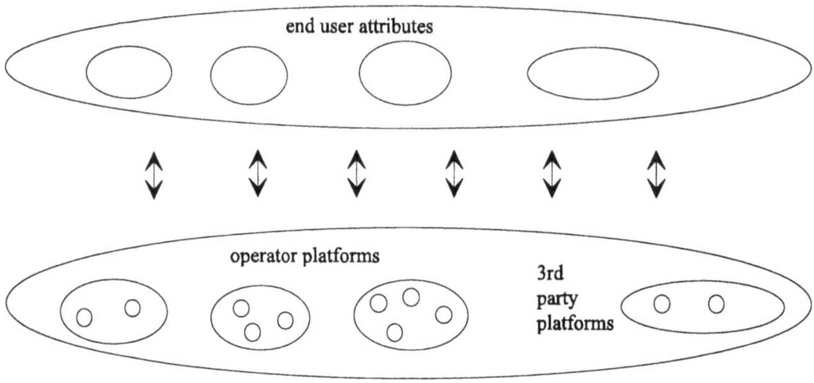

Figure 11. A model of 9700-service with 3rd party service providers as platforms.

A comparison of the results of the two approaches would provide interesting results about the pricing of the services. Necessarily the customer value would not be divided "correctly" between the 3rd party service provider and the telecommunications operator. Possible reasons for this are the competitive situation determining the price levels or the telecommunications industry structure in general. The results would also be interesting considering cooperation opportunities between the companies. Information could be obtained about efficiency improvement, cooperation, and task division possibilities. Currently there is no research data available for this kind of analysis.

6 DISCUSSION AND CONCLUSIONS

The growth of the demand for advanced telecommunications services provides opportunities for IN-based differentiated value-added services. As the competition increases in the telecommunications industry the customer oriented service development provides possibilities for competitive advantage. Therefore, a systematic approach to analyse customer needs and values in the service development process is important, because it clarifies concepts and changes the emphasis in service differentiation from technology to customers. In this paper we have formulated the framework that serves these objectives.

The framework connects customer needs and values with the value activities of the service provider. Definitions and meanings of customer value are discussed in the context of service differentiation. We analysed customer preferences and the service provider's operations using the concept of value network. What is new in our approach is that the framework developed in this paper connects the customer value network and the service provider value network. The framework also provides new possibilities to analyse interdependencies between the service attributes and features.

In the paper we described the methods to perform customer value analyses. Telecom Finland has undertaken several customer research projects which have provided valuable information about customer preferences and the customer value network. The cash flow model developed in Telecom Finland is used in analysing the value network of the company. The cash flow model is useful in intertemporal strategic analysis, service portfolio analysis, and scenario calculations. The company value network is easier to analyse than the customer value network since more information is available, the subject is less ambiguous, and methods have been developed for a longer time.

We presented an example of the customer value creation in 9700-Telemarket service. The service utilises the IN-platform of Telecom Finland in its service creation. The example describes the logic of customer value hierarchy. Customer value structures of the service features and platforms were presented.

The framework can be used with several related research subjects. The efficiencies and synergies in platforms can be analysed with the model. The scope of the analysis can be expanded to include a whole company or several companies that form a value network. The results could help to explain the formation of network relationships and facilitate cooperation between companies. An example of such analysis is the inclusion of telecommunications equipment manufacturers in the value network.

Telecom Finland will use the framework in its service development projects in the future. More empirical data about customer preferences and attributes will be gathered and the effect of service features on the value of attributes will be analysed thoroughly. The methods and results will be utilised in service development pilot projects. The object of pilot projects is to develop succesful differentiated service concepts based on the technological know-how of Telecom Finland.

ACKNOWLEDGEMENTS

The authors express their sincere appreciation for helpful support they obtained from colleagues, especially from Mr. Sakari Luukkainen, Mr. Kari Koivisto, Dr. Arto Lahti, Dr. Liisa Uusitalo, Mr. Janne Yli-Äyhö, and Mr. Olli Ranta.

REFERENCES

1. Akao: Quality Function Deployment: Integrating Customer Requirements into Product Design, Productivity Press, Cambridge, Massachusetts, 1990.

2. Bunn: *Applied decision analysis*, McGraw-Hill, New York, 1984.

3. Gerpott, Schefczuck, and Pospischil: "International efficiency comparisons of Deutsche Telekom", *Communications & Strategies*, no. 10, 2nd quarter 1993.

4. Glahe and Lee: *Microeconomics, theory and applications*, Harcourt Brace Jovanovich, New York, 1981.

5. Green and Srinisavan: Conjoint analysis in consumer research: issues and outlook, *Journal of Consumer Research*, vol. 5, September 1978.

6. Infosino, Parker, and Unger: "Market analysis and product design for telecommunications equipment and services", *AT&T Technical Journal*, March&April 1991.

7. Kotler: *Marketing management: analysis, planning, implementation, and ontrol*, Prentice-Hall, Englewood Cliffs, New Jersey, 1988.

8. Lahti and Martikainen: *Value creation in networks*, Helsinki School of Economics series, 1994.

9. Peter and Olson: *Consumer behavior and marketing strategy*, Irwin, Homewood, IL, 1990.

10. Porter: *Competitive advantage*, The Free Press, New York, 1985.

11. Wilkie: *Consumer behavior*, John Wiley, New York, NY, 1990.

12. Von Winterfeldt and Edwards: *Decision analysis and behavioral research*, Cambridge University Press, Cambridge, 1986.

16

Value creation for multimedia services on broadband networks

Sakari Luukkainen, Christer Englund
VTT Information Technology
Tekniikantie 4 B
02044 VTT Finland
Tel. 90-4561 Fax. 90-456 7028

1. INTRODUCTION

This paper deals with the requirements of the market and with the marketing strategies of multimedia in realtime business communications, in particular at the prevailing early stage of the market. The targets in the covered area are: the qualifications for the takeoff of the services, marketing strategies, promotional marketing actions and policies needed to accelerate the startup. This is a summary of the results, consisting of the essential documentary material produced by the CEPT Telecom CAC Project Team Picture Services. The work has been coordinated by CEPT Telecom CAC BUS. The work started in March 1991 and ended in May 1992. Some material is also taken from work done in RACE EuroBridge -project.

The projects focused on the market requirements of realtime videocommunications in business communications, the main target being such applications, where the business user can amortize the investment already during a short period of time, regardless of the high terminal prices. Another aim of the work has been to find out, what kind of marketing policy should be adopted to accelerate the development of the critical mass. The market requirements have been tackled through case-studies as the most feasible way at this instance to research the market.

The terminal prices are expected to decline, but they may still exceed, for a while, the level that would be necessary for home market penetration. Like in most teleservices, business communications is also expected to be the driving force for the proliferation of multimedia communications. Therefore, the studies are in this context restricted to the prime business communication market segments.

The promotion of multimedia services needs and also deserves a lot of research and profound insight about the user needs before we can expect the services to take off. Our view is that the results can be used in the design and introduction of all new multimedia services.

2. DEFINITIONS

There are a lot of confusion concerning the use of term "multimedia". Whilst users see it as a solution to their needs, regardless of what the underlying technology may be, suppliers see it as a combination of enabling technologies that may help them to break into new markets. It is also used to describe a wide variety of products. Many people would like to replace the term with something else [Jeff-92].

Multimedia is defined to be in our context as an *adjective* characterizing such a *technology* that enables *computer* platforms based systems to make use of several digital media types like audio, still image or video. This paper is restricted on *studying multimedia only in the telecommunication area.*

A *telecommunication service* is defined to be that part of the system, which implements OSI layers 1-7 (bearer service OSI 1-3, teleservice OSI 4-7) and provides this functionality through *application programmer interface (API)* to another part of the system (application) using it. An *application* can be then defined as that part of the system, which implements a set of related tasks, for which business users have goals (e.g. document editing, medical diagnosis) and uses telecommunication services through API [Byer90]. A service will be *generic*, when it can be used by different systems as an element through its modularity. RCO-90 has defined generic applications as applications of advanced communications, which are widespread, and which provide a market opportunity for product and service providers by addressing users common application needs. A *market segment* is defined to be a group of users which have similar needs and requirements for the application and the whole system. A *platform* is a limited set of generic service software and hardware components or tools from which the system can be build.

3. SUMMARY OF THE CASE-STUDIES

The following cases were studied and analysed in depth:

- remote troubleshooting and maintenance of heavy earth moving machinery
- remote consulting at unmanned power transmission stations
- remote inspection of damaged cars
- desktop videocommunications in distant education
- remote consulting using videotelephony in the customer services of banking

The cases and other experiences indicate that remote consulting, which means geographical extension of various expertise, will be the most important application in business communication. The consulting may relate to a wide range of different application areas and market segments.

In installation, troubleshooting and maintenance remote consulting may be given by foremen, a group of experts etc. for the guidance and instructions of the field operations. Image information is necessary for the experts to get a correct view of the problem. The expert can use eg. document camera to present graphical and other image based information for the field personnel. The benefit is that the experts can supervise more field operations without the need of making a trip to the target. Besides, the throughput times will be reduced as a result of increased efficiency. The time of an expert tends to be extremely precious.

The Finnish Company Rotator, importing heavy earth moving machinery, has exploited videotelephony with great success in after sales operations of the machinery. The top experts are located in the head office of the company in the city of Vantaa, next to Helsinki. The company has 4 branch offices with service personnel carrying out the maintenance and repair operations. All the operations are conducted from the head office. The service personnel at the field use videotelephony to show a problem eg. a broken part for the expert group. They can give direct instructions, how to tackle the problem and what spare parts are needed. As a result, no foremen are needed at the branch offices, and the repair time has been minimized. For the entrepreneur this is essential, because there may be some 20 lorries awaiting, when an excavator stops. The money wasted in every minute of interruption is huge. Rotator has assessed that the investment already paid off during the first year of operation.

The degree of automatization of power transmission and the distribution network is getting higher and higher. The network is run by fewer operational people being in charge of the supervision and maintenance of the network. In many cases, when the power supply is interrupted, the losses for both the power company and the end customers are huge. Hence, if videocommunication facilitates a timely dealing with the problem, the money needed for the investment would pay off in a short time frame. This application demands a mobile terminal or a mobile link, because the power station covers an area of roughly 200 x 200 metres. The access to the terrestrial network is provided only in the station house, usually located at the centre of the station. The videocommunication could be used in an equal manner also as a support for the operational activities of telecom operators. When the traffic flow is interrupted, a large amount of money is lost. Hence, there is a clearly indicated potential to use videocommunication for minimizing the time needed to remove the fault.

The remote inspection applications seem to be a very promising area for desktop videocommunications. In Finland, a sum of about 400 MECUs is used for the repair of damaged cars. If the insurance companies could inspect a car straight from their office, without the need of travelling to the garage, a considerable amount of money and time could be saved. Besides, the insurance companies are now planning a joint venture to establish a inspection station network based on the use of videotelephony, covering the whole country. This enables the companies to ask for tenders from several local garages, which is expected to induce huge savings in repair costs as a result of proper function of the market mechanisms.

In distant education, desktop videocommunications will have a future. The case study indicates that this application is a very demanding one, both as far as the audio and video quality and the functionalities are concerned. The best choice is desktop videoconference service, based on FCIF resolution and multipoint capabilities. This application will remain a system solution application, where the service offers only one element for the solution.

In France, the bank CIC has made experiments with France Telecom in using videotelephony for advisory consulting for stock investments and other financial operations. The top experts of the bank are located in the head office in Paris. The customers can contact the experts from branch offices without the need to make appointments and a trip to the head office site. The videocommunication seems to create the feeling of confidence, which is extremely important in money affairs. Besides, the customers have felt being important for the bank. The experts can serve more customers than otherwise would be the case. It remains to be seen, whether the number of contracts have increased as a result of the trial.

4. MARKETING OF MULTIMEDIA COMMUNICATION SERVICES

4.1 General

The essence and requirements of the market have been studied by analyzing different application projects. The adopted approach has been considered a much more reliable way of exploring the market than by interviewing the users before they have not had any concrete idea or experience of multimedia services.

The first commercial videophone products have indicated that the traditional marketing channels in the telecommunication area are not good enough (Fig. 2.). The customers have had difficulties in noticing the value that the use of e.g. videophones might bring to them with regard to the high costs of purchase and usage. The market has been addressed with mass products, when the market has not been mature for it. It has been discovered that at the beginning of the life cycle there is only demand for special applications, in which videocommunication has been integrated as a part of a larger system. Under these circumstances, *the modularity of the system is essential*, enabling the tailoring to varying customer needs.

It can be recognized in the long run that broadband communication, using optical fibre network with digital broadband swithches and the corresponding terminals will become a widespread standard in business communication. However under the conditions to be expected up to the year 2000 (avaibility, cost), there has to be concentrated on special applications which lead to *strategic organizational advantages*. During this time span the market potential will mainly be determined by the group of highly qualified information professions, i.e. managers, products developers, researchers, consultants, experts, marketing specialists and instructors. For these professional groups access to existing knowledge and collected experience can be improved or made possible using advanced services for cooperative solution of problems, cooperative design of products, and cooperative preparation for decisions can be speeded up and secured, and training can be intensified and brought to the workplace [Bier-91].

When offering new technology to new markets the risks are very high. When implementing multimedia technology as an extension to existing services or products like e-mail, EDI, databases, text based CSCW, CAE the markets are already available. Completely new services like videotelephony do not have any existing markets. It can not yet be used in the beginning of its lifecycle as an extension of voice telephony. In the case of Video On the Demand services, there is already a customer need when offering centralised VCR and video renting services. VOD service can be offered to the recidental market also in the beginning of its lifecycle in order to have fast economies of scale. However in many other services different marketing strategy should be adopted. The added value in VOD comes from the fast and comfortable delivery of videomaterial and improved picture quality. By integrating IN nodes to broadband networks also more added value features can be created in VOD services. This service will still then be very cost sensitive.

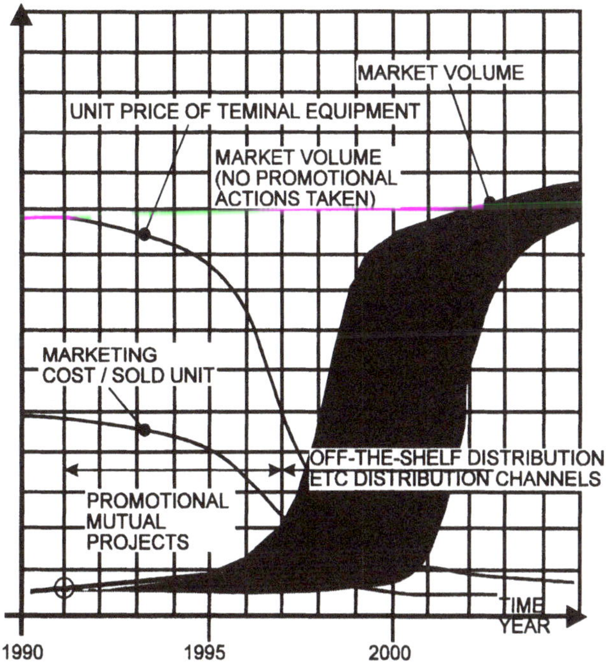

Figure 1. S-curve for the penetration of multimedia communication services.

The evolution of new services follow the shape of the wellknown S-curve (Fig. 1.). The incubation period i.e. the initial phase of the curve can be rather long before it matures up, shifting to the take off. The experiences gained so far indicate that it is very difficult to predict the demand of a new service. Nevertheless, the launch of the service usually requires huge investments, which imply big risks. To ensure profitable business activity, the new service has to meet the user needs as well as possible and concurrently sustain the competitive edge with regard to the old ones. *In other words, the new service has to offer distinct added value for the user in terms of cost.* The new service may enhance an old one i.e. stimulate the customer satisfaction or activate new latent needs.

4.2 Marketing strategies to achieve the critical mass

The marketing strategy of a telecommunication service should be chosen, taking into account its position on the S-curve (Fig. 1.). In order to achieve *economies of scale* and a *mass market* a lot of marketing efforts has to be done. The successful strategies can be completely different in different phases of the life cycle of a given product or a service. At the initial phase, the cost of the marketing efforts with regard to the proceeds is high, as the market is being promoted. *The experiences have confirmed the fact that, under these circumstances, the products and services pertaining to multimedia especially videocommunication are system products that cannot be marketed separately from the larger entity they are a part of.* It seems to be that in the beginning of the market only certain *components* of the system can achieve economies of scale and mass market. *The most profitable business is in system integration near endusers* (Fig. 2.).

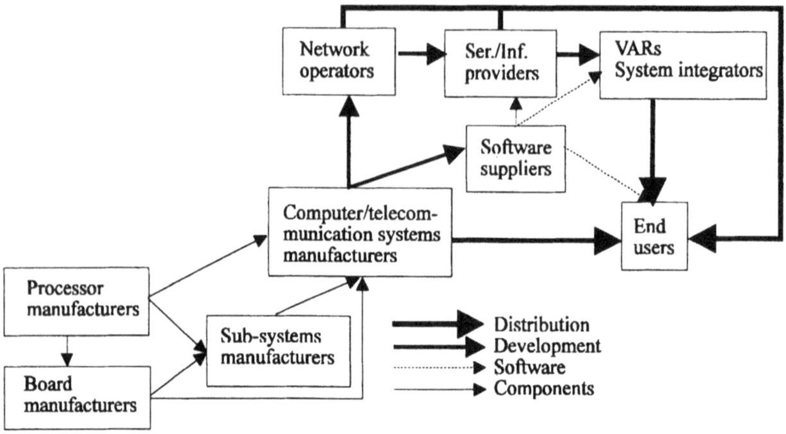

Figure 2. Value-added chain in the multimedia communication market [Jeff-92].

During the lifecycle there happends a shift that changes the characteristic of product from investement commodity to consumer electronics. This is now happening e.g. in the mobile phone market. In the beginning of the mobile market the terminal and transmission prices were high and the service were used only at some niche applications in the business market. Now the prices have dropped and the customers do not more require so high value of the service and the number of recidental users grow quickly.

The first experiences obtained from the marketing of videotelephony indicate that at this stage, the marketing schemes, based on normal distribution channels, do not give satisfactory results (Fig. 2.). In Finland, where the installed base of proprietary videotelephone terminals is more than 100, the *normal distribution channels have proved* ineffective, contributing to only a fraction of the sales. It is obvious that the promotion of desktop videocommunication requires, in the introductory phase of the services, particular measures and a marketing strategy that imply a new approach. The marketing strategies to be adopted do not necessarily differ significantly from those of other new multimedia services, which offer a new type of media with a lot of unfamiliar novel features for the end user. Under those circumstances, the customers first want to experiment and test the usability of a new service in a specific application before taking further actions.

Another important point is that the operators and terminal manufacturers have, for the time being, very little knowledge about where and how the end users could utilize multimedia in their business activities in such a way that the investment would pay off in a reasonably short time. The service providers cannot expect to make much progress before they possess this kind of marketing knowhow.

The services, supporting multimedia communications, are expected to constitute subsequently the essential part of the teleservices of the public telecommunication network domain.

Accordingly, the traffic generated by them is going to be massive for years to come and bring a substantial part of the revenues to the network operator. Therefore, the large market potential and long life cycle of the services will weigh up the large promotional marketing investments. The significance of multimedia communications justifies a preferential treatment with regard to other new teleservices. The promotional marketing efforts are likely to be rewarded by a much shorter maturing period, preceding the takeoff, than otherwise would be the case. The ruled area in Fig. 1, restricted by the two curves, is proportional to the proceeds of the sales. The aim is that it will balance the deficit of the initial period with a reasonable interest for the invested capital. As noted above, the available market at the beginning of the S-curve is explicitly different from the circumstances following the takeoff. After the takeoff has started, marketing of the services can be gradually transferred to the customary distribution networks and channels. *At that stage, the critical mass has been created, the service has gained wide knowledge and recognition among users, and the service providers are familiar with the utilization of the service.*

At the beginning, mainly such applications, in which *the value for the end user measure up the high initial cost*, can be successful. The potential customers and applications should be sought on this basis. *Cost versus benefit analyses* of the first experiences have proved that, even at the present price level, the end user can in some applications amortize the investment in quite a short time, taking into account saved labour cost, travel cost etc. and the benefits of increased efficiency. Moreover, *new business structures* may be needed at the end user level. A good example is the remote inspection of damaged cars, where there is a need for the insurance companies to establish a common enterprise running a network of common inspection stations with advanced visual communication capabilities. In some cases, subsidizing may be needed like in RACE ACE projects, especially in cost sensitive applications, having a large long term potential. In Finland state has supported the development of sign language systems.

Application software are typically system products, which cannot be sold on off-the-shelf basis through distributors for a long time. *The users want to have new services interworking with their existing systems* and they do not want to invest in new equipments only because of the multimedia extensions [Delp-92].

Good way to lower the barrier, encountered by the users, is to establish mutual experimental projects, consisting of the service providers, vendors, system integrators, and the enduser companies willing to cooperate. In this way application oriented technical and marketing know-how can be obtained. The terminal equipment can be rented for a given period of time on favourable terms. For instance, the customer should be entitled to deduction of the paid rent in the purchase price of the equipment involved. As noted before, this kind of approach requires quite a large quantity of manpower and high initial marketing cost. For the time being, the lack of marketing people with sufficient insight on multimedia communications is a problem. Consequently, the network operators should immediately take necessary actions to acquire such know-how so that qualified resources are available in due time.

The interviews of the end users have shown that there exist pioneeringminded companies, willing to join the experiments. They are naturally interested in gaining full value for their money. Therefore, the use of multimedia communication has to be customized to their specific business environment. This implies that the service providers, vendors, system integrators and VARs (value-added resellers) have to establish *a very close cooperative relationship with the customers* (Fig 2.). The companies do not want to take further steps and make the investments on new technology and services unless they are sure that the application will work and will pay off in a reasonable time span.

Behind the concept of a system product is the idea that the end user is interested in a complete solution only for a given communication problem. The ability to implement *costeffectively* complete system solutions, tailored to the user needs, requires *system integration capabilities and modularity.*

Besides the basic service, various *value-added features play an important role* when the service is aligned with the requirements posed by a specific application. *The basic service can be characterized as the platform covered by layers of additional functionalities, which may vary considerably depending on the application.* The vendors and service providers should be prepared to offer terminal equipment, based on a modular concept. At the same time when the markets require more flexibility of the products, the costs have to be kept down. In Mar-94 has been defined a concept of a modern product, which optimizes synergetic combinations of product elements as platforms. To the generic platform it is important that the tailoring costs can be kept as minimal as possible. When creating cost effective generic elements and elements creting differentation to the system it is possible to answear to the market demand cost effectively.

The tailoring can be supported by a PC/Unix based terminal environment, preferably with the windowing option and a API. API enables a flexible design and configuration of the user interface, facilitating the system integration in accordance with the needs of a given application. *The widespread of objectoriented programming makes it possible to create* ***modular service component library***. The *reusability* of software increases the productivity and enables costeffective tailoring.

The integration of the various elements into a complete solution requires wide system integration knowhow, which is neither available in the operator domain, nor among the terminal manufacturers. For the time being, the system integration business activity is concentrated mainly on rollabout and on studio videoconferencing, mainly in USA. Outside USA, there still are quite few VARs, having the capability for system integration. In due course, this business activity is expected to be diffused to Europe and to multimedia communications also. At the very beginning, the system solutions are customised to the internal use of companies, taking into account the requirements of the end user's business environment. The insight on these requirements is important for all the parties involved to promote the market.

The provisions presented above are necessary, but not yet adequate. Moreover, *the timing is a crucial factor for success.* The significance of the timing will be stressed under such circumstances, when a new product or service is launched in a new market. The first videophones were introduced too early to the market, the market being immature to utilize the product, among other things, because of deficiencies in the infrastructure. The launch ended in disaster to the terminal manufacturer. However, thanks to those enterprises, we now have valuable experiences that have been highlighted in this paper.

4.3 Proposals for the promotion of the multimedia communication to the market

An important thing in offering multimedia technology is *the integration to the existing systems.* As mentioned previously, joint venture trials, consisting of the network operator, service provider, the terminal manufacturer, software supplier, system integrator and the enduser are recommendable. That is also the main goal for RACE ACE (Advanced communication experiments) projects. The trials and experiments should be carried out in such a way that the end users are offered the possibility to test the feasibility of new technical solutions without any obligation to commit themselves to the purchase of the equipment concerned. The equipment may be rented for a given period and after that the user can buy it off on agreed terms. It is essential that the end user is given the choice to first make sure that the investment is beneficial and meets his needs.

Through the trials, the knowledge of how to multimedia communication, in various fields of business activities, can be obtained. The tailoring to the user's needs requires modularity and standard interfaces, through which the modules are linked together. The most profitable business is then the system integration using standard components. The promotional actions to be taken are as follows:

- recognition of the pioneeringminded customers, who want to be at the top of their own business, and who are willing to gain experience in multimedia communication in the business activities on their own
- analysis of their business environment and selection of a suitable application for a trial
- analysis of the requirements, which the application is placing on the equipment and the functionalities
- offering of the equipment and services on profitable terms for the field trial by redressing the paid rent, in case the customer is making a purchase
- subsidizing the rent of the equipment and the service charges, if necessary
- follow-up of the trial and gathering of the experiences already during the trial

The competitive edge, gained by the pioneering users, is a *technical advance*, because the competitors are likely to follow up. Besides the technical advance, the pioneering company may have the opportunity for *cream skimming*. Another important point is that, *with the aid of new services, the end users might establish firmer connections with their customers.*

5. STANDARDIZATION OF MULTIMEDIA SERVICES

The significance of offering also multimedia services on an open basis is of great importance. There has to be created also an open market for terminals and applications. Therefore the aim of launching quickly new customised multimedia services in the market poses new requirements for service standardization also. The old 3-stage ITU-T model I.130 for service standardization should be replaced with a novel service definition framework, which is based on the use of a standard set of *modular reusable service components* and a limited set of generic service tasks. This kind of approach enables the use of existing building blocks for new service also in the standardization area. This will result to a substantial reduction of work, as well as timely introduction of the desired standards, terminals and applications. ITU-T SG1 has recently initiated this kind of action, which is intented to be accomplished by the end of 1995.

6. CONCLUSION

The dominating trend of competition and reduced lifecycles for products is likely diffusing to services also, in particular for broadband multimedia services. The added value may be created either as a IN-capability in the network node or in an embedded intelligent terminal with dedicated sw-packages or as a combination of these two alternatives. It is propable that *the market for the software supporting customized value creation in the terminals will increase* radically in the future. Cost-effective provision of customized applications requires a standard set of modular reusable platforms with open interfaces. The embedded intelligent terminal environment seems most flexible solution in a timely manner to the needs of various customized applications. The service provision should adapt to the requirements of *customized mass production and fast service launch*. Stardardization work of teleservices should directed to support this trend.

It can be recognized in the long run that broadband multimedia communication, using optical fibre network with digital broadband swithches and the corresponding terminals will become a widespread standard in business communication. However under the conditions to be expected up to the year 2000 (avaibility, cost), there has to be concentrated on special applications which lead to *strategic organizational advantages*. Like in most teleservices, business communications is also expected to be the driving force for the proliferation of multimedia communications. During this time span the market potential will mainly be determined by the group of highly qualified information professions, i.e. managers, products developers, researchers, consultants, experts, marketing specialists and instructors. The promotion of this kind of services needs and also deserves a lot of research and profound insight about the user needs before we can expect the services to take off.

The marketing strategy of a telecommunication service should be chosen, taking into account its position on the S-curve. In order to achieve *economies of scale* and a *mass market* a lot of marketing efforts has to be done. The successful strategies can be completely different in different phases of the life cycle of a given product or a service. During the lifecycle there happends a shift that changes the characteristic of product from investement commodity to consumer electronics.

At the initial phase of the lifecycle, the cost of the marketing efforts with regard to the proceeds is high, as the market is being promoted. *The experiences have confirmed the fact that, under these circumstances, the products and services pertaining to multimedia especially videocommunication are* ***system products*** *that cannot be marketed separately from the larger entity they are a part of.* It seems to be that in the beginning of the market only certain *components* of the system can achieve economies of scale and mass market.

Besides the basic service, various value-added features play an important role when the service is aligned with the requirements posed by a specific application. The basic service can be characterized as the *platform* covered by layers of additional functionalities, which may vary considerably depending on the application. The vendors and service providers should be prepared to offer terminal equipment, based on a *modular concept*. A concept of a modern product *optimizes synergetic combinations of product elements as platforms*. To the generic platform it is important that the tailoring costs can be kept as minimal as possible. When creating cost effective generic elements and elements creating differentation to the system it is possible to answear to the market demand cost effectively.

Application software are typically system products, which cannot be sold on off-the-shelf basis through distributors for a long time. ***The users want to have new services working on their existing systems*** and they do not want to invest in new equipments only because of the multimedia extensions. They want also to experiment and test the usability of a new service in a specific application before taking further actions.

Good way to lower the barrier, encountered by the users, is to establish mutual experimental projects, consisting of the service providers, vendors, system integrators, and the enduser companies willing to cooperate. In this way *application oriented* technical and marketing knowhow can be obtained.

REFERENCES

Bier-91 Bierhals R., Nippa M., Setzen J.: Market Potential for the Future Use of Digital Broadband Networks, Berlin, 1991

Byer-90 Byerley P.: Enabling States: a new approach to usability, Proc. of 13 th Internaational Symposium on Human Factors in Telecommunication, Torino, 1990

Mar-94 Martikainen O., Lahti A.: Value Creation in Networks, Helsinki School of Economics Series, 1994

Delp-92 Delpho H., Eyett, E, Mahling, D.: Multimedia Systems & Services, A Survey on User´s Attitude and Requirements and Market Developments in Western Europe and USA up to 1995 and beyond, Basel, 1992

Jeff-92 Jeffcoate J., Templeton A.,: Multimedia: Strategies for the Business Market, Ovum Ltd, London, 1992

RCO-90 RACE Central Office: Business Opportunities Using Advanced Communications, A Report from RACE Workshop, Commission of the European Communities, Brussel, 1990

17

Use of INtelligent networks in the Universal Mobile Telecommunication System (UMTS)

Håkan Mitts
VTT Information technology, Telecommunication
P.O.Box 1202
02044 VTT
Finland
email: Hakan.Mitts@vtt.fi

1. INTRODUCTION

UMTS (Universal Mobile Telecommunication System) is a third generation mobile communication system currently being developed in Europe. UMTS related activities are lead by research conducted within the RACE II program and standardisation activities within the European Telecommunication Standards Institute (ETSI). ETSI SMG5 has the overall responsibility for UMTS related activities. More information on UMTS work in ETSI can be found in [SMG5]. For IN related matters also NA6 is involved. The ETSI working group responsible for Broadband ISDN, NA5, has created a special working group called Advanced Network Architectures (ANA). One of the objectives of ANA is to study UMTS impact on fixed networks.

Within RACE, the Mobile Project Line Assembly (MPLA), consisting of several RACE projects, is responsible for UMTS related activities. The major projects involved in UMTS development are MONET (R2066, responsible for fixed network aspects of UMTS including the relation between UMTS, IN and UPT), ATDMA (R2084, developing a radio interface for UMTS based on Time Division Multiple Access, TDMA), CODIT (R2020, developing a radio interface based on Code Division Multiple Access, CDMA), MBS (R2067, developing a broadband, ~ 100 Mbit/s, mobile access) and MAVT (R2072, studying aspects of a mobile audio and video terminal for UMTS). A RACE view of UMTS can be found in [BB92].

An effort similar to UMTS is underway in ITU under the name of FPLMTS (Future Public Land Mobile Telecommunication System), lately renamed to the more catchy IMT-2000 (International Mobile Telecommunication 2000). It is expected that UMTS and IMT-2000 will be compatible so as to provide global roaming but it is too early yet to say whether this goal will eventually be attained. ITU work on UMTS is described in [ITU].

This paper will present the use of Intelligent Networks in UMTS with an emphasis on the work done in RACE. To give some background, the UMTS system is first briefly described. Next the role and use of IN in UMTS are discussed. Finally some major areas will be presented, where significant enhancements to IN are expected.

2. A BRIEF DESCRIPTION OF UMTS

As for all mobile systems, UMTS provides wireless communication services to users on the move. The system supports roaming, that is the UMTS users can be reached and they can place calls where ever they are located within the coverage area of UMTS.

UMTS should be clearly distinguished from Universal Personal Telecommunication (UPT). UPT will offer discrete mobility, i.e. the possibility for a UPT user to register for incoming or outgoing services at any terminal. While moving between terminals, the UPT user is not registered and cannot therefore be reached. If a UPT user registers onto a wireless system like UMTS, the user also benefits from the wireless capabilities of the hosting system. One of the main design goals of UMTS is to allow UPT users to make use of UMTS in this way.

The distinction between UMTS and UPT has been made less clear by the fact that UMTS is expected to provide native support for UPT. Native support means that a UPT user (i.e. a user that only has a valid UPT subscription but with no UMTS subscription) could register onto a UMTS terminal without a prior subscription to UMTS. Such a mode of using UMTS would probably be more costly than using a normal UMTS subscription and thus UMTS users that are not UPT users will also exist. The resulting relation between UMTS and UPT, assuming native support for UPT in UMTS, is shown in figure 1 below.

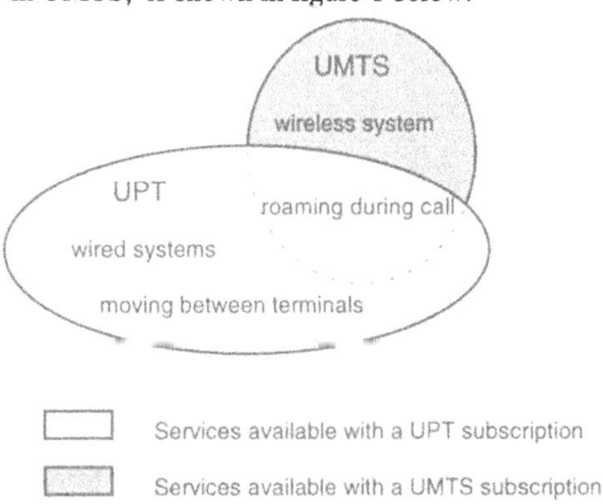

Figure 1. Relation between UMTS and UPT

UMTS is expected to satisfy a future demand for mobile multimedia services at transmission speeds up to a maximum of 2 Mbit/s. The services will include high speed data services and video services like mobile videophony. The ability to provide mobile multimedia - and in particular video services - at an acceptable cost and with good quality is expected to be one of the main features distinguishing UMTS from second generation mobile systems like GSM.

Other important features of UMTS will be the ability to support multicomponent calls as well as multiple simultaneous calls. As an example, a user might be engaged in a video conference that uses both a speech and a video component and at the same time initiate an interactive computer session from a personal computer connected to the UMTS terminal. In many respects the goals of UMTS are very similar to those of Broadband ISDN (B-ISDN) if one disregards the fact that the maximum transmission speed in UMTS will be much lower.

An other important aspect of UMTS is expected to be the support for various operating environments (public, domestic and business) using a single terminal. Global roaming between these environments will also be possible. To give an example, a user could use a UMTS mobile phone as a cordless phone at home. Next, without performing any explicit actions, the user could move outside the home to become reachable over the public network. Finally the user moves into an office and automatically becomes a wireless user of the PABX equipment in the office. It is even expected that the user could have maintained a call while moving from home through the public environment to the office.

3. STRUCTURE OF THE UMTS NETWORK

One of the main characteristics of UMTS is that it - in contrast to existing mobile systems like GSM - is planned for integration into a fixed network infrastructure and not as a separate overlay network. To achieve the integration, UMTS has been divided into three different parts as shown in figure 2.

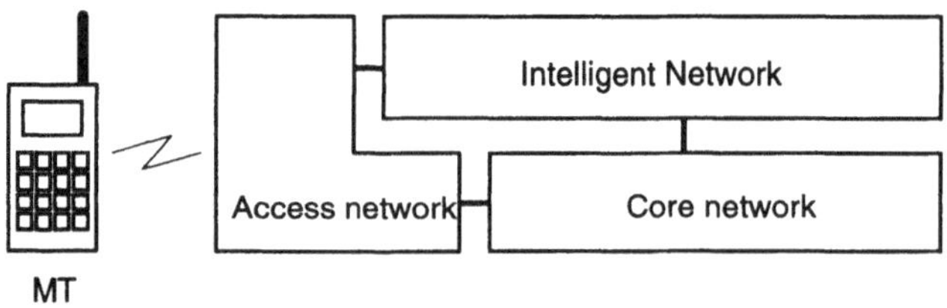

Figure 2. Main components of UMTS

The UMTS Access network is responsible mainly for radio related matters, it provides the radio access needed for wireless operation. Many important functions of a mobile system, e.g. handovers, are to a large extent implemented in the Access network.

The Core network provides the switching and transmission functions of UMTS. It will also provide interworking with fixed users and existing networks. Broadband ISDN is viewed as the likely candidate to act as the Core network of UMTS as it is the only fixed network technology that provides the required functions and the kind of flexibility needed to support the advanced UMTS functions.

Finally, the mobility specific control functions of UMTS are foreseen to be provided by an Intelligent Network (IN) based on the IN Capability Set 3 (CS-3)[1]. Some of the functions implemented using IN are:

1 IN Capability Set 2 is planned to support many of the functions required for UMTS but CS-3 is the first Capability Set that is expected to be able to support a fully functional UMTS network.

- database functions to store and retrieve globally accessible UMTS data; most UMTS IN functions update or access this data
- location management functions needed to track the user while roaming
- user registration functions allowing users to register for services at different UMTS terminals
- security functions providing user identification and access control
- support for certain aspects of handovers, e.g. handovers occurring between Local Exchanges

The goal of UMTS development is that, of the three components above, only the Access network will be UMTS specific. To achieve this, a number of requirements on the reused components, i.e. the fixed network technology acting as the UMTS Core network and IN, have been identified. If these requirements are rolled into the base standards of the respective technologies - as currently seems possible - UMTS can be realised reusing major fixed network components to provide a highly integrated network supporting both fixed and mobile communication.

4. THE ROLE OF IN IN UMTS

Next we will examine in some detail how UMTS is expected to use IN. The functions needed for which UMTS is expected to use IN are very different from those targeted by IN CS-1. The main benefit of IN still remains the same in both cases: IN provides a separation between basic switching functions and the additional intelligence needed to support of mobile specific features.

This model is very useful for UMTS, as it offers a way to separate the UMTS specific control functions from the switching functions provided by the Core network. This in turn is essential to enable the (re)use of an existing fixed network switching fabric as the Core network of UMTS and thus fulfil the UMTS vision of a single, integrated network providing both fixed and mobile communication. It is an additional benefit that IN also could be used to provide value added services in UMTS. This could make interactions between value added services and mobility much easier to manage. The study of IN for UMTS has however largely disregarded this aspect of IN. As a consequence, this paper only discusses the use of IN to provide mobile functions to a fixed network.

To illustrate the use of IN for separating mobile functions from the functions of the Core network, the Network architecture of UMTS is shown in figure 3. Each node shown has been indicated as (mainly) performing fixed network switching functions, IN-type mobility related functions or radio access related functions. As can be seen, the architecture allows adding mobile functions to a fixed network by simply plugging in a number on IN nodes that support the mobile functions[2]. Figure 3 also shows the extent of the UMTS Access network.

---

2 Note that the Core network assumed here has been enhanced with some limited functions required to support the 'plug and play' mobility.

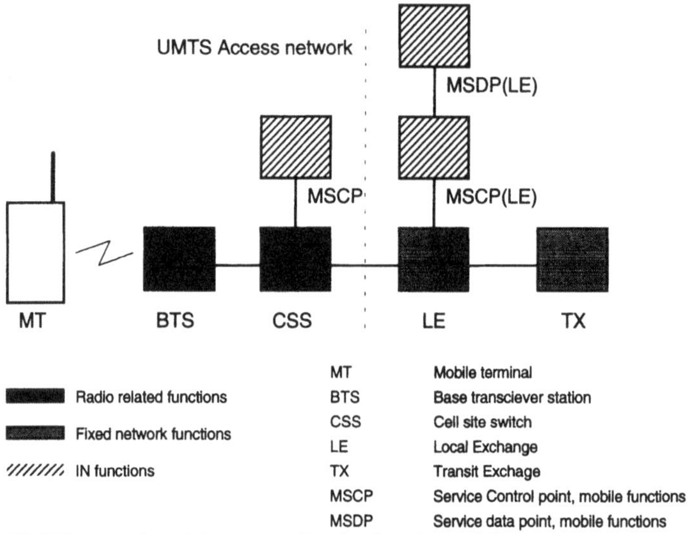

Figure 3. UMTS network architecture and major functions of the nodes

The MSCP is the UMTS node hosting the IN Service control function (SCF) with necessary provisions for the support of mobile users. Similarly, the MSDP is the UMTS node that hosts the IN Service data function (SDF). Figure 3 shows a slightly simplified view of UMTS in that the MSCP in the Access network could have an MSDP associated to it in large CPNs. Also the basic UMTS model identifies MSCPs and associated MSDPs controlling the Transit Exchanges but these have been omitted as they so far have very little functionality.

The UMTS network architecture shows maximal distribution in that the MSCP is assumed to host only the SCF-type functions of IN. Similarly the MSDP only contains the SDF functions. Nodes can of course be combined into bigger entities as required in an eventual implementation. It should also be reminded that the MSCP and MSDP are foreseen to be standard IN Entities and could therefore also be used to support any other IN-based services.

5. UMTS REQUIREMENTS ON IN

In this section we will identify a number of aspects of IN where improvements are critical to ensure the usability of IN for UMTS. Other requirements can be found in [BK&94]. Requirements on B-ISDN can be found in [KMS94].

In its current form, CS-1 is mainly seen as a vehicle to quickly develop and deploy new value added services for the fixed telephone network, e.g. Freephone or Premium Rate services. This is in contrast to the mobile functions needed for UMTS that are well defined and will not change significantly over time. On the other hand, the performance requirements on the mobile related functions, and thus the IN components implementing them as well, will be very high and the distributed aspects of control will be much more pronounced than in CS-1. This implies that IN for UMTS will be a more stable but also more efficient system than IN CS-1 performing a well-defined task in an optimised way. This might eventually result in implementations where specialised IN nodes, responsible only for mobile functions, might occur.

5.1 Distribution of UMTS functions

Perhaps the main modification required to IN is the ability to support much more distributed operations. While current IN systems are based on a very centralised mode of operation, UMTS will require that various control functions are distributed so that they will reside close to the end user in order to provide adequate performance. This also implies that the same functions exist in a large number of UMTS/IN entities, each serving a small geographical area.

It is also foreseen that different UMTS/IN functions are distributed in different ways. Thus for one IN-based function, execution speed might not be very critical. This function could then be implemented in a fairly centralised way. Other functions might have much more pronounced efficiency requirements and might consequently be much more distributed. This is shown in figure 4 where the UMTS FunctionA is provided by a single MSCP for all areas shown while FunctionB is distributed and thus requires an MSCP for each area.

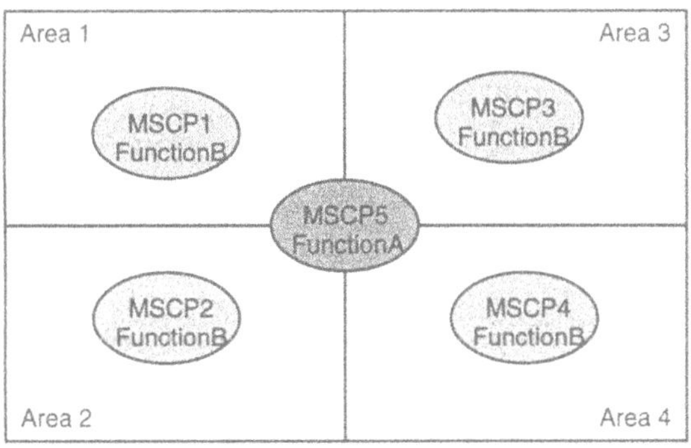

Figure 4. UMTS areas served by specialised MSCPs

Handovers (the actions required to support the movement of terminals when a call is active) bring yet other requirements on the distribution aspects of IN. As a terminal moves from one area to another, functions related to the terminal must be very quickly moved between entities responsible for the various areas visited. This implies a new, dynamic form of interaction between MSCPs not required by fixed network functions. This concept can be illustrated by assuming that a mobile moves from Area 1 to Area 2. If the mobile makes use of the UMTS/IN features implemented by FunctionB, the execution of the function needs to be moved from MSCP1 to MSCP2. This not only implies direct signalling between the MSCPs but also the existence of mechanisms to execute the transfer of control related to one mobile terminal between the MSCPs.

One final aspect of distribution is that IN functions are also foreseen to be moved to the UMTS access network (the MSCP connected to the CSS in figure 2) and onto customer premises. The interface between the UMTS LE (because the reuse of the B-ISDN LE is assumed) is the B-ISDN UNI including Q.2931, the B-ISDN UNI protocol, over a Signalling ATM Adaptation Layer (SAAL). This implies that the Signalling System 7, currently the main method of transporting signalling between IN entities in CS-1, is not available to interconnect the MSCPs in the access network to those deeper down in the network. This implies that IN must be updated to support new underlying signalling transport mechanisms in addition to Signalling System 7[3].

Distributing IN functions to the access network also results in moving IN functions onto Customer Premises Networks (CPNs). When a CPN is used to provide UMTS services in either a domestic and business environment, the UMTS/IN functions pertaining to the access network actually reside in the CPN. Thus IN must also be updated to support interactions between customer operated IN entities and those of the UMTS network operators. This has implications mainly on security related matters in the communication between customer entities and network operator entities. Requirements related to authentication, etc. must be fulfilled to provide adequate security for the signalling functions when customer operated IN entities are involved.

5.2 Handling of non-call related events

Another important aspect of mobile systems that is not present in current IN specifications is the need to support events and actions taking place outside a call. Current IN systems only act on events initiated by triggers detected during the basic call processing in the SSP. This results in a tight coupling of the IN Basic Call Process (BCP) to all IN CS-1 functions.

In mobile system, including UMTS, many control procedures foreseen to use IN for their implementation take place outside a call. An example of such a function is Location updating whereby the location of a UMTS terminal is stored in the UMTS database to enable calls to be routed to the terminal as it moves. Location updates in UMTS can take place both during and between calls. Whenever a mobile terminal detects a change in its Location area (evident from data constantly broadcasted over the radio interface), the terminal reports its new location to the MSCP responsible for managing terminal locations in the new area. The MSCP then updates the information into the UMTS database (in the MSDP) to be globally accessible to any other entity that might require the information.

As Location updates a more frequent than calls, most Location updates take place when no call is present. It is certainly not acceptable to initiate a call just to execute the Location update mechanism, this would be far too costly in terms of overall system performance, in particular as the location update does not require any processing in the Local Exchange (LE).

Due to the above, it is not appropriate to model all IN related events in UMTS within a call state model. To support mobile systems, IN must be enhanced to support interactions between 'user entities' (here used in a very broad meaning) and IN entities without involving the basic call processing. To perform this efficiently requires that IN models and protocols be extended to provide direct 'user'-IN interactions. In this context both the UMTS mobile terminal and IN entities in the access network and CPNs could be viewed as users of the main IN functions located in conjunction with the UMTS Core network.

Recognising the problems above, a joint experts group was set up by ETSI SMG5 and NA6. Based on the work of this group, a new functional model for IN has been proposed. The

---

3 This is actually mentioned in IN standards, but not actually supported.

functional model attempts to solve some of the problems related to non-call related functions [NA6]. The model is shown below in figure 5.

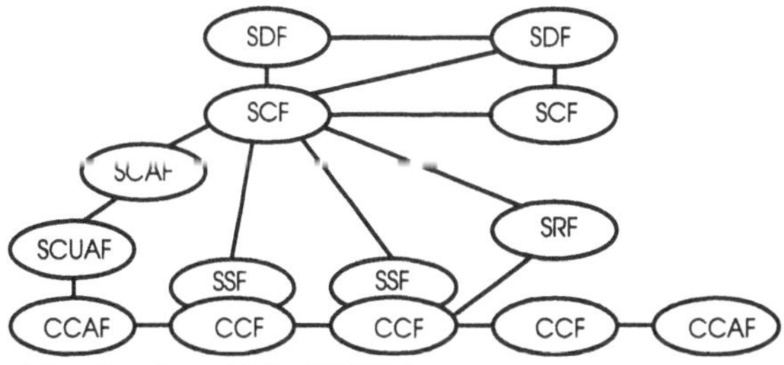

Figure 5. Proposed functional model for IN CS-2

In the model, two new Functional Entities are introduced, the Service Control Agent Functions (SCAF) and the Service Control User Agent Function (SCUAF). The SCUAF function provides service control access for users. It is the interface between user and network for service control. The SCAF provides the set of functions required for access and interaction between the SCUAF and a SCF. The SCAF and SCUAF are responsible for non-call related functions. They could well be used to model the non-call related functions required for UMTS. It should also be noted that the SCAF could be used to model the functions performed in the MSCP in the access network (see figure 2).

5.3 Distributed database

Data storage and retrieval are also a very important aspect of UMTS. While IN functions in support of IN CS-1 mainly read relatively static data from the IN database, UMTS functions frequently update the database as well. Further, access delay to the data is of crucial importance and the number of accesses per user will be very high. Further, due to the ability of users to roam, the data required to support a certain user might not be stored near the location where it is used.

To achieve the required processing speed for the UMTS data base and to manage the complexities related to locating the data for a given user, a number of mechanisms can be used. To maintain a simple interface between the UMTS entities using the data base and the data base itself, all UMTS data is stored in a single logical database distributed over all UMTS networks. The goal is to isolate the UMTS entities processing the data from the complexities related to the database implementations. This allows UMTS functions to access the required data without the need to know details related to the data base like the current location of the data for a given user. It also allows the data base to be modified and improved without changing the way it is used.

As a consequence, even though the MSDPs are shown as separate entities, they all provide access to a single global database. The main functions of this database are:

- the basic functions of providing read and write access to the data
- locate the requested data within the database
- move data to locations close to the user when the user has roamed to provide fast access to the data where it is frequently being used

- provide mechanisms that ensure the integrity and reliability of the data even when it is copied between locations and possibly duplicated as a consequence
- ensure that the data is only access by entities that have been authorised to do so

Many aspects of the UMTS database can be found in the X.500 specification. X.500 has also been used as the basis when defining the improved access interface between SCF and SDF in the ETSI Core INAP specification [Core]. It therefore seems likely that the UMTS distributed database will resemble the X.500 database even though X.500 cannot directly be used.

6. CONCLUSIONS

Evidently, due to the gap between IN CS-1 and the requirements imposed on IN by UMTS, a number of modifications and improvements to IN are necessary before UMTS can be implemented as planned. If these enhancements are rolled into standard IN - as now seems to be possible mainly in the CS-3 time frame - standard IN can be used to support mobile control functions in UMTS and thus fulfil the goal of reusing existing systems to implement UMTS.

As significant changes to IN are required, one can ask whether it has been correct to choose IN as the starting point for UMTS developments. Perhaps a better and more economic solution could have been found by starting from a clean slate? While this question certainly is valid, the advantages of the chosen path must not be forgotten. At the time when UMTS developments where started, no other scheme existed that provided the crucial division between basic (fixed network) switching functionality and the processing intensive functions to be added to provide the mobile functions.

Using IN is also an important step in enabling the integration between UPT and UMTS. As UPT will use IN, integrating the two technologies is expected to be much easier when UMTS is based on IN. If the current vision of enabling the use of UMTS terminals with only a UPT subscription becomes reality, mobile communications have taken a great step forward towards enabling ubiquitous communication.

It should also be pointed out that even though a number of requirements stemming from UMTS have been pointed out, they are not unique to UMTS. For instance UPT has many similar requirements, as have other advanced functions. By looking to all supported services in an integrated way, duplication of functionality and effort can certainly be avoided.

Finally it should be mentioned that other recent developments to provide advanced service processing separate from the basic switching functions, notably the TINA (Telecommunication Information Network Architecture) initiative [Nat93], have goals that are well aligned with the long term goals of IN. Thus basing UMTS development on IN is not in contradiction with other emerging technologies but evidently efforts are needed to realise a merging if it becomes desirable.

ACKNOWLEDGEMENTS

This paper is based on work done in the RACE II project MONET (R2066) where the author is working. The views presented are however those of the author and do not necessarily represent those of the project as a whole. The work has been partly funded by the Technology Development Centre in Finland (TEKES).

REFERENCES

[BB92] Hans de Boer and Evert Buitenwerf, *Networks for 3rd generation mobile telecommunication systems*, Proceedings 1992 IEEE International Conference on selected topics in wireless communications, 25-26 June, 1992, Vancouver.

[BK&94] Wouter van den Broek, Eric Kuisch, Chris van Maastricht, Toon Norp, *Impact of UMTS on IN Developments*, 3rd International Conference on Intelligence in Networks, Bourdeaux, France, October 11-13, 1994 (to be published)

[Core] ETSI ETS 300374-1, *Core Intelligent Network Application Protocol (INAP)*, 1994, European Telecommunications Standards Institute, Sophia Antipolis, France.

[ITU] ITU-RS TG8/1, *FPLMTS - Future Public Land Mobile Telecommunication Systems*, ITU document 8-1/TEMP/173.

[KMS94] John Korinthios, Håkan Mitts, Efstathios Sykas, *Scenarios for integration of the Universal Mobile Telecommunication System (UMTS) into Broadband ISDN*, 44th Vehicular Technology Conference, Stockholm, Sweden, June 7-11, 1994.

[NA6] ETSI NA6, *Enhancements of the IN distributed functional architecture*, TC-TR NA 60401, Version 2.0.0, European Telecommunications Standards Institute, Sophia Antipolis, France.

[Nat93] N. Natarjan, M. Ferro *The Distributed Processing Framework in the INA Architecture*, The 4th TINA workshop (TINA'93), September 27-30, 1993, L'Aguila, Italy.

[SMG5] ETSI SMG5, *Work programme for the standardization of the Universal Mobile Telecommunications System (UMTS)*, DTR/SMG-05 00-01, European Telecommunications Standards Institute, Sophia Antipolis, France.

18

On location of service control

Heikki Hämmäinen and Pekka Lahtinen
Nokia Research Center
P.O.Box 45, FIN-00211 Helsinki, Finland
heikki.hammainen@research.nokia.fi,
pekka.lahtinen@research.nokia.fi

Abstract

This paper discusses the requirements imposed by broadband services on location of service control. Discussion is based on the early results of TINA consortium. TINA proposes a modular object-oriented control architecture which allows services to be designed with little consideration on the final deployed configuration. This flexibility promotes "free competition" between locations of service control, between CPE and telecom platforms in particular.

1. INTRODUCTION

TINA[1] consortium is specifying the TINA software architecture in order to speed up the harmonization of computing and telecommunications technologies [TINA-RM, 1993].

So far TINA has contributed significantly in the areas of connection management [TINA-CMA, 1993] and distributed processing environment (DPE [TINA-DPE, 1994]). TINA connection management is independent of transport network technology but aims at a standards impact on ATM signaling phase 3 based on separation of call and connection control. TINA DPE extends the concepts of CORBA [OMG, 1994] by for instance integrating message-based and stream-based object interfaces as well as connectionless and connection-oriented communications.

An important part of TINA is the proper organization of service control, i.e. distributed processing and storage of service logic and data. The wide range of existing telephony-based services and the difficulty of predicting the future broadband service features suggest to have functional flexibility as a high priority design goal. TINA adds flexibility for instance by pursuing location independence and decentralized control of services.

This paper proceeds bottom-up. The general technical requirements of location independent object interactions are first elaborated. Then the essence of service control objects and service

[1] TINA-C stands for Telecommunication Information Networking Architecture Consortium. The circa 30 member companies of TINA-C maintain an international group of 40 researchers at Bellcore in Red Bank, New Jersey, USA. The author of this paper worked in the group during 1993.

control architecture are reviewed. Finally some problems of flexible location of service control are discussed.

2. LOCATION INDEPENDENCE

By location independence we mean that potentially active objects (i.e. code and data) can be moved from one location (i.e. computing environment) to another with minimal technical and managerial overhead. External interfaces and semantics of the affected service are preserved. Only the non-functional quality of service aspects may change. This is an ambitious goal that requires common agreements on

1. processing platform
2. object language
3. common object services
4. service design guidelines.

TINA DPE platform is based on the POSIX standard and maps the application programming interfaces onto the underlying native operating systems: threads, local system calls, remote procedure calls, stubs, kernel transport service, etc. DPE facilitates portability of application software modules written in C and C++.

TINA ODL (Object Definition Language) extends the OMG IDL by making a distinction between an object and its interfaces and by supporting the specification of both. ODL facilitates interoperability and portability of object and interface templates across DPE platforms. So far ODL does not, however, sufficiently cover the specification of object behavior. Therefore TINA Preprocessing Language (TPL) was developed to provide a temporary solution for easy port of TINA conformant object templates between DPE platforms. TPL is an extension to C++. TPL preprocessor takes TPL as an input and generates C++. ODL and TPL generate C++ stubs which support location and access transparencies [ODP, 1992]. Location transparency, i.e. late binding of object interactions, is based on removing hardcoded object references by using traders to maintain and mediate information on object locations [ODP, 1992]. Access transparency, i.e. interworking across heterogeneous object implementations, is based on automatic conversion between network's message transfer syntax and the local object's internal message representation.

TINA has produced draft specifications for a service infrastructure, i.e. a set of common object services. Object portability is significantly enhanced if an object template can assume that the same or corresponding common object services are available to it in any new location. Examples of such services are naming service, event service, object factory service, terminal mobility service, personal mobility service, etc. [TINA-SA, 1993, OMG, 1994].

TINA provides design guidelines which promote location independence in several ways. For instance, adhering to a service design process based on the ODP viewpoint model [ODP, 1992] postpones decisions on object locations up to the deployment phase. Service objects are designed and implemented on the ODP computational level with minimal assumptions on their physical locations.

3. SERVICE CONTROL OBJECTS

The term service control denotes here the user perceived on-line control aspects of networking services. Examples of such services and service components are directory searches, call forwarding, tailored billing, authentication, authorization, video conference bridging, voice mail, and quality of service negotiation.

Control is implemented as logic and data. Distinction between logic and data is, however, vague and sometimes unimportant. In practice, logic can be implemented for instance as interpreted scripts, rule sets, executable binary programs, or dynamic link modules. Data may appear for instance as rows of relational tables, state variables of object instances, or less structured global variables.

TINA, like other object-driven approaches, encapsulates logic and data in objects. Logic is mostly embedded in object templates and data in object instances. It should be noted that an object instance can be activated only in an environment where its behavior description, e.g. ODL/TPL/C++ object template, is available in executable form. Further, the behavior description needs its execution environment, e.g. ODL interpreter and DPE-extended operating system, to be available and activated.

In IN terms (e.g. IN CS1 [CCITT, 1991]), service logic can be represented as a script running on Service Logic Interpreter (SLI) in Service Logic Execution Environment (SLEE) in Service Control Point (SCP). Scripts assume that SLEE contains a predefined set of functional Service Independent Building Blocks (SIB). If the script language and SIB set are standardized, scripts can be easily moved across SCP platforms. Service data is represented as (1) volatile service-specific variables called Service Instance Data belonging to a script, (2) SIB-specific managed parameters called Service Support Data, and (3) service-specific data stored in a database potentially residing on a separate Service Data Point (SDP) platform. If the representation of database schemas and query language are standardized, service data can be easily moved across SDPs.

4. SERVICE CONTROL ARCHITECTURE

In the IN architecture, service logic runs in servers (i.e. in SCPs) whereas access references reside in switches (i.e. in SSP trigger tables). That is, switches know where the service logic is located and how to activate it. Activation happens as a "side-effect" of calling to a given number. In the Internet architecture, terminals (i.e. application client modules) know how to find the servers running the service logic (i.e. application server modules). Some service logic may run in the terminal side, but the network (of routers) remains as a provider of plain bit pipes.

As IN takes the service control software and databases out of switches, TINA does the same to IN call processing software and trigger tables. Switches are seen as a simple switching function controlled by objects running on DPE compliant terminals and servers. Distinction between terminals and servers can be vaguely made based on differences in degree of DPE capability, processing capacity, reliability, mobility, set of stubs and objects for common object services, application objects, scope of node management, and ownership.

Designers of applications and new services can assume the existence of a set of common object services. This includes services for controling connection-oriented communication sessions such as person-to-person audio. According to TMN layered abstractions (see

Figure 1), Element Layer consisting of network elements (e.g. switches) is controlled by Connection Managers (CM) residing on the Element Management Layer. CMs on Network Management Layer are responsible of link-by-link control thru the Element Management Layer. Session managers (SM) residing on the TMN Service Layer facilitate end-to-end negotiation between Service Access Functions (SAF) before reserving SAF-to-SAF transport resources from the lower layers. SAF is further decomposed into Terminal Agent and User Agent which represent the interests and capabilities of individual terminals (i.e. owners) and users.

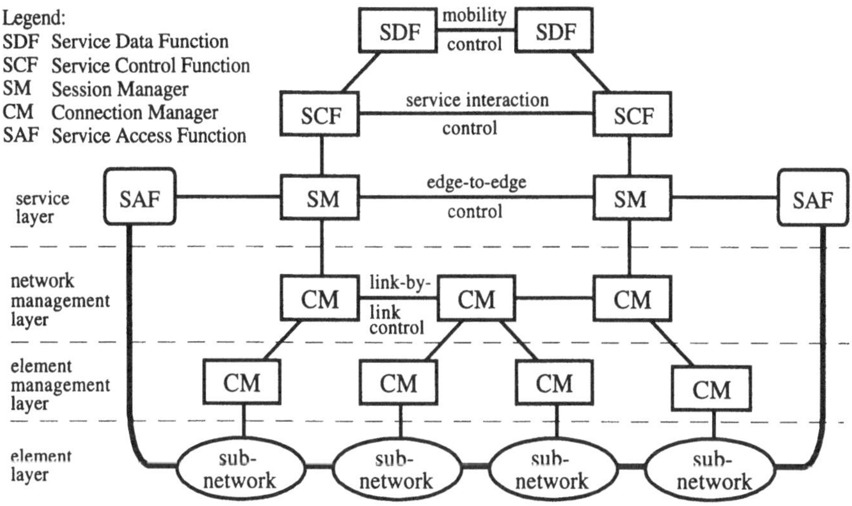

Figure 1. IN/TMN based reference model for service control.

Connections are represented as Connection Graphs (CG) where vertices represent switches, edges represent connections between switches, and ports represent end points of edges [TINA-CMA, 1993]. Each CG has a level-specific synchronized representation on each level of abstraction (i.e. Element Layer through Service Layer). For example, user initiated changes to CG (e.g. add party to conference) are propagated top-down from Service Layer to Element Layer, and network-initiated changes to CG (e.g. bulldozer cut a wire) are propagated bottom-up from Element Layer to Service Layer if necessary. State changes of a CG need to be propagated horizontally between SMs if the participants of a CG reside in different SM domains.

CG provides a multiparty mechanism for users, or application programs, to negotiate on usage of network resources. In addition to switches, more specialized resources such as video bridges and signal converters can be controlled thru CGs. In an opportunistic scenario new types of resources can be deployed with little extensions to CG objects. The CG-based approach to session control has got positive feedback from small scale experiments performed on the Touring Machine platform [Coan et al., 1993].

5. EXAMPLE

Traditionally, two kinds of disjoint network and terminal scenarios have flourished:

- dumb terminals (like ordinary telephones) connected to an intelligent network (like an ISDN with IN features), network providing the services, and
- smart terminals (like computers and workstations with TCP/IP protocols and networking capabilities) connected to a dumb network (like the fixed connections onto which the Internet is based), terminals and terminal-like servers providing the services.

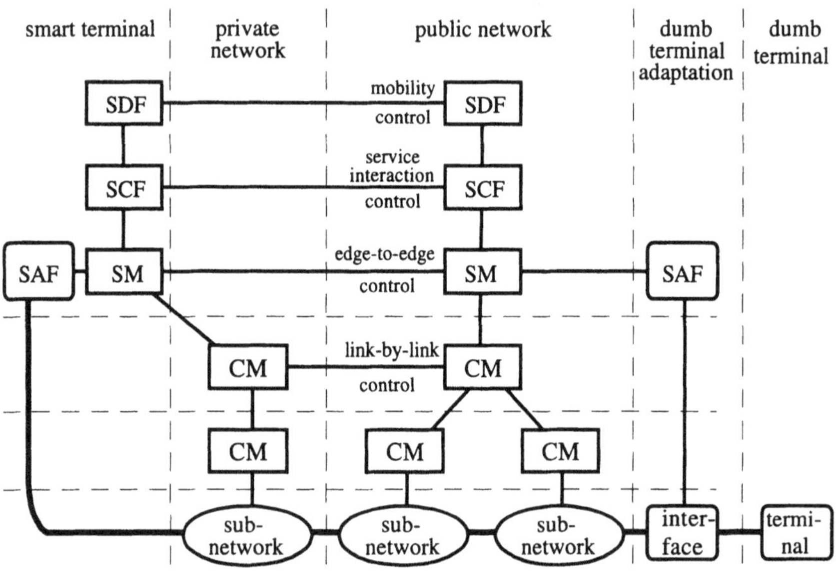

Figure 2. An example of the assignment of functionalities onto organizations and network components.

In the future, this cannot be the case any more, but the same network and terminal scenario must support various assignments of functionalities onto the terminals, network components, and participating organizations. **Figure 2** illustrates one example of such a scenario.

One cannot expect that all terminals were capable of supporting any functionalities, not even protocol-based negotiation of services. Therefore the dumb terminals must be adapted to a future telecommunications architecture by including on the network side as an adaptation interface some functionalities which logically reside on the terminal itself. This is illustrated on the right side of Figure 2.

On the other hand, a smart terminal - illustrated on the left of Figure 2 - may itself perform session, service and data management (SM, SCF, SDF) for itself, in addition to the actual Service Access Funtion (SAF). Nevertheless intelligent and dumb terminals can freely interwork with each other within the same service, thanks to the adaptation - it does not matter where the functionalities are actually placed, as long as they are available for the terminals.

Figure 2 also illustrates the fact that the telecommunications network of the future will consist of public parts (owned by telecoms operators) and private parts (owned and operated by their customer organizations), and the control and management implemented by the various participating organizations must co-operate in order to jointly provide the services.

6. DISCUSSION

IN currently supports telephony services based on single point of control which means that at most one service logic script in a single SCP can be active at a time for a call. TINA distributes the functionality into terminals and specialized servers. For instance, users may have personal scripts for controling incoming and outgoing calls and users may access several services potentially at the same time. This implies dealing with multiple points of control, or distributed peer-to-peer control, since it is likely that even a basic person-to-person call involves two potentially conflicting scripts.

The conventional routing and billing logic of IN is distributed into the SAF and SM functions in TINA. Both SAF and SM use available distributed common object services for various purposes. SM typically runs on shared DPE servers whereas SAF can partly run on DPE capable terminals. SAF has a modular object structure including user and terminal agents, which allows some flexibility in physical configuration of SAF objects between terminal and server platforms. Exploiting the location independence of object interactions, on-demand downloading of stubs, instances, and even templates of SAF objects is possible. In some cases, this may happen per service instantiation. For instance, a mobile call involves downloading the mobile user's personal profile from his remote home location register to the local visitor location register, or even to the mobile terminal, for local processing. It has been suggested that on-demand transfer of executable application and service logic, e.g. object templates, becomes desirable along with more powerful workstations and terminals [Hämmäinen, 1991].

Flexible location of service logic and data brings information networking services onto a new level of dynamics. This raises difficult issues concerning feature interactions, data security, service reliability, service introduction, billing, etc. For instance, the problem of unwanted feature interactions emerging in IN, see e.g. [Cameron et al., 1993], gets more complex in the case of multiple points of control. Preparing for future service features when designing the current one is difficult. Within a single server such as IN SCP it is possible to analyze a new feature against existing features and make changes if necessary. In an open network containing multiple servers managed by different service providers this is not possible. Independently designed service features are likely to interact, often with harmful consequences. Introduction of object-oriented techniques, generic negotiation protocols between servers, and user-adjustable interaction attributes are however expected to alleviate the feature interaction problem in the long term [Griffeth et al., 1992]. The fine tuning of service features must be left to users because they finally decide whether a feature interaction is harmful or not.

7. CONCLUSION

We have described how TINA architecture adds flexibility to the location management of service control. This flexibility is necessary to facilitate deployment and usage of alternative location schemes. The underlying assumption is that location flexibility leads into a more optimal location scheme for each particular usage situation. Optimal can mean for instance more efficient exploitation of distributed computing resources or more rapid introduction of new service features.

The inevitable downside of location flexibility is the increased complexity. Feature interactions are problematic in the relatively simple single point of control architecture of IN. Applying location flexibility for multiple points of control in TINA services is likely to be significantly more complex. Much experimentation and trialing is needed to make this part of TINA mature enough for commercial deployment.

8. REFERENCES

Cameron J., Velthuijsen H.: Feature Interactions in Telecommunications Systems: A Software Perspective, IEEE Communications Magazine, August 1993

Coan B., Gopal G., Herman G., Leland W., Mak V., Sekar R., Vecchi M., Weinrib A., Wuu S.: The Touring Machine System (Ver. 3): An Open Distributed Platform for Information Networking Applications, Proc. of TINA Workshop, L'Aquila, Italy, September 27-30, 1993

CCITT: CCITT Study Group XI, draft recommendation Q.1200 series "Intelligent Networks", Geneva, September 1991

Griffeth N., Velthuijsen H.: The negotiating agent model for rapid feature development, proc. of Software Engineering for Telecommunications Systems and Services, Florence, Italy, March/April 1992

Hämmäinen H.: Form-based Approach to Distributed Cooperative Work, PhD thesis, Faculty of Information Technology, Helsinki University of Technology, 1991

ODP: Basic Reference Model of Open Distributed Processing - Part 3: Prescriptive Model, DR X.903, ISO/IEC JTC1/SC21/WG7, November 1992

OMG: Common Object Services Specification, ed. John Siegel, The Object Management Group, March 1994

Ryan R.: The SCAI Standard, proc. of GlobeCom '93, Houston, Texas, December 1993

TINA-CMA: Connection Management Architecture, TINA-C Core Team, Red Bank, December 1993

TINA-DPE: Distributed Processing Environment, Version 0.1, TINA-C Core Team, Red Bank, January 1994

TINA-RM: TINA Architecture Rooadmap, TINA-C Core Team, Red Bank, December 1993

TINA-SA: Service Architecture, TINA-C Core Team, Red Bank, December 1993

19

Comparison of broadband intelligent network signalling architectures

Olli Martikainen[a], Tapani Karttunen[b], Valeri Naoumov[c], Konstantin Samouylov[d]

[a]*Telecom Finland, Helsinki*

[b]*Lappeenranta University of Technology, Lappeenranta*

[c]*Institute for Problems of Information Transmission, Russian Academy of Sciences, Moscow*

[d]*Russian Peoples' Friendship University, Moscow*

Abstract

In this paper we compare different distributed intelligent network signalling architectures based on broadband transmission and switching. As a result of the comparison we choose one example architecture, where we specify the broadband equivalents of Service Switching Point *(B-SSP)* and Service Control Point *(B-SCP)*. The protocols for user signalling and broadband application part as well as the service distribution architecture are discussed. With the proposed Broadband Intelligent Network Signalling Architecture an example interactive multimedia service is concepted and analyzed.

1. INTRODUCTION

In modern telecommunications deeply influential changes are taking place, caused by the emerging competitive telecommunications services market and the new technological breakthroughs. The market changes are due to the integration of telecommunications and information technology, which makes it possible to develop data intensive telecommunications services. There are two types of application data involved: the customer and service management data, and service content data. Typical examples of the former are the mobile and value added network *(VANS)* services, where customer location data and customer personal service data are used for network control. With broadband networks media service applications bring interactive real time video and multimedia contents available to users. Examples of media services are digital interactive TV, video on demand services *(VOD)*, electronic press and publishing. The service portfolio will move from Interconnection services towards Mobile, VANS and Media services (Figure 1-1).

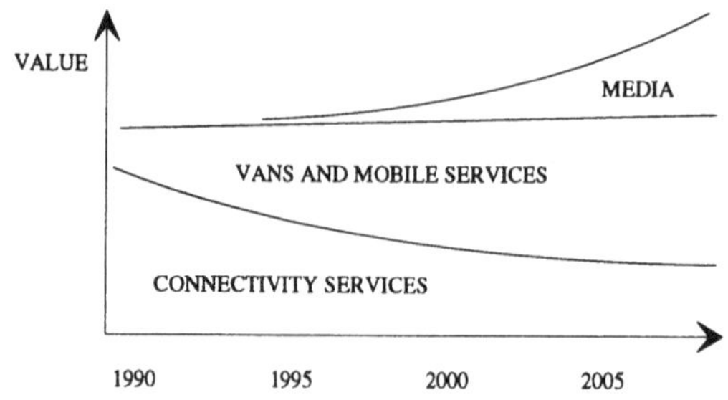

Figure 1-1: Service Portfolio Chances

The technological requirements for these new services are cost effective broadband transmission and access technologies, flexible computer based management and control of networks, switching and service applications and the support of mobility.

In technology substantial new breakthroughs are going on. The introduction of cellular radio networks and mobility is probably the most influential one in the next few years. The broadband transmission and switching technology is also maturing and will provide a cost effective platform for service provision. When the broadband customer access will be available, interactive business and consumer services based on video and multimedia will become possible. Common to all these developments will be the computer controlled structure of modern telecommunications, where protocols, application technology and resource management are key factors.

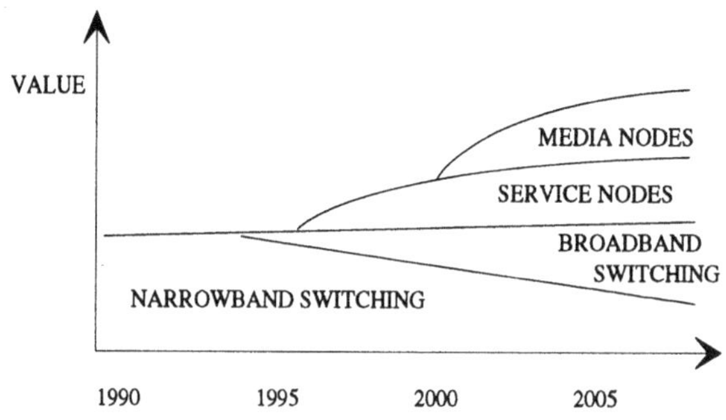

Figure 1-2: Switching and Service Nodes

In modern telephony the most promising computer control and management concept is the Intelligent Network *(IN)*. It facilitates the development and management of new services by service providers. Intelligent networks can be considered as the bridging technology between the customer based service management, the service application execution and the connection control in the networks. Considering the mobile and broadband technologies it is a natural question, how this concept of intelligence will develop with these new networks. It seems to be clear that the service applications will be distributed over terminals and different **service** control and **media** provider **nodes** (Figure 1-2). Examples of distributed intelligence can be found from Internet World Wide Web *(WWW)* and the specifications of Universal Mobile Telecommunications Services *(UMTS)*.

2. MOTIVATION FOR DISTRIBUTED INTELLIGENCE

The new broadband and mobile infratructures will enable new service types and more versatile roles for service providers. We believe that in broadband and media services the integrated service combinations will grow more important. This is because there are advantages of scale in basic interconnection and basic media production. Hence the basic production has to be done in a large scale, preferably by global network or service providers. On the other hand, the specialized knowhow depending on segmented markets creates specialized enterprises, that use their expertise locally or even in global scale. The generic and specialized services considered as product elements can be combined to synergic concepts [ValC]. The key question is the ability to exploit strong synergies of different basic and specialized product elements so that the customer perceived value can be maximized (Figure 2-1).

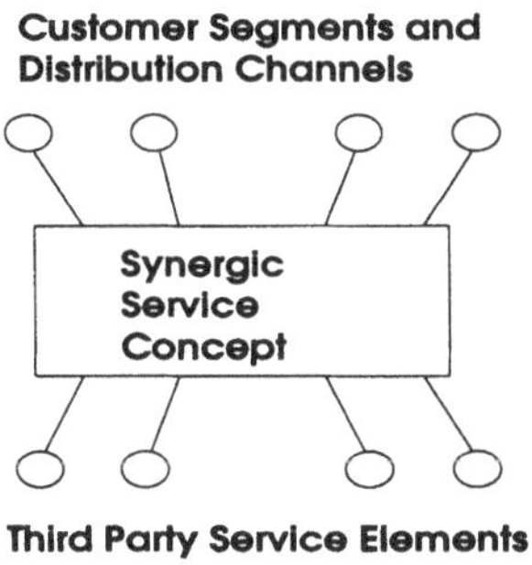

Figure 2-1. Service product consisting of several elements

To facilitate the provision of new services there are demands for open service environments and common service architectures. How these demands can be taken into consideration in the evolution of Intelligent Networks is an important question. Let us first consider the demands in more detail.

2.1. New Service Types

The Intelligent Network Architecture *(INA)* enables rapid development and deployment of advanced services on the current telephony network. Its main principle is the separation of the supplementary services control aspects from the connection control aspects. When the broadband transmission, switching and access capabilities become available in the basic network, the scope and scale of supplementary services will grow remarkably. Different supplementary service types as well as different service provider roles will be present. The following service types can be identified:

- Voice and media services provided in special service nodes by network operators or service providers.

- Intelligent network services provided in intelligent switches or separate service nodes usually by network operators.
- Mobility services provided today in intelligent switches by mobile network operators.

Voice and media services are controlled by the user with signalling sequences and the service data is resident in databases, which should be interfaced to the network. A promising user and service signalling system for them is the ISDN D-channel signalling, which is separated from the ISDN B-channels. Intelligent Network architecture separates the connection control and service logic. However, the narrowband IN user signalling is still the basic telephony signalling, which is in the speech channel in the access network. In mobile services signalling must be separated from calls because roaming and handovers take place independently of calls. Hence in GSM and DECT the ISDN type signalling protocols have been developed.

2.2. Different Roles of Enterprises

In telecommunications the enterprises can be categorized as different stakeholders with respect to the following roles:

- Service users and service subscribers which use the network and service capabilities.
- Network provider that provides the access, transport and switching capacity.
- Service provider that providers services using the network interconnectivity.
- Service manager that manages the operation of services and maintains the normal condition of the network.
- Service broker that subscribes to the service and provides it to another subscriber.
- Others such as network and service designers.

We would like to introduce also the following role:

- Concept provider that integrates different services as synergic combinations and provides these combinations for the service users. Concept provider works in an enterprise network using other service providers as subcontractors and network providers as distribution channels.

How the service logic programs are distributed and interfaced with respect to the different roles is a question that has to be studied.

2.3. Service Architectures

Since the service types and the organizational roles of service providers will be more fragmented, the management and interworking of networks and services should be quaranteed. For this purpose unified service architectures will be needed. There are several standardization and research inputs for these kinds of architectures, such as:

- The ISDN Supplementary Services Framework [Q932], where call and supplementary service control messages can be separated from the actual data stream. A lot of pioneering work for supplementary services specification has been done in ISDN recommendations [Q950].

- IN, Intelligent Network architecture [Q1200], which specifies the IN Service Processing Model separating call control and service logic, and the INCM Conceptual Model for service specification and distribution for current telephony networks.
- ROSA, the RACE Open Service Architecture [R1093], which defines a set of concepts and rules for building services that are open in time, space and technology. The ROSA conceptual model divides a service in terms of access, core, management and transport capabilities.
- CASSIOPEIA is a RACE project which has continued the ROSA work in defining an Open Services Architecture (OSA).
- INA, the Bellcore Information Network Architecture [Bellc], which separates the architecture into user, data and processing parts and classifies the service management functions into fault, configuration, accounting, performance and security management.
- ODP, the ISO work on Open Distributed Processing [N7524] defines a reference architecture for distributed systems using the enterprise, information, computation, engineering and technology viewpoints and associated languages.
- TINA, Telecommunications Information Networking Architecture [TINA], is a service architecture defined by TINA-C consortium. TINA has adopted ODP reference architecture and applies Object Modelling Technique [OMT], GDMO [X722] and General Relationship Model [X725] specifications in modelling the information concepts. The computation modelling concepts are based on structures and interactions between computational objects, such as operational and stream interfaces, traders and servers. The engineering modelling concepts are based on Distributed Processing Environment (DPE) infrastructure, that provides distribution transparencies by kernel, server and stub -parts.

In ATM networks the control signalling can be independent of the connections, as in ISDN networks, which gives new possibilities in the introduction of intelligence. Since the Intelligent Network Architecture is dominant in the present day networks, there is a clear need to analyse its possible developments with emerging broadband and mobile networks. In this study also the different roles of service users and providers are to be taken into account.

3. SWITCHING AND INTELLIGENCE MODELLING

The IN Service Processing Model is designed for modularization of the network functions and standardization of the communication interfaces between network functions [Q1201].

The three main elements of the Service Processing Model are the basic call processes, the "hooks" that allow the basic call processes to interact with IN service logic, and IN service logic that can be "programmed" to implement new supplementary services (Figure 3-1).

The main principles of the Service Processing Model are:

- The basic call process is available all over the network and is designed to support, with optimal performance, network services that do not require special features.
- "Hooks" are to be added to the basic call process forming the links to the service logic in the Service Switching Points *(SSP)*. The "hooks" are trigger points that are able to start an interaction session with the IN service logic. For this it should continuously check the basic call process for the occurrence of conditions on which an interaction session with IN service logic should be started. During an interaction session the basic call process can be temporarily suspended.

- IN service logic uses a programmable software environment in Service Control Points *(SCP)*. New supplementary services can be created by means of "programs" containing IN service logic. The IN service logic is able, via the "hooks", to interact with the basic call process. In this way IN service logic can control the basic call process.

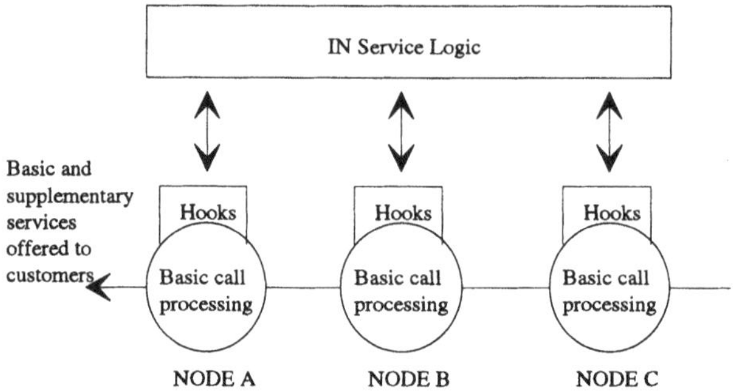

Figure 3-1: IN Service Processing Model

Thus, by changing logic at the service control point and modifying network data, a new service that uses existing network capabilities can readily be implemented.

In addition IN service logic can decide to terminate an interaction session with the basic call process. The basic call process will then resume its execution as specified by the IN service logic. In order to allow fast service implementation, the IN service logic should have a logical view of the network resources that constitute the basic call process and additional (specialized) network functions.

In present day implementations the service logic is usually centralized and controls the centralized Service Switching Point (Figure 3-2).

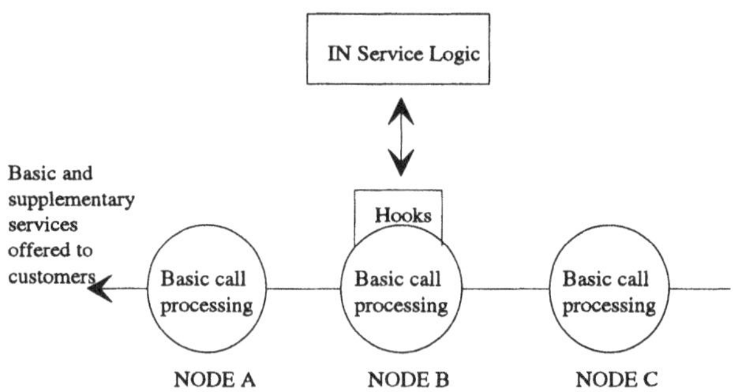

Figure 3-2: Centralized Service Logic

If the service logic is distributed so that its different parts reside in different network nodes, one has to specify the relevant service logic and network node types. Let us here choose the nodes according to the following roles (Figure 3-3):

- Local Node *(LN)* used by the Service Subscriber
- Switch operated by the Network Operator

- Control of Service Combinations operated by Concept Provider
- Service Node *(SN)* operated by the Service Provider.

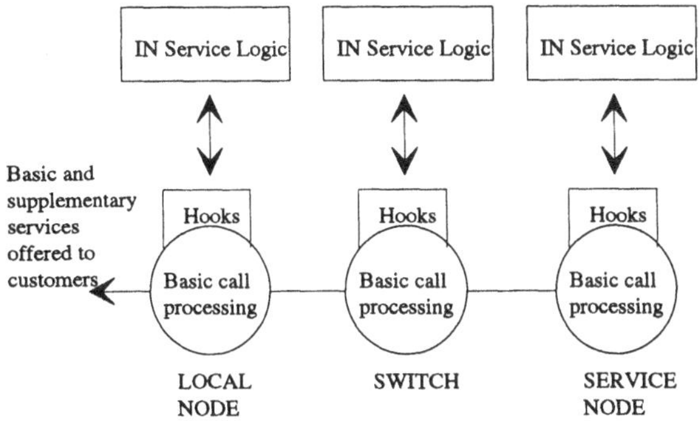

Figure 3-3: Distributed Service Logic

4. ATM SWITCHING

4.1 ATM Networks

A simplified example for the structure of an ATM network is shown in Figure 4-1. The

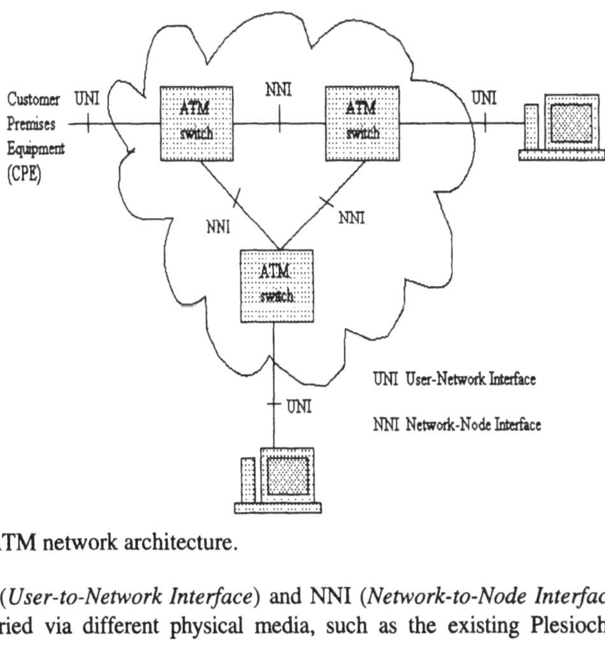

Figure 4-1. ATM network architecture.

various UNI (*User-to-Network Interface*) and NNI (*Network-to-Node Interface*) connections could be carried via different physical media, such as the existing Plesiochronous Digital

Hierarchy (*PDH*) or the new Synchronous Digital Hierarchy *(SDH)*. Several standards have been defined on how to interface the physical layers, and work is continuing to specify additional physical layers to be used to transport ATM cells. [Forum]

4.2 Virtual Channels and Virtual Paths

The concepts Virtual Channel (*VC*) and Virtual Path (*VP*) are applied when ATM cells are transported through the entire network (Figure 4-2). Virtual Channel Connections (*VCCs*) are set up between any source and any destination in the ATM network, regardless of the way it is being routed across the network. Fundamentally, ATM is a connection-oriented technology. The way the network sets up the connection is therefore by means of signalling, i.e. by transmitting a SETUP request which passes across the network to the destination. If the destination agrees to form a connection, the VCC is set up between the two end-systems. A mapping is defined between the Virtual Channel Identifiers (*VCI*)/ Virtual Path Identifiers (*VPI*) of both UNIs, and between the appropriate input link and the corresponding output link of all intermediate switches. [Forum]

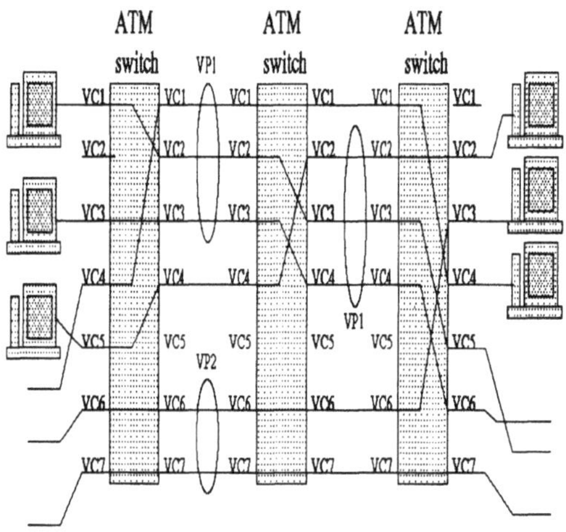

Figure 4-2. Virtual Channels and Virtual Paths.

A VCC is a connection between two communicating ATM end entities, possibly consisting of a concatenation of several ATM VC links. All communication proceeds along this same VCC which preserves cell sequence and provides a certain Quality Of Service (*QOS*). The Virtual Channel Identifier in the ATM cell header is assigned per network entity-to-entity link, and may change across the network within the same VCC. Each connection is identified by VCC and Virtual Path Connection Identifier *(VPCI)* consisting of VPI and InterfaceID.

4.3 ATM Signalling

The user signalling for call control is specified in [Q2931] for UNI. For point-to-point access configurations VCI=5 is used as the Signalling Virtual Channel *(SVC)*. In Q.2931 the following message types for call and connection control are defined: ALERTING, CALL PROCEEDING, CONNECT, CONNECT ACKNOWLEDGE, SETUP, RELEASE, RELEASE COMPLETE, NOTIFY, STATUS and STATUS ENQUIRY.

In B-ISDN there are no additional signalling message types yet to support other features. In ISDN the additional message types to Q.931 defined in Q.932 are: HOLD, HOLD ACKNOWLEDGE, HOLD, REJECT, RETRIEVE, RETRIEVE ACKNOWLEDGE, RETRIEVE REJECT for call related control, and FACILITY and REGISTER for non call related control [Q931, Q932]. The facility messages can convey service dependent information.

Since the SVC is a special VC, the call and facility control type messages can also be used in other VCs. So, they can be used also to control other services than the basic switching. However, these messages may not be sufficient for advanced service control such as the INAP operations between SCP and SSP.

5. DISTRIBUTION OF INTELLIGENCE IN BROADBAND NETWORKS

The enhanced broadband network services will be networked applications and the implementations may include several nodes external to the transmission network. If several applications control the service, there is a need of additional functions as compared to the basic call control.

The networked applications may consist of the following: Users that have their own service profiles (own control data), Network that provides the routing and QoS and the Service Providers which offer services and have their own rules how to use the services. Synergic combination of services, called as Service Palettes, can be differentiated with respect to QoS, service types and charging rules.

The ATM switching control is evolving towards ISDN type signalling [Q2931]. Its specification has broadband specific extensions concerning mainly the reservation of the virtual channels and the compatibility with narrowband ISDN. The networked services put new requirements for the protocols to be used. When the services may be controlled from several points, the specified signalling may not be capable enough to deliver control information from one node to another. How Q.2931 and the related protocols could be utilised for this purpose?

Another question is, how to organise service control into network in such way, that the upcoming needs of new services can be fulfilled in a sufficiently flexible way.

5.1 Control in Existing Systems

To be able to specify signalling architectures satisfying the demands for distributed intelligence it is helpful to analyse example services. Three existing services could be analysed for this purpose: Mobility services, Intelligent Network services and Media services.

The mobility services are seen as user having portable communication facilities. An example of this is the mobile phone solutions (GSM, DECT). The control of mobility services is handled near the Service Subscriber in terminals and in access networks.

The media services are offered from media service nodes by Service Providers. These MS's in turn have large databases containing the programs or multimedia material. The rules to access and to manipulate this information are applied in the server. In other words, the control is placed near the Service Provider.

In traditional Intelligent Network the service is controlled inside the network in Service Control Points. The IN control is focused to control network resources The user or service provider control is more or less added into the concept in a proprietary way.

5.2 Example Service in Broadband Network

We shall use the Video on demand (VOD) service as a common example describing the need for broadband network services. VOD service offers television programs and movies to the customers at the time they want to. The main difference to ordinary TV broadcasting is that the service is offered to one customer (unicasting) .

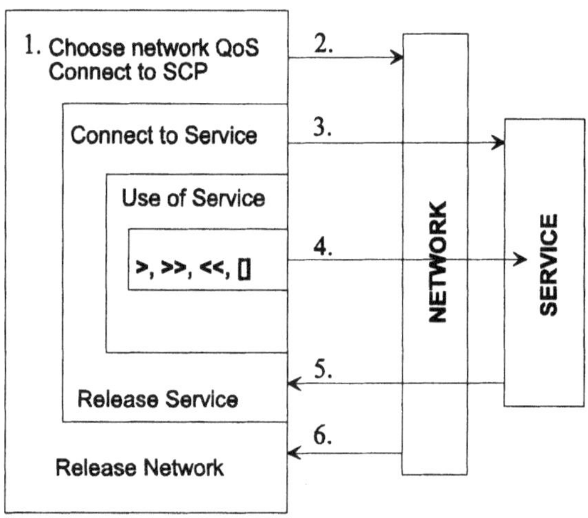

Figure 5-1. VOD Service Phases

The VOD service implementation in broadband networks (on ATM) could be the following (Figure 5-1):

1. the service is selected with service identifier (service number),
2. the service identifier is recognised by the network, the routing and QoS information is retrieved,
3. user is connected to video server,
4. user is controlling video server by the controls offered by server,
5. user is disconnected from video server and
6. network releases the network resources.

5.3 Analysis of VOD Service Control

The VOD service presented in previous chapter can be implemented in a real network at least in four different ways: complete control in subscriber terminal, complete control in network controller, complete control in video server and the control is distributed among these three instances. Let us analyse these different approaches.

5.3.1 Subscriber Terminal Based Service Control

The VOD service control could be placed to the subscriber terminal. In this scenario subscriber terminal (LN, local node) takes care each new connection establishment. The control of the networked service (e.g. how to utilise different connections) is in the user application. In Figure 5-2 the call phases are: a) the original call setup is placed on VCI=5 to connect SN 1, b) the 1st connection is created to SN 1, c) the second call setup is placed on VCI=5 to connect SN 2 and d) the 2nd connection is created to SN 2.

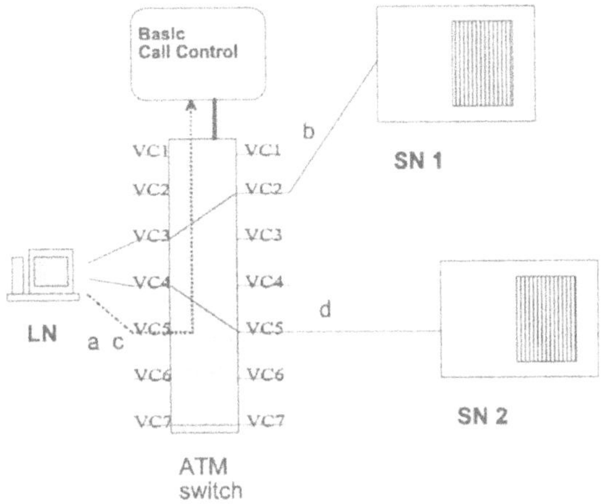

Figure 5-2. The LN based service control

The following consequences can be pointed out:

- the terminal software will be service and furthermore VOD server specific,
- the terminal software will be complex and therefore will rise the cost of the terminal,
- network resource control may be difficult to arrange and
- terminal based security is vulnerable.

This signalling architecture may be optimal for limited service types. It is difficult to support service element or equipment reuse in this architecture.

5.3.2 Network Based Service Control

In the next alternative the VOD service control is in the network (either in the switch or in the centralised Service Control Point) (Figure 5-3). In this scenario network controller takes care of each new connection establisment. The control of the networked service, e.g. how to utilise different connections, is in the network control application. The connection creation phases are: a) the original call setup is placed on VCI=5 to connect SN, b) the destination is consulted from SCP, c) SCP responds the routing address, d) the connection leg is created to LN and e) the connection is continued to SN. The possible followup can be created into SCP, when the first connection is dropped by SN. LN is consulted for next operation (e.g. the next SN address).

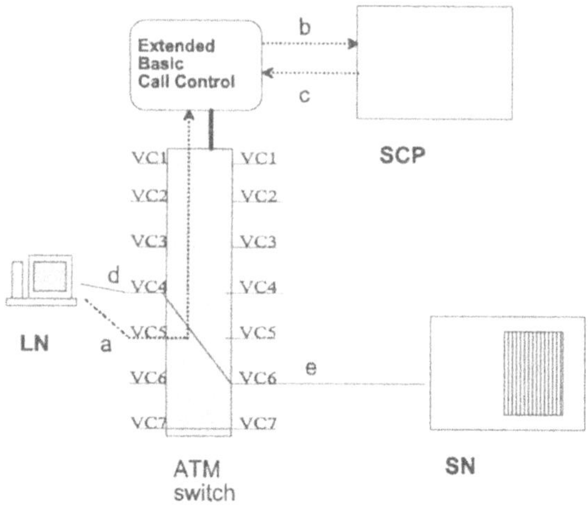

Figure 5-3. The network based service control. The configuration includes SCP. The corresponding scenario could be implemented also with plain switch with its own logic.

The following can be determined from this centralised architecture:

- the service control traffic may cause severe problems in centralised control node,
- service provider specific applications may be difficult to implement,
- service provider and user specific management may be difficult to arrange.
- complex software in centralised control point and
- applications in centralised systems are typically proprietary.

This scenario is similar to the existing Intelligent Network solutions. When looking the existing solutions critically, the efficient service creation has not become true and the external service provider applications cannot be implemented. Furthermore, the management solutions are inflexible and they usually cannot be offered outside network providers due to security problems.

5.3.3 VOD Server Based Service Control

The VOD service control can also be placed in the Service Node. In this scenario VOD server takes care each new connection establisment. The server routes forward each new needed connection. This can be seen from the phases in Figure 5-4: a) original call setup is placed by LN, b)1st connection is established to SN 1, c) SN 1 requests new connection to SN 2 and d) route to SN 2 is established. In this scenario SN 1 routes the communication between LN and SN 2.

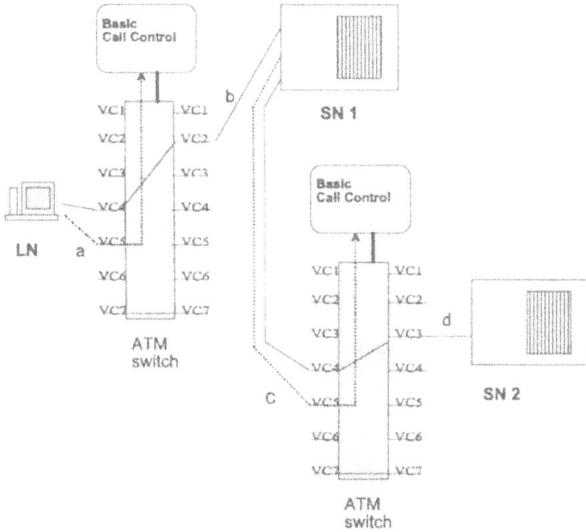

Figure 5-4. The service node (SN) has control of the service. The new connection is routed by original server.

The following can be determined from this scenario:

- the control software will be VOD server and terminal specific,
- the terminal may not be of general purpose,
- inefficient use of network resources may result and
- network QoS cannot be set dynamically.

Basically the VOD service itself is controlled in optimal way in this scenario (server has the control). Anyhow, the network reources cannot be configured dynamically and this causes inefficiency in the network. Also the user profiles are difficult to manage in a VOD server and the service management can control only this special service.

5.3.4 Distributed Control Between User Terminal, Network and VOD Server

The control can be distributed in several ways, we present two of the most obvious ones.

1. The external controller has an external control access (separate protocol like CS-1 INAP [Q1218]) (Figure 5-5) or
2. An external controller has an extended D-channel signalling (extension of DSS2 [Q932]) (Figure 5-6).

In both cases several controlling points can be added in the network and these Service Nodes (SN) can control the VC creation.

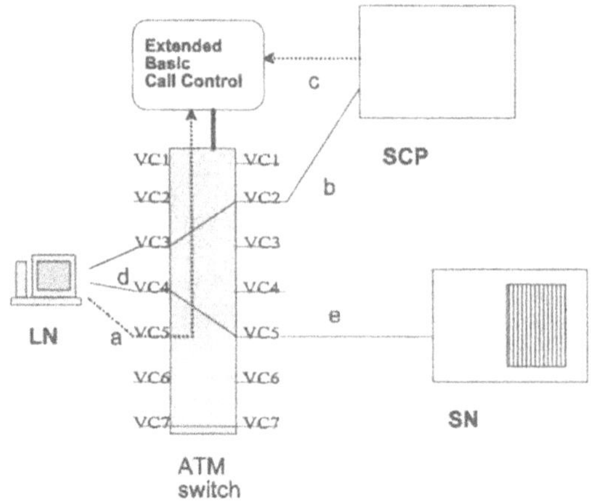

Figure 5-5. The service is distributed. The service access is controlled by SCP and the service itself is controlled by both LN and SN (scenario 1).

The scenario 1 describes the usage of external controller with external control access. The call establishment phases are: a) original call setup to SCP, b) call is established to SCP, SCP is requested to connect the service, c) SCP commands through command interface the switch to establish connection between LN and SN, and finally d) and e) connection legs are established for a service.

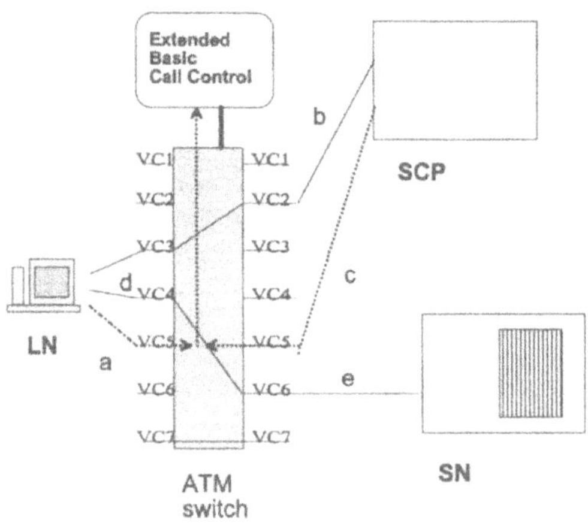

Figure 5-6. The service is distributed. The service access is controlled by SCP and service itself is controlled by both LN and SN. The extension of Q.2931 is utilised as external command channel (scenario 2).

The scenario 2 describes the usage of external controller with Q.2931 extension access. The call establishment phases are: a) original call setup to SCP, b) call is established to SCP, SCP is requested to connect the service, c) SCP commands through Q.2931 interface the switch to establish connection between LN and SN, and finally d) and e) connection legs are established for a service.

When distributing the service control into three separate instances, the following properties are gained:

- optimal use of network resources controlled by network provider,
- network resource usage can be negotiated dynamically,
- reusability of terminal software (service specific modules offered by the VOD server),
- QoS can be defined by user or by service provider,
- each entity can have its own security management,
- user profile configuration can be managed locally and
- service profile management can be implemented in the VOD server.

This distributed alternative seems to be flexible enough to be used for networked applications. The distribution of the control gives the freedom to configure instance specific matters by each stakeholder. The architecture gives the modular solution and the modules can be reused for new services.

The solution creates requirements for the signalling. The protocols have to support dirstributed computing and control passing. Therefore the proposed protocol has to be analysed, how and if this distribution can be handled. This work is extensive and needs a lot of research of service examples. We shall only give some outlines here.

If the extension of DSS2 signalling is utilised the straightforward approach is to specify DSS2 supplementary services extension. This can be adopted from N-ISDN specification Q.932 facility messages. The question is, if this message set offers enough features to implement efficient call and charging control.

The other possibility is to have separate protocol and this can be approached from IN technology. The ITU-T Q.1214 and Q.1218 Recommendations specify the CS-1 protocols. Their broadband extensions would be needed and these extensions have been discussed to be added into IN Capability Set 3.

It is important to remember that control distribution extends requirements for both DSS and network control signalling. An important issue which may affect a great deal on how to select architecture is the different roles of service provider and network provider. Furthermore an important open issue is how to arrange broadband network charging. Is it completely a task of network providers or is it divided between service providers and network providers?

6. CONCLUDING EXAMPLE

Let us finally present a VOD service with signalling and distributed service logic according to the scenarios given in Figures 5-5 and 5-6. The organizational roles within the VOD service are Service Subscriber (SS), Network Operator (NO), Concept Provider (CP) and Service Provider (SP). We believe that these roles are substantial for broadband VANS and Media services. In present media terminology we can say that the CP acts like a newspaper publisher and the SP acts like an advertiser or news editor. These different types of stakeholders should have different privileges to control the services. The CP should have almost similar rights as the network operator in network and service control, the control of the capacity allocation and charging should be allowed. The CP needs to utilize service control interfaces with NO and SP. The SP should control communication with SS and furthermore SP should be able to deliver assisting information to CP, so that more capacity or new connection to next server (in a pool) can be requested. The INAP type external control is used by the Concept Provider's Service Node (CP-SN) and the extension of the Q.2931 control is used by the Service Provider's Service Nodes (SP-SN1).

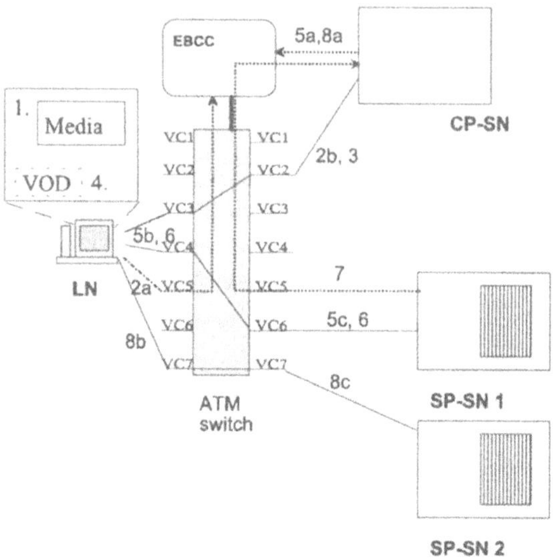

Figure 6-1. The VOD service example with distributed applications.

The Concept Provider's Service Node controls combinations of several services from different Service Provider's Service Nodes. Since the control of services is in the CP-SN, the interfaces, access rights and service parameters can be managed. CP-SN could also be an environment to manage the feature interactions of the services.

The service invocation and use sequence could be the following:

1. initial Service Icon is in LN,
2. icon click activates the connection to CP Service Node (CP-SN),
3. a new icon palette is given to LN,
4. LN activates a service icon,
5. the connection to SP Service Node (SP-SN1) is created,
6. the service is accessed and controlled via control channel between LN and SP-SN1,
7. SP-SN1 requests new network resource,
8. LN connection is rerouted by CP-SN to new destination (SP-SN2) and finally
9. the service is terminated.

This service logic distribution was presented first in [Wint]. The Icon concept as the service stub in the Local Node and the Service Palette concept as the service directory and control library in the Concept Provider Service Node acting as SCP are presented in [BINP]. In that study the ATM switch which supports this kind of Extended Basic Call Control was called as B-SSP and the Concept Provider Service Node was called as B-SCP. The synergic combinations of services provided by the Concept Provider were called Service Palettes.

REFERENCES

[N7524] ISO/IEC JTC1/SC21 N7524, ISO Committee Draft on Basic reference Model of Open Distributed Processing - Part 2: Descriptive Model, 1992

[Q931] CCITT Recommendation Q.931, ISDN User-Network Interface Layer 3 Specificatiom for Basic Call Control (DSS1)

[Q932] CCITT Recommendation Q.932, Generic Procedures for the Control of ISDN Supplementary Services

[Q12XY] CCITT Recommendation Series Q.1200 on Intelligent Networks

[Q2931] ITU-T Recommendation Q.2931, B-ISDN, Digital Subscriber Signalling No. 2 (DSS2), User Network Interface Layer 3, Specification for Basic Call Connection Control

[R1093] RACE Project R.1093 (ROSA) Deliverable 93/BTL/DNR/DS/A/005/b1, The RACE Architecture, Rel. Two, Version 2, RACE, 1992

[X722] CCITT Recommendation X.722, Guidelines for the Definition of managed Objects (GDMO)

[X725] CCITT Recommendation X.725, General Relationship Model

[Bellc] Bellcore SR-NWT-002268, Cycle 1 Specification for Information Networking Architecture (INA), Issue 2, 1992

[BINP] K. Molin, O. Martikainen, Broadband Intelligent Network Project, Workshop on Intelligent Networks, Lappeenranta University of Technology, August 1994

[Forum] ATM Forum, Introduction to ATM, The ATM Forum, 1993

[OMT] J.Rumbaugh, M.Blaha, W.Premerlani, F.Eddy and W.Lorensen, Object-Oriented Modeling and Design, Prentice Hall, N.J., 1991

[TINA] P. Leydekkers, V. Gay, Multimedia Services in TINA and RM-ODP, IN'94 Workshop, Heidelberg, 24-26.5.1994

[ValC] O. Martikainen, A. Lahti, Value Creation in Networks, Helsinki School of Economics, Report W-67, February 1994

[Wint] K. Molin, O. Martikainen, Intelligent Network Tutorial, The Second Winterschool on Telecommunications, Helsinki, Telecom Finland, March 1994

20

Broadband intelligent network project

Olli Martikainen,
Telecom Finland,
P.O.BOX 106, 00511 Helsinki
Tel: +358 2040 3513, Fax: +358 2040 3251

Kim Molin,
Lappeenranta University of Technology,
Telecommunications laboratory,
Laserkatu 6, 53850 Lappeenranta
Fax: +358 53 574 3650,
E-mail: Kim.Molin@lut.fi

Abstract

This paper presents a case study on a service architecture called Broadband Intelligent Network (*BIN*). BIN consists of Broadband Service Subscribers (*BSS*), Broadband Service Switching Point (*BSSP* or ATM-switch), Broadband Service Control Point (*BSCP*), Broadband Service Management Point (*BSMP*) and Broadband Service Providers (*BSP*).

Unlike in conventional IN, BIN service logic programs are distributed into each active component of BIN. BSS and BSCP contain the basic service logic programs with the basic functions such as controlling the icons, the hypermedia documents, QoS and charging. In the project a sophisticated application protocol was designed, called BINAP (*Broadband INAP*). BINAP is ends-oriented being used as an application protocol between the components of BIN. BINAP includes messages for security, performance, Quality of Service (*QoS*), managing customer service palettes and controlling the multimedia flow between the BSS, BSCP and BSP.

Services provided in BIN are based on the transfer of multimedia data from BSP to BSS, known also as the fast multimedia data stream. The control of the services is provided by BSCP together with the distributed service logic.

Keywords: ATM, Broadband ISDN, BIN, IN, TMN

1. INTRODUCTION

Telecom Finland and Lappeenranta University of Technology (*LUT*) have started a project to research broadband, ATM-based, Intelligent Network called the Broadband Intelligent Network *(BIN)*. The research considers the future broadband multimedia services and their implementation. BIN has not been standardized or even implemented anywhere, being now in the prototype design phase. It has been only a name for the project, which was started in March 1994 at Telecom Finland /4/.

The project refers to the conventional narrowband IN /1/ with fast service introduction and support for external broadband service providers. The main objectives concerning the architecture have been centralized customer and service management and multiservice offerings at single service points.

2. BIN ARCHITECTURE

2.1 Components

BIN components are BSSs, BSPs, BSCPs, BSMPs and BSSPs (*Figure 1*). BSSs, BSPs and BSCPs are connected to BSSPs, which form the ATM-network. BSMP is connected to BSCP and is not tied to any physical implementation technique.

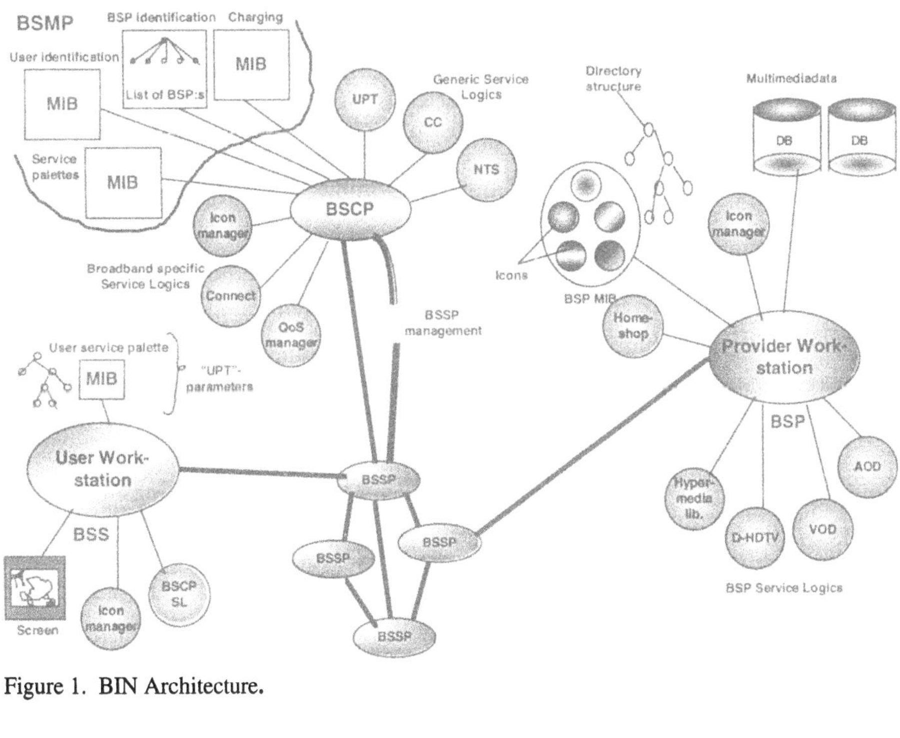

Figure 1. BIN Architecture.

Several service providers, which would use different protocols and managing architecture, could be designed. If done this way, every service provider should implement its own mechanisms of charging and managing services. One advantage of this architecture would be quite simple interface to the end user, and one disadvantage quite complex service introduction. Intelligent Networks have centralized service management architecture, whereby components can be distributed from each other. The objective of this project is to have a centralized model, where the users (BSS) could use services provided by BSPs and have a centralized service management system.

BSMP contains customer identification and charging information, a list of BSPs known to it and possibly a customer service palette. BSCP contains the general Service Logics (*SL*) also existing in conventional IN, which can be used in several services, such as CC (*Credit Calling*), NTS (*Number Translation Service*), and UPT (*Universal Personal Telecommunications*). In addition, broadband services have SLs for controlling icons and QoS (*Quality of Service*) -parameters, and SL for a connection initializer. BSS has possibly its own customer service palette, icon manager, SL for BSCP and a screen handler for showing the multimedia data. In BSP lies the actual multimedia databases (*DB*) and icon databases. Also the SLs for broadband services are located there.

2.2 BIN and IN

In conventional narrowband IN the user has a simple interface to the network, i.e. either from the SSP (*Service Switching Point*) or the NAP (*Network Access Point*). The highly simplified function of CS1 (*Capability Set 1*) IN is that the user is connected via SSP or NAP to the SS7 (*Signalling System No. 7*) network which forms the signalling network of the IN. The user dials a phone number, which is then generated into a message and transferred to the nearest SSP. If the number begins with a 0700 or 0800, SSP knows that the user chose an IN number. In this case, SSP triggers an IN call, otherwise the call setup procedure is just as for a normal call. SSP forms the intelligent part of conventional IN. SSP triggers the IN call and forwards an INAP (*Intelligent Network Application Protocol*) message to the SCP (*Service Control Point*) via the SS7 network using the services of TCAP (*Transaction Capabilities Application Part*). SCP then has control on the next step e.g. sends a control message to SSP.

In BIN the function of BSCP is different from SCP. BSSP is in point of fact a simple high-speed switch architecture, e.g. ATM-switch. The functions of such a switch is to route the 53-byte cell from the input to output end according to the Virtual Circuit Identifiers (*VCI*) and Virtual Path Identifiers (*VPI*) and the information that has been configured in its routing tables. The switch is not intelligent like conventional IN switch, because it does not trigger an IN call. It handles any data in the cells transparently. Compared to the SS7 network's 64 Kbit/s capacity ATM-network forwards cells at a much higher rate i.e. ≥155 Mbit/s.

SSP solutions are mainly based on hardware solutions. They can not be programmed as easily as computers. While BSSP being simple, the BSCP has to be quite complex. The end nodes, BSS and BSP, are thus also quite complicated. This means that application layer protocol BINAP is tranferred between all the end nodes unlike in conventional IN, where INAP is mainly used between SSP and SCP.

2.3 Broadband services

In fact the few services shown in *table 1* are in a way quite similar to each other. They could be grouped into three different categories: controlled file transfer[1] based AOD and VOD, hypermedia database[2] based hypermedia library and homeshopping, and single- or multi-party[3] calls. By doing this controlling of the services can be done in a similar way. There does not have to be different controlling mechanisms for every service provided.

Table 1. Some broadband services

AOD	❶	AOD (*Audio On Demand*) is a service that corresponds to a CD- (*Compact Disk*) player. The service includes therefore PLAY and STOP functions.
VOD		VOD (*Video On Demand*) is similar to AOD, but in addition to voice also moving picture is transferred. In future, it might be competing with the video rent activity.
Hypermedia library	❷	An interactive service, where the user can browse hypermedia documents over the network, e.g. real-time Internet Mosaic.
Homeshopping		An interactive homeshopping service that is based on hypermedia documents.
Videotelephony	❸	Videotelephony is a conventional two-party telephone, where in addition to voice also moving video picture is transferred.
Video conferencing		Videoconferencing is quite similar to videotelephony, but it enables multiparty instances.

2.4 Functioning of BIN architecture

The main idea of BIN architecture is that the user does not directly communicate with the service provider with BINAP messages. BSCP, which forms the controlling part of the BIN, processes the BINAP message sent to it, makes statistics, and controls other points by sending BINAP messages to them. The advantage of this kind of architecture is that the end systems (BSS and BSP) do not have to be as complex as the controlling systems. The main intelligence of a service lies in BSCP and BSP, where are the BSLs (*Broadband Service Logic*). BSS has mainly the intelligence of requesting a service and interpretating BINAP messages sent to it.

BIN consists of two data streams: the management data streams and multimedia data streams. BSS communicates only with BSCP, but BSCP is responsible for communicating with all the other components. The transport layer below BSS does communicate with the BSP, but it gives upwards only data indications from BSP. (*Figure 2*)

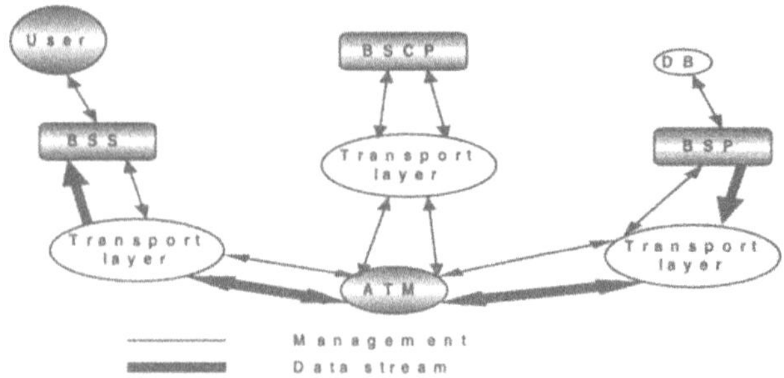

Figure 2. Management and data streams separated.

2.5 Requirements of ATM network

The signalling protocol itself is not yet developed to the level where it could be used to implement BIN architecture using ATM. The BSS uses the ATM signalling protocol (Q.2931) to set up the path to the destination, which is BSCP, or it could use the PVCs that are available to BSCP. However, the BIN architecture suggests to use BSCP to establish the path between BSP and BSS. The Q.2931 signalling protocol does not define these kind of functions, but similar extension as Q.932 to Q.931 in ISDN is needed /2,3/. Q.2931 UNI version 3.0 does not define a third parties connection setup. UNI 4.0, which is going to be introduced in late 1994, should contain the third party connection setup function defined.

2.6 Course of BIN events

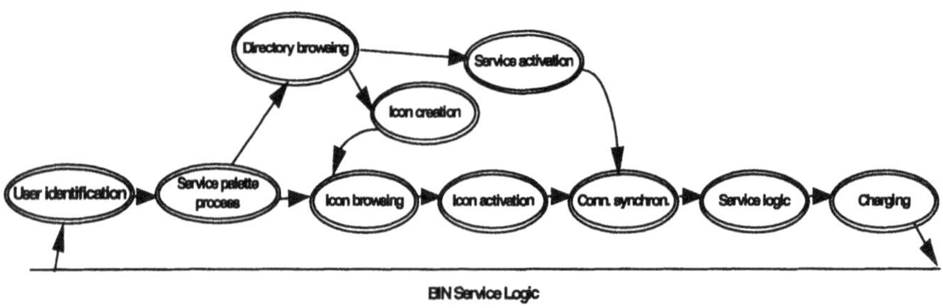

Figure 3. Course of BIN events.

Let us consider the event sequences when applying BIN services (Figure 3).

2.6.1 Service request phase

User identification:

The user sends a BINAP message to BSCP and gives sufficient identification information of him/herself. The user has to know the ATM-address (CCITT NSAP (*Network Service Access Point*), 20 bytes) of BSCP. BSCP then fetches more accurate information about the user, whereby the location of the customer service palette is also found. If the user information is not found in the current BSMP, the user must give the address of his/her Home BSCP (*HBSCP*). This enables usage of broadband services from mobile stations.

Directory browsing:
In case of a new user, requesting of a service is proceeded via directory browsing. BSCP knows one or more BSPs. BSCP sends a BINAP message to BSS, which contains a hypermedia document with links to BSPs. (*Figure 4*) The first level of the hypermedia document contains the BSPs (located at BSCP) and next levels contain all the information the BSP is able to offer, which are fetched by BSCP from the BSP in case. BSCP does not have to know all the services that every BSP can offer, just knowing the addresses of the BSPs is sufficient. By using the identification information of the user, the BSCP may filter the information given to the user.

The hypermedia documents residing in BSPs contain structured information about the type of service. The types are controlled file transfer, hypermedia database or single- and multiparty calls.

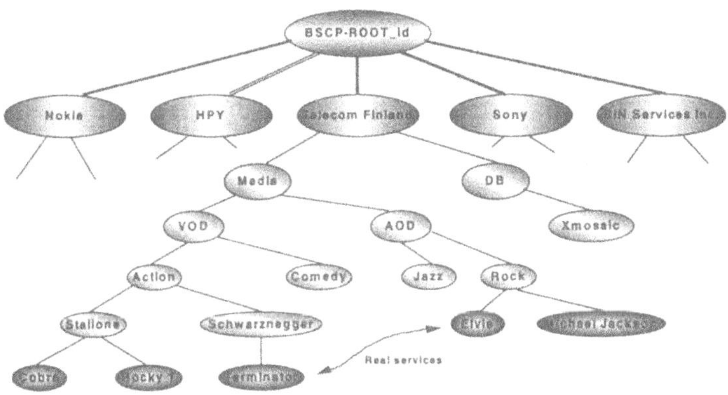

Figure 4. The structure of directory tree.

Service palette process:
When the user has chosen the 'real' service, an icon can be created to the customer service palette either to BSMP or BSS. When such a service palette exists containing icons and an accurate description of the parameters of the service, the service may be executed via icon browsing and activation. This enables much simpler and faster usage of broadband services, because of icons' graphical presentation and short-way execution process.

2.6.2 Service activation phase

After activation of the icon or 'real' service, BSS informs BSCP, which sends BSP a BINAP message containing sufficient information about the icon or 'real' service. In BSP the BSL for the service is executed. In BSCP the QoS-manager is initialized for this connection, which has received the QoS-parameters from the BSS. The QoS-parameters are also negotiated with BSP and if the user has too little network capacity, the BSP might reject the service usage.

2.6.3 Service usage phase

This phase is highly dependent of the type of service requested. For example, controlled file transfer type of service could have the functions of PLAY and STOP. During the service execution phase the QoS-manager is responsible for the quality parameters of the service. The parties inform BSCP of the changes in service quality, which then tries to restore the values.

2.6.4 After-usage phase

After the service has been used, the user should inform BSCP of the connection closing. BSCP then starts the controlled connection close phase. BSCP has stored the necessary information about the service usage, e.g. actual usage time and ability to perform with the requested QoS-parameters. The actual time here means for example in controlled file transfer the time between PLAY and STOP functions summed up in the entire session. The charging information is then added to the MIB (*Management Information Base*) in the BSMP of the user.

3. BINAP

3.1 BINAP-messages

BIN uses BINAP application protocol to communicate with the external parties of BSCP. BINAP-messages have been categorized into the following subclasses:

Initialization:
- User identification
- Customer service palette
- Directory handling

Service usage:
- Service type dependent control messages during the service usage
- QoS-messages

Service close:
- Controlled closing messages
- Customer charging

4. CUSTOMER SERVICE PALETTE

4.1 BIN conceptual model

The BIN conceptual model does not correspond to the conventional IN conceptual model and it is divided into to three planes: user, network and service planes (Figure 5).

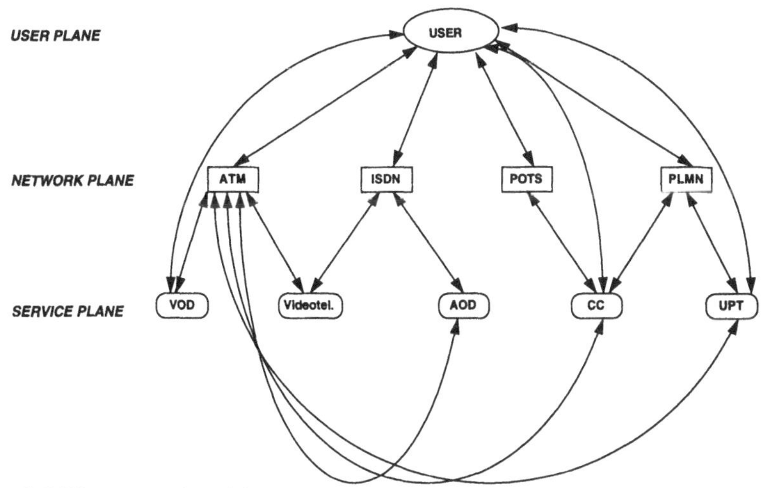

Figure 5. BIN conceptual model.

The user plane contains the information about the user. The network plane defines all the possible networks that the user system can interface with. The service plane indicates all the possible services that the user can make use of. The BIN conceptual model shows the correlations of the three planes. The correlation between user and network planes is such that the user has a number of accessible networks. The user and service plane correllation defines the services that the user have subscribed or installed. The network and service plane correlation defines the services that can be provided in the networks that the user is allowed to access.

4.2 BIN MIB

BIN MIB is the customer personal service palette and it should be structured according to BIN conceptual model. (Figure 6) The location of such a database can be either in BSS itself or in HBSCP's BSMP.

Actually the icons form the basis of this BIN MIB framework. They contain all the information of the service that has to be known by BSCP in order to be able to execute the service. They include for example in VOD, the BSP and its address, used picture formats and used network. On the other hand, the allowed networks are also listed with necessary parameters, e.g. transmission speed.

BIN MIB is designed to be managed with TMN (*Telecommunications Management Network*) architecture. An example considering UPT management can be found in /5/.

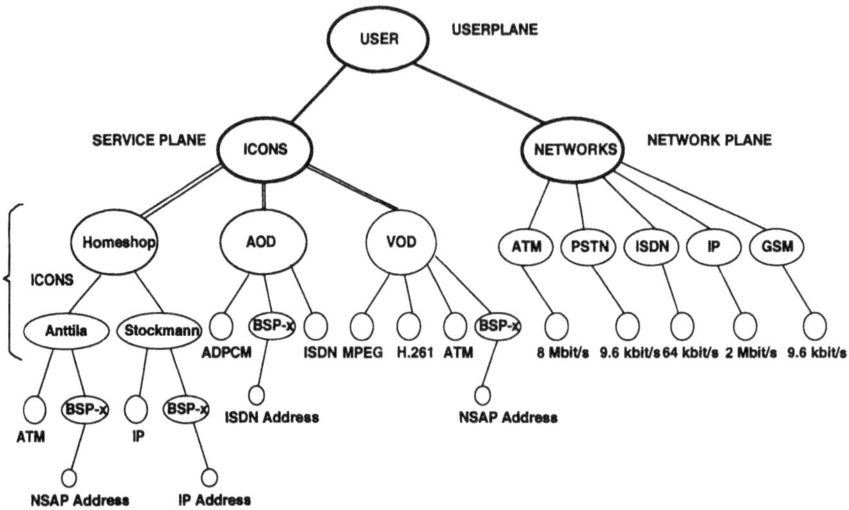

Figure 6. A framework of BIN MIB.

4.3 TMN and BIN

In the previous sections the Broadband IN services were discussed. The services were static services which could not be configured by the customer. The meaning of this stage was just to have a view of BIN and its possibilities. The next step is to have a remotely configurable service database where the customer could remotely configure for instance his VOD service table and get the true VOD service capabilities. The aim of this stage is to have a TMN configurable BIN service parameters. The customer's configuration would affect the BSCP database (MIB) and naturally also the BSMP, because of the charging. (*Figure 7*)

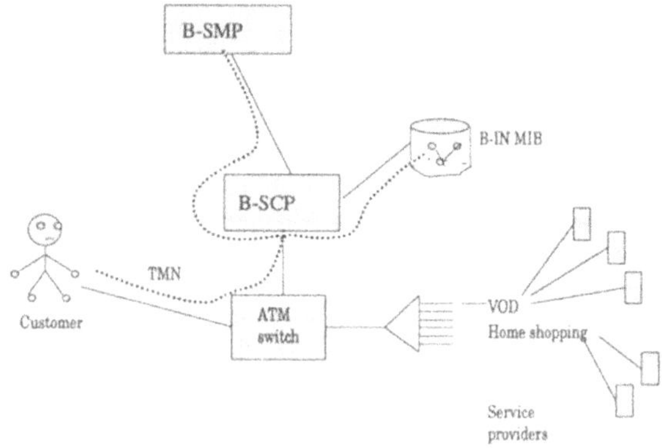

Figure 7. The use of TMN in BIN.

5. CONCLUSION

A new architecture for managing Broadband Intelligent Network services has been presented. The BIN architecture is based on ATM networks which would allow also mobile station access. It makes also use of TMN architecture and QoS-related approaches, which are still under design phase.

REFERENCES

/1/ CCITT Recommendations Q.1200 series, Intelligent Network

/2/ CCITT Recommendation Q.931, ISDN User-Network Interface Layer 3, Specification for Basic Call Control

/3/ ITU-TS recommendation Q.2931, B-ISDN, Digital Subscriber Signalling No. 2 (DSS2), User Network Interface Layer 3, Specification for Basic Call/Connection Control

/4/ K. Molin, O. Martikainen, Intelligent Network Tutorial, The Second Winterschool on Telecommunications, Helsinki, Telecom Finland, March 1994

/5/ J. Airaksinen, O. Martikainen, J. Sonninen, H. Töhönen, UPT Service Management, Proceedings of the Workshop on Intelligent Networks, Lappeenranta University of Technology, 1993

21

Which intelligence for future networks?

D. Gaïti and G. Pujolle***

**Columbia University*
Center for Telecommunications Research
New York, NY 10027-6699, USA
E-mail: gaiti@ctr.columbia.edu

***University of Versailles*
Laboratoire PRiSM
45, avenue des Etats-Unis
78035 Versailles Cedex - France
E-mail: Guy.Pujolle@prism.uvsq.fr

Abstract

Service providers should be able to satisfy the demand a new services, to improve the quality of service, to reduce the cost of network service operations and maintenance, to control the performance and to adapt the network to the user demands. In other words, it seems essential to investigate the ways we have in mind to perform such tasks. Three ways may be determined:

- Add a new architectural concept, the Intelligent Network, on the network to be controlled.
- Add an intelligent architecture able to provide a control on the network.
- Introduce intelligence in the nodes of the network to control the protocols.

In this paper we describe these three solutions. First, we present the Intelligent Network solution whose aim is to allow the inclusion of additional capabilities to facilitate provisioning of service, independent of the service and network implementation in a multi-vendor environment. Then, we show that the architecture proposed for the Intelligent Network is very similar to what could be an intelligent architecture. Finally, we propose the introduction of intelligent agents coming from Distributed Artificial Intelligence concepts. These agents could be supported by predetermined nodes of the network.

INTRODUCTION

It has been envisaged that network infrastructures will become highly complex as the evolution of telecommunication toward B-ISDN (Broadband Integrated Services Data Network) take place. Growth in new service deployment and equipment provisioning is expected to mean that very high volumes of information will have to be processed under stringent response-time constraints. Today's users are demanding customization from their telecommunication services. To obtain satisfaction, they are obliged to ask for changes in the switching system of the operator. So, a new service becomes available only when all the switching systems are updated. And, each time a new service is requested, some changes have to be done to introduce new capabilities.

It turns out that the main problem in the future is to control and manage the network and adapt the network to user demands.

Three possibilities are available:

- Add a new network, the Intelligent Network, over the network to be controlled.
- Add an intelligent architecture able to provide a control on the network.
- Introduce intelligence in the nodes of the network to control the protocols.

This paper is divided into 3 sections corresponding to each possibility to introduce the intelligence in future networks. Section 1 presents the Intelligent Network solution. In section 2, we show that the architecture provided for the Intelligent Network is very similar to an intelligent architecture (as TINA and ODP). In the last part, we propose the introduction of intelligent agents coming from Distributed Artificial Intelligent concepts. Finally, in the conclusion we propose a general solution for introducing intelligence in networks.

1. THE INTELLIGENT NETWORK

Intelligent Networks (IN) should promote a very rapid introduction of any new service through a flexible network architecture. This flexibility should be reached using standard interfaces. Intelligent Network is an architectural concept for the operation and provision of new services which is characterized by (see recommendation I.312/Q.1201):

- extensive use of information processing techniques;
- efficient use of network resources;
- modularity and reusability of network functions;
- integrated service creations and implementation by means of the modularized reusable network functions;
- flexible allocation of network functions among physical entities;
- standardized communication between network functions via service independent interfaces;
- service subscriber control of some subscriber-specific service attributes;
- service user control of some user-specific service attributes;
- standardized management of service logic.

To provide these capabilities, it is necessary to define IN functional requirements. Two types of requirements have been defined:

- Service requirements corresponding to the customer needs;

- Network requirements corresponding to the network operator needs.

Five areas have been identified for service and network requirements that are defined in recommendation I.312/Q.1201.

- Service creation: An activity whereby supplementary services are brought into being through specification phase, development phase and verification phase.
- Service management: An activity to support the proper operation of a service and the administration of information relating to the user/customer and/or the network operator.
- Network management: An activity to support the proper operation of an IN-structured network.
- Service processing: consist of basic call and supplementary service processing which are the serial and/or parallel executions of network functions in a co-ordinated way, such that basic and supplementary services are provided to the customers.
- Network interworking: a process through which several networks (IN to In or IN to non-IN) co-operate to provide a service.

These network capabilities in IN-structured network are shown in Figure 1.

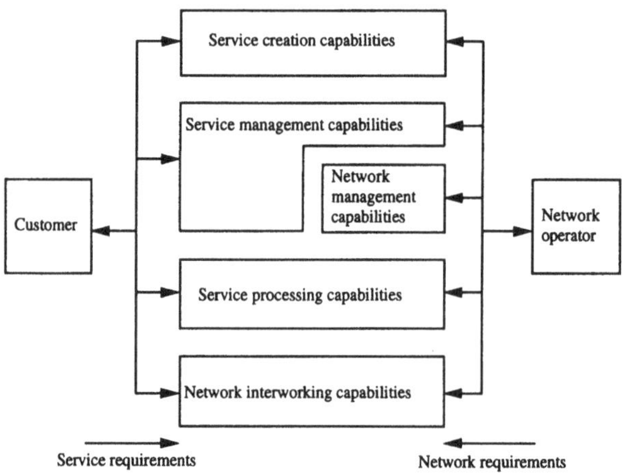

Figure 1. Network capabilities in IN-structured network.

The ITU-T has defined a methodology called INCM (Intelligent Network Conceptual Model) for specifying the IN architecture. It is not an architecture in itself, but a framework for the design and the description of the IN architecture [1] [2].

The INCM consists of four planes, where each plane represents a different abstract view of the capabilities provided by an IN-structured network (Figure 2). The planes represent the service view, the global functional view, the distributed functional view and the physical view of IN.

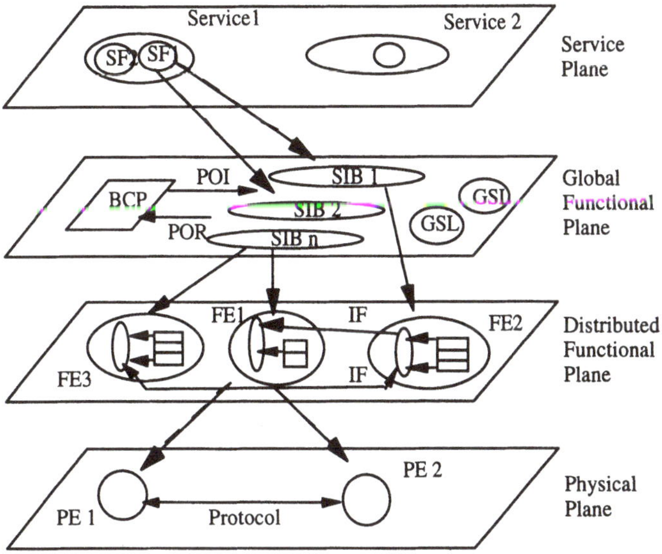

SF: Service Feature
SIB: Service Independent Building block
FE: Functional Entity
PE: Physical Entity
POI: Point Of Initiation
POR: Point Of Return
IF: Information Flows
BCP: Basic Call process
GSL: Global Service Logic

Figure 2. The INCM.

The service plane gives an abstract view to the end-users on the services provided by an IN-structured network. A service can be decomposed into one or more Service Features. A Service Feature is built from several Service Independent Building blocks.

The global functional plane defines the Service Independent Building blocks (SIBs) with the following principle: the network is viewed as a single virtual machine. So the distribution is transparent at that level. At this time, 14 SIBs have been defined by ITU-T. This number is insufficient and has to be enlarged by new capabilities. Actually, we have SIBs as "charge", "compare", "distribute", "limit", "log call information", "screen", "status notification", "translate", "user interaction", "verify", "queue", "algorithm", "GSL", "service data management".

In this plane, the Basic Call Process (BCP) and the Global Service Logic (GSL) are defined too.

- The BCP is a specialized SIB that provides the basic call capabilities. SIBs are used to build Global Service Logic (GSL).
- A GSL may be seen as a particular Service Independent Building block and corresponds to a service processing scheme. A GSL is shown in Figure 3. It is presently assumed that the GSL for service processing has no parallelism. The GSL describes the order in which the SIBs can be chained together to accomplish services.

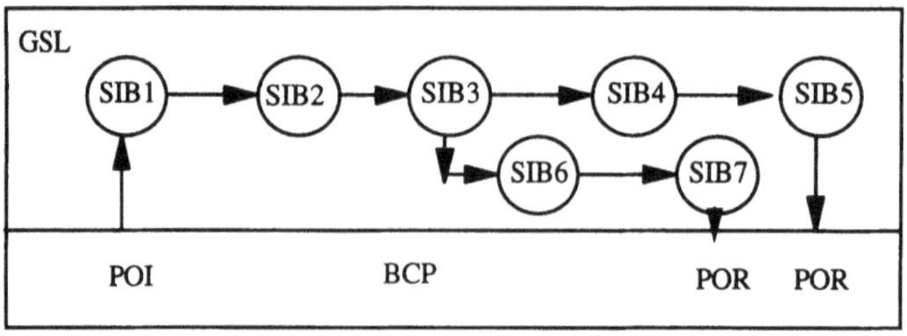

Figure 3. The Global Service Logic.

Two BCP functional interfaces to the GSL have been identified: the Point Of Initiation (POI) and the Point Of Return (POR).

The distributed functional plane models a distributed view of an IN-structured network. This view consists of Functional Entities (FEs). A FE is a unique group of functions in a single location and a subset of the total set of functions required to provide the Service Feature performed by the SIB. A FE is built with Elementary Functions (EFs) and a Functional Entity Action (FEA) as described in Figure 4.

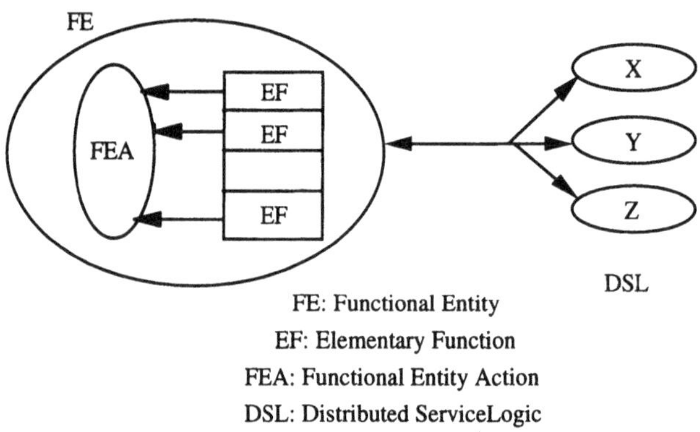

Figure 4. A Functional Entity.

A group of centralized or distributed FEs is mapped onto a SIB. Then a SIB is realized by the interaction between corresponding distributed FEs. These interactions are provided by Information Flows (IF), which are short messages. The advantage of this representation is to have reusable FEs, i.e., a FE can be used by several SIBs.

Moreover, concerning the service logic, there is one set of distributed service logic (DSL) per SIB and it uses FEAs and information flows (see Figure 4).

The physical plane identifies the different Physical Entities (PEs). A Physical Entity consists of one or more Functional Entities. One or more FEs may be mapped onto the same PE, but a FE cannot be split between two PEs. This plane takes into account physical considerations, optimization aspects, and protocols.

The goal of the IN architecture is to significantly reduce the new service introduction interval with the ability to keep the same communication infrastructure. The IN logical organization [3] may be seen as described in Figure 5.

Two kinds of interfaces have been defined:

- a high-level logical programming interface that defines the access through the interface to logical modules named SIBs (Service Independent Building blocks).
- a resource-control interface that allows the IN infrastructure to control the physical resources. These physical resources can be machines from different vendors.

An important aspect of the objectives of an IN architecture is to come up with an evolutionary perspective for defining new capability sets. The reasoning behind the capability sets is that they serve to address incremental sets of functionalities, achievable within reasonable time frames and with measurable impact on the existing network.

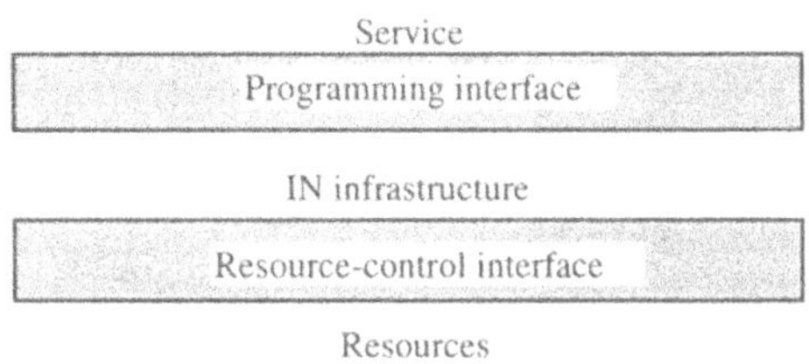

Figure 5. The logical organization of IN.

A number of significant issues remain to be addressed for introducing a complete intelligence:

- Evolution towards broadband, multi ended, multimedia services. This is expected to include co-operative and high throughput requirements.
- Performance issues. Due to the increase of service traffic, performance objectives need to be placed in an overall IN context.
- Address OAM (Operations And Maintenance) functionalities. The significant impact of OAM issues on network and service operation is critical for the life of future broadband-based networks.

2. INTELLIGENT ARCHITECTURE

In the previous section we described the architecture of the Intelligent Network. Indeed, this architecture is very similar to other architectures able to manage a network. The work to define the IN architecture was performed in parallel with two other initiatives directed towards the definition of flexible architectures: one coming from ISO and the other one from operators and carriers.

The ISO contribution comes from the ODP (Open Distributed Processing) normalization [4] [5]. The purpose of the ODP standardization is to facilitate systems integration where there is distributed processing and heterogeneity. The conceptual architecture is composed of five viewpoints:

- enterprise — the enterprise viewpoint studies the role of a computer system within an organisation and the interaction between the system and the people within the system
- information — the information viewpoint addresses information issues that are influenced by distribution and that affect distribution.
- computational — the computational viewpoint models a distributed system as a collection of programs that interacts with each other. The specification of an interaction in this viewpoint includesthe distribution transparencies used in the interaction.
- engineering — the engineering viewpoint studies the mechanisms needed for providing different kinds of transparencies.
- technology — the technology viewpoint studies how a distributed system can be built using existing technologies including hardware, software and standards.

This architecture is quite identical to the architecture proposed by the TINA (Telecommunication Information Networking Architecture) initiative [6]. This initiative was set up by the major Telecom operator in 1990 to encourage the development of architectural techniques to offer better control in the telecommunication area.

The general TINA Reference Model is shown in Figure 6 [7].

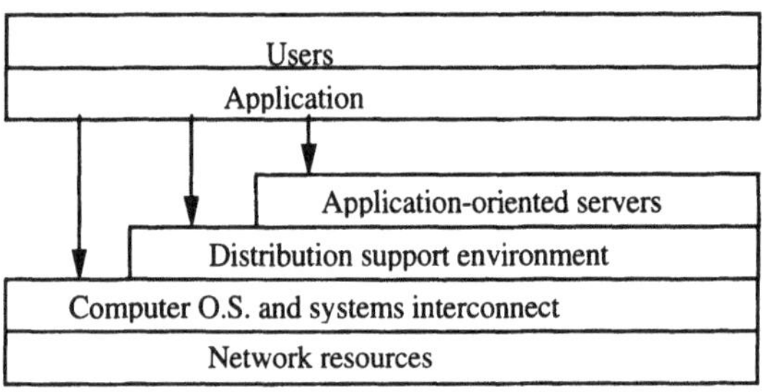

Figure 6. The TINA architecture.

Between the network resources and the applications, the TINA platform offers three layers. The two first ones may be bypassed to access directly the computer operating system.

These three architectures (IN - ODP - TINA) can be compared as follows (Figure 7). They are structured in three main parts:

- An upper part, which is user-oriented. The role of this part is to take into account the users' requirements to provide them with well-suited services.
- A lower part, which is a physical one. It represents the components (hardware and software) on which the system is implemented.
- A medium part, which makes the link between the two others. It realizes the distribution transparencies on one hand and provides an environment for the development of applications on the other hand.

Using the different viewpoints to examine architecture issues encourages a clear separation of concerns, which in turn leads to a better understanding of where to add intelligence:

- Description of the definition of services independently of the way in which that role is automated.

- Description of the information content of a system independently of the way in which the information is stored or manipulated.

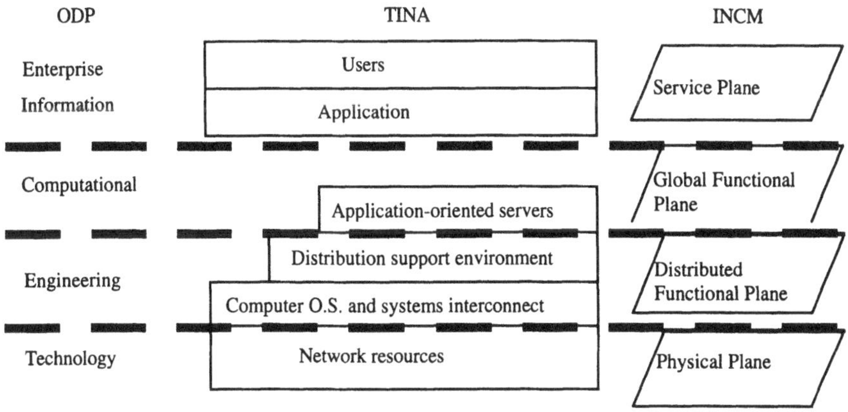

Figure 7. The mapping of the different architectures.

- Description of the application programming environment independently of the way in which that environment is built.
- Description of the components, mechanisms and techniques used to build the system independently of the machines on which they will have to run.
- Description of the system hardware and software independently of the role it plays in the system.

3. THE DISTRIBUTED ARTIFICIAL INTELLIGENCE APPROACH

Another way to make the network intelligent is to introduce intelligence in the node through a distributed multi agent system able to control the resources of the network [8]. Distributed Artificial Intelligence concepts give an interesting way to support shared intelligence [9]. This third solution to introduce intelligence in the network can be considered as the only one really "intelligent" in the sense of human intelligence.

The Distributed Artificial Intelligence (DAI) field is often divided into two sub-areas: Distributed Problem Solving (DPS) and multi agent systems. DPS considers how the work of solving a particular problem can be divided among a number of modules that co-operate by dividing and sharing knowledge about the problem and the developing solution [10]. Multi-agent systems include intelligent behavior among a collection of possibly pre-existing, autonomous, intelligent "agents", when they can co-ordinate their knowledge, goals, skills, and plans to take action or solve problems [11]. The agents in a multi agent system may be working towards a single global goal or separate individual but interacting goals [12]. Agents in multi agent and DPS systems must share knowledge about problems they face and solutions they reach, but they must also reason about the co-ordination processes among the agents.

DAI can be useful in domains in which action, perception, and/or control are naturally distributed. It is also a way to reproduce the human behavior when a group of experts work together to realize a task. For these reasons, multi agent systems are envisaged to control future networks.

We propose a methodology to solve any kind of problems that can appear in the control of networks. Our system is generic, even if some specific problems (diagnosis, performance or security management) correspond better to the proposed system. We consider each task of the network control as a distributed problem solving [13].

We propose a global architecture following the principle of a blackboard architecture [14]. This architecture is composed of a collection of agents under the supervision of a control system. The agents co-operate through a knowledge base in writing and reading data corresponding to a given problem. The agent itself has a knowledge part that is controlled to define the problem solving strategy and choose the next action to perform. The agents communicate through an interpreter of messages; they all have a model of the others agents in their environment. The global organization and the agents are described in [15] §3.2.

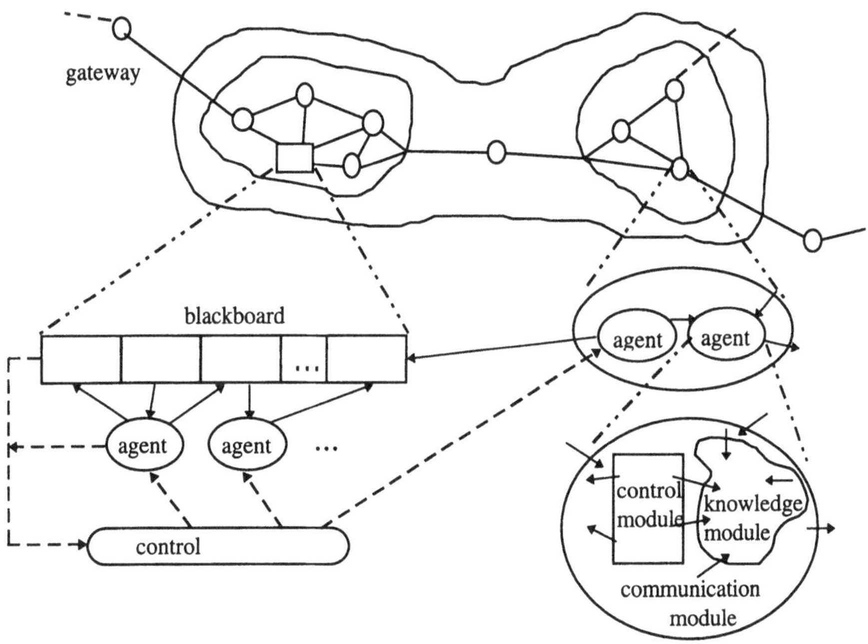

Figure 8. The multi agent system supported by the network.

The integration of this blackboard architecture into the network can be seen as shown Figure 8. One node can have one or several agents depending of the size of the node. We choose one particular node to be responsible for the blackboard.

The blackboard will be more or less used depending on the task we want to achieve. For example, a global management task will take place in the blackboard with the co-ordination and the collaboration of several agents distributed in different nodes.

But, if we are interested in the control of a node, the agent responsible for this node will solve the task alone with its own knowledge about the task and with its knowledge of the state of the environment. In that case, the agents will be almost totally independent and autonomous. When the network is not too loaded, the agents may communicate each others to update their global view of the state of the network. The blackboard is used in writing to receive information from the nodes through the associated agent when the node is not loaded; this is performed to permit the other agents to update their knowledge of the global state of the system. It is used in reading when an agent of a node wants to update its knowledge. Real time is reached by reducing at the minimum communication requirements. We use this approach for flow control in ATM networks [16] and for an intelligent management in IN [15] [17].

CONCLUSION

Control and management capabilities play a key role in the viability and success of the emerging broadband networks. Substantial progress has been performed in introducing new technologies in control and management software. An interesting way is to introduce intelligence in operations functions.

We proposed three solutions to introduce some intelligence in the control of future broadband networks. Indeed, these three solutions are based on very different concepts and could be superposed in future networks.

A nice solution would be the introduction of an intelligent architecture (ODP, INCM or TINA) controlled by co-operative agents for problem solving processes. This architecture would be based on concepts developed in IN, distributed networking and distributed artificial intelligence. As a conclusion, we would like to give a sketch of what could be done to introduce simultaneously the three previous types of intelligence.

Let us assume we want to manage and control an ATM network. First, we may adopt an ODP architecture on the global system. On this architecture we implement a distributed model following the concepts introduced in [18]. This architecture is composed of domains that is a first principle for organizing the resources. A domain is a set constituted from resources to be managed in a distributed system and a specific management policy. The second organizing principle concerns the Management Process (MP) that contains two components, the MP view, which is the abstract representation of the resources and of the activities under control and the MP core that executes the management process. It manages and activates the resources through the MP view. At the MP core may be associated the intelligent architecture developed in this paper. Capabilities described in the first part of the paper are provided through the Intelligent Network associated to the ATM network. This architecture may be seen as a platform to support open services.

REFERENCES

[1] ITU-T, Recommendation I.312/Q.1201, Principles of Intelligent Network Architecture, October 1992, Geneva.

[2] José M. Duran and John Visser, International Standards for Intelligent Networks, IEEE Communications Magazine, February 1992.

[3] R. Kung, Rationale for Intelligent Networks, Proc. of the International IFIP Workshop on Open Distributed Processing, Berlin, October 1991.

[4] Basic Reference Model of Open Distributed Processing, Part 2: Descriptive Model - ISO 10746-2, 1994.

[5] Basic Reference Model of Open Distributed Processing, Part 3: Prescriptive Model - ISO 10746-3, 1994..

[6] T. Boyd, Telecommunication Information Networking Architecture Initiative, Proc. of the International IFIP Workshop on Open Distributed Processing, Berlin, October 1991.

[7] W. Barr, T. Boyd, Y. Inoue, The TINA Initiative, IEEE Communication Magazine, vol. 31, no 3, pp 70-77, march 1993.

[8] I. Rahali and D. Gaïti, A multi-agent system for network management, proceedings of the Second International Symposium on Integrated Network Management, Washington, April 1991.

[9] D. Gaïti and M.P. Gervais, Artificial Intelligence Environment for Intelligent Network Management, Proc. of the IEEE International Conference on Communication Technology, ICCT'92, pp. 30.05.1-30.05.4, Beijing, China, September 1992.

[10] Alan H. Bond and L. Gasser, eds., Readings in Distributed Artificial Intelligence, San Mateo, Calif.: Morgan Kaufmann, 1988.

[11] E. Werner, The Design of Multi-Agent Systems, Decentralized A.I.-3, Elsevier Science Publishers B.V., 1992.

[12] L. Gasser, Distributed Artificial Intelligence, AI Expert, July 89.

[13] P. Lebouc and P.E. Stern, Distributed Problem Solving in Broadband Telecommunication Network Management, Proc. of the Specialized Conference on Artificial Intelligence, Telecommunications and Computer Systems, 11th International Conference Expert Systems and their Applications, Avignon, 1991.

[14] R. Engelmore et al., Blackboards systems, Reading, Mass.: Addison-Wesley, 1988.

[15] D. Gaïti, An Advanced Management Architecture for IN, International workshop on Intelligent Networks, IFIP, Lappeenranta, Finland, August 1994.

[16] G. Pujolle and D. Gaïti, ATM Flow Control Schemes through a Multi-agent System, Proc. of SICON/ICIE'93, pp. 455-459, IEEE Computer Society Press, Singapour, September 1993.

[17] D. Gaïti, A Proposal for Integrating Intelligent Management in the Intelligent Network Conceptual Model, à paraître dans International Journal on Computer Networks and ISDN Systems, 1994.

[18] D. Gaïti, I2NMA: An Intelligent Integrated Network Management Architecture, International Journal on Network Management, vol 4, 3, 1994.

INDEX OF CONTRIBUTORS

Bihain, A. 68
Bisseling, H. 34

Englund, C. 225

Gaïti, D. 137, 280
Granö, K. 77

Haitao, T. 126
Hämmäinen, H. 246
Harju, J. 77
den Hartog, J. 34
Herrigel, A. 149

Järvinen, T. 77
Jormakka, J. 97

Kärkkäinen, E. 126
Karttunen, T. 77, 253
Knight, C. 47

Lahtinen, P. 246
Lai, X. 149
Larikka, T. 77
Lehtinen, P. 61
Luukkainen, S. 225

Martikainen, O. 208, 253, 270
Mitts, H. 236
Molin, K. 270

Naoumov, V. 253
Nørgaard, J. 159

Paakki, J. 77
Pujolle, G. 280
Puro, P. 111

Raatikainen, K. 173
Ranasinghe, D. 159
Ruusunen, J. 208

Samouylov, K. 253
Sonninen, J. 111

Taina, J. 194
Takala, J. 208

Warsta, M. 61
White, J. 68

CPSIA information can be obtained
at www.ICGtesting.com
Printed in the USA
LVHW081111200420
654119LV00010B/690

9 780412 717307